# Speciation and Patterns of Diversity

Bringing together the viewpoints of leading ecologists concerned with the processes that generate patterns of diversity, and evolutionary biologists who focus on mechanisms of speciation, this book opens up discussion in order to broaden understanding of how speciation affects patterns of biological diversity, especially the uneven distribution of diversity across time, space and taxa studied by macroecologists. The contributors discuss questions such as: Are species equivalent units, providing meaningful measures of diversity? To what extent do mechanisms of speciation affect the functional nature and distribution of species diversity? How can speciation rates be measured using molecular phylogenies or data from the fossil record? What are the factors that explain variation in rates? Written for graduate students and academic researchers, the book promotes a more complete understanding of the interaction between mechanisms and rates of speciation and these patterns in biological diversity.

ROGER BUTLIN is Professor of Evolutionary Biology, Animal and Plant Sciences, at the University of Sheffield. He has held a prestigious Royal Society Research Fellowship at the University of Cardiff and his work has been recognized by honorary fellowships at the Natural History Museum, Zoological Society of London, and Royal Belgian Institute of Natural Sciences.

JON BRIDLE is Lecturer in Biology, in the School of Biological Sciences, University of Bristol. Since completing his PhD in Evolutionary Genetics in 1998, he has conducted research in quantitative genetics and evolutionary biology at University College London, University of Cardiff, Universidad Autónoma de Madrid, and the Institute of Zoology, London. He was awarded a fellowship at the Zoological Society of London in 2002.

DOLPH SCHLUTER is Professor and Canada Research Chair, Biodiversity Research Centre and Zoology Department, at the University of British Columbia. He is a former President of the Society for the Study of Evolution and recipient of the Sewall Wright Award from the American Society of Naturalists. He is a Fellow of the Royal Society of London and of Canada.

*Ecological Reviews* will publish books at the cutting edge of modern ecology, providing a forum for volumes that discuss topics that are focal points of current activity and are likely to be of long-term importance to the progress of the field. The series will be an invaluable source of ideas and inspiration for ecologists at all levels from graduate students to more-established researchers and professionals. The series will be developed jointly by the British Ecological Society and Cambridge University Press and will encompass the Society's Symposia as appropriate.

*Biotic Interactions in the Tropics: Their Role in the Maintenance of Species Diversity*
Edited by David F. R. P. Burslem, Michelle A. Pinard and Sue E. Hartley

*Biological Diversity and Function in Soils*
Edited by Richard Bardgett, Michael Usher and David Hopkins

*Island Colonization: The Origin and Development of Island Communities*
by Ian Thornton
Edited by Tim New

*Scaling Biodiversity*
Edited by David Storch, Pablo Margnet and James Brown

*Body Size: The Structure and Function of Aquatic Ecosystems*
Edited by Alan G. Hildrew, David G. Raffaelli and Ronni Edmonds-Brown.

# Speciation and Patterns of Diversity

Edited by

ROGER K. BUTLIN
*University of Sheffield*

JON R. BRIDLE
*University of Bristol*

DOLPH SCHLUTER
*University of British Columbia*

CAMBRIDGE UNIVERSITY PRESS
Cambridge, New York, Melbourne, Madrid, Cape Town, Singapore, São Paulo,
Delhi, Dubai, Tokyo

Cambridge University Press
The Edinburgh Building, Cambridge CB2 8RU, UK

Published in the United States of America by Cambridge University Press, New York

www.cambridge.org
Information on this title: www.cambridge.org/9780521883184

First published 2009
Reprinted 2010

Printed in the United Kingdom at the University Press, Cambridge

*A catalogue record for this publication is available from the British Library*

*Library of Congress Cataloguing in Publication data*

Speciation and patterns of diversity / edited by Roger Butlin, Jon Bridle and Dolph Schluter.
   p.   cm. — (Ecological reviews)
Includes bibliographical references.
1. Species diversity.  2. Biodiversity.  I. Butlin, Roger, 1955–  II. Bridle, Jon.  III. Schluter,
Dolph.  IV. Series.
QH541.15.S64S62   2008
577—dc22

                                                                2008025507

ISBN 978-0-521-88318-4 hardback
ISBN 978-0-521-70963-7 paperback

# Contents

The colour plate section is placed between pages 180 and 181

# Contributors

JOHN ALROY
The Paleobiology Database, University of California

TIMOTHY G. BARRACLOUGH
Division of Biology, Imperial College London

GRAHAM BELL
Biology Department, McGill University

JON R. BRIDLE
School of Biological Sciences, University of Bristol

ROGER K. BUTLIN
Animal and Plant Sciences, University of Sheffield

THOMAS P. CURTIS
School of Civil Engineering and Geosciences, University of Newcastle upon Tyne

DIEGO FONTANETO
Division of Biology, Imperial College London

DANIEL J. FUNK
Department of Biological Sciences, Vanderbilt University

SERGEY GAVRILETS
Department of Ecology and Evolutionary Biology and Department of Mathematics, University of Tennessee

LUKE HARMON
Department of Biological Sciences, University of Idaho

ELISABETH A. HERNIOU
Division of Biology, Imperial College London

JODY HEY
Department of Genetics, Rutgers, the State University of New Jersey

JAMES MALLET
Galton Laboratory, Department of Biology, University College London

PATRIK NOSIL
Zoology Department and Biodiversity Research Centre, University of British Columbia

C. DAVID L. ORME
Division of Biology, Imperial College London

PAUL N. PEARSON
School of Earth, Ocean and Planetary Sciences, Cardiff University

ALBERT B. PHILLIMORE
NERC Centre for Population Biology and
Division of Biology, Imperial College
London

JITKA POLECHOVÁ
Biomathematics & Statistics Scotland

TREVOR D. PRICE
Department of Ecology and Evolution,
University of Chicago

ANDY PURVIS
Division of Biology, Imperial College
London

CLAUDIA RICCI
Università di Milano, Dipartimento di
Biologia

ROBERT E. RICKLEFS
Department of Biology, University of
Missouri – St. Louis

DOUGLAS W. SCHEMSKE
Department of Plant Biology and W. K.
Kellogg Biological Station, Michigan State
University

DOLPH SCHLUTER
Zoology Department and Biodiversity
Research Centre, University of British
Columbia

OLE SEEHAUSEN
Centre of Ecology, Evolution and
Biogeochemistry, EAWAG

WILLIAM T. SLOAN
Department of Civil Engineering,
University of Glasgow

NICOLA H. TOOMEY
Division of Biology, Imperial College
London

TIM H. VINES
Centre d'Ecologie Fonctionelle et Evolutive
Montpellier

AARON VOSE
Department of Ecology and Evolutionary
Biology and Department of Computer
Science, University of Tennessee

NIGEL C. WALLBRIDGE
Nomad Digital Limited

# Preface

This volume is derived from the Annual Symposium of the British Ecological Society on 'Speciation and Ecology' which was held at the University of Sheffield, 28–30 March 2007. The idea for this Symposium arose during a previous meeting in the series, the 2002 'Macroecology: Concepts and Consequences' meeting organized by Tim Blackburn and Kevin Gaston. The 2002 meeting concentrated on large-scale diversity patterns. Many speakers acknowledged the role of speciation in generating diversity and influencing patterns of diversity. Although there was some discussion of the factors that determine rates of speciation, it was striking how little contact there seemed to be between the discipline of macroecology and the large and active field of research into mechanisms of adaptive divergence and speciation. 'Ecological speciation' has been an area of research growth in recent years, asking how ecological drivers influence the speciation process. However, the opposite direction of effect, how speciation processes impact on ecological patterns, has been studied less. Therefore, we proposed a meeting whose central objective was to foster dialogue between these two fields.

The meeting had an unusual mix of participants but we hope that they managed to communicate effectively with one another! The chapters in this book reflect the range of topics discussed and we hope that they will help to continue the conversations that were started in Sheffield. In our introduction, we try to set the scene by considering mechanisms of speciation and their potential impacts on biodiversity, both in terms of species' geographical distributions, and their interactions within ecological communities. In particular, speciation mechanisms can be divided into those that generate ecologically distinct species and those that do not, for example because reproductive isolation is driven by sexual conflict. These two classes of speciation mechanism may contribute to diversity in different ways, adding to either local diversity or beta diversity, for example, or affecting levels of within-species functional diversity. We also introduce the problems surrounding the estimation of speciation rates, an essential preliminary to understanding the factors that influence those rates.

We have not subdivided the book because we are reluctant to separate parts of a spectrum. However, the chapters are ordered in a progression, similar to that used at the Symposium, which begins with the nature of species, how we count the units of biological diversity and some of the limits on diversity; considers mechanisms of speciation and adaptation; and then looks at rates of speciation both from the perspective of molecular phylogenetics and that of the fossil record. We hope that readers will see some logic in this arrangement and will find the book stimulating and enjoyable.

# Acknowledgements

We are very grateful to the British Ecological Society for the opportunity to organize the Symposium on 'Speciation and Ecology' and especially to Hefin Jones, Hazel Norman and Richard English for their excellent support in the organizational stages and at the meeting. The Department of Animal and Plant Sciences and the Conference Office of the University of Sheffield also helped to make the meeting a success. As with any meeting, the interest and excitement of the event was dependent on the contributors: we thank those who gave oral presentations, those who brought posters and all participants for making the meeting go so well.

During the preparation of this volume, our contributors have done a great job of providing stimulating chapters more or less on schedule, as well as providing very valuable assistance with the review process, and responding constructively to reviewers' comments. We have had strong support both from the British Ecological Society, Hefin Jones and Lindsay Haddon, and from Cambridge University Press, Dominic Lewis and Alison Evans. Alan Crowden was also a great help in the early stages.

# CHAPTER ONE

# Speciation and patterns of biodiversity

ROGER K. BUTLIN, JON R. BRIDLE
AND DOLPH SCHLUTER

There are many more species of insects (>850 000) than of their putative sister taxon (Entognatha, 7500 species) (Mayhew 2002). More than 1600 species of birds have been recorded near the Equator in the New World compared with 300–400 species at latitudes around 40° North or South (Gaston & Blackburn 2000). Mammalian families with average body sizes around 10 g have nearly 10 times as many species as those with average body sizes around 3 kg (Purvis *et al.* 2003). In a catch of 15 609 moths of 240 species over 4 years of light trapping at Rothamsted, England, the majority of species (180) were represented by 50 individuals or less (Fisher *et al.* 1943). These observations illustrate the highly uneven distribution of the world's biological diversity. They are examples of four well-known patterns: species richness varies among clades; it varies spatially, with the latitudinal gradient being a classic example; it is higher in small animals than large ones; and rare species are more numerous than common ones. Documenting and explaining such patterns is a major enterprise of ecology (Gaston & Blackburn 2000).

In their introduction to a previous British Ecological Society (BES) Symposium Volume, Blackburn and Gaston (2003) identified three evolutionary processes that underlie large-scale patterns of biodiversity: speciation, extinction and range changes. Anagenetic change might also contribute to some patterns, for example if there is a general tendency for size increase among mammalian lineages (Alroy 1998). The 2002 BES Symposium dealt mainly with rates of extinction and the ecological processes limiting the distributional ranges of species, but had relatively little to say about speciation. Therefore, our intention in this volume is to examine the thesis that mechanisms and rates of speciation are key determinants of biodiversity patterns. Since ecological factors, including the current diversity of a community, may influence speciation and this in turn may alter ecological relationships, there is a rich web of interactions to explore. This volume, like the Symposium from which it is derived, aims to foster communication among the various disciplines that can contribute to this exploration, especially evolutionary ecologists interested in the speciation process and macroecologists interested in

*Speciation and Patterns of Diversity*, ed. Roger K. Butlin, Jon R. Bridle and Dolph Schluter. Published by Cambridge University Press. © British Ecological Society 2009.

explanations for patterns of diversity. In this first chapter we highlight some of the key issues.

## How many species are there?

Diversity is measured by counting species. This is not a straightforward process for several reasons, and the complexities should be kept in mind when analysing patterns of diversity. Counting species is relatively easy when there is a tight association between species identity and easily measured traits such as plumage or floral morphology. This is one reason why many studies of the distribution of biodiversity focus on the best-known groups, such as birds (Gaston & Blackburn 2000; Phillimore *et al.*, this volume; Ricklefs, this volume), mammals (Alroy, this volume; Purvis, this volume) and angiosperms (Schemske, this volume). However, even in these taxa, it is likely that northern temperate diversity is better documented than tropical diversity.

Analyses of biodiversity patterns make the necessary simplification that species in all taxa, whether prokaryote or eukaryote, plant or animal, can be treated as equivalent units and counted. One reason why this is inadequate relates to the genetic diversity contained within species. This can be highly variable across taxa, partly because some speciation mechanisms generate very unequal products. One could argue that a highly polymorphic species represents more diversity than a monomorphic one (Tregenza & Butlin 1999). This is akin to the argument for weighting the conservation value of species according to phylogenetic distinctiveness (Mace *et al.* 2003) but has been less-widely discussed.

Establishing the diversity of microorganisms is a special challenge. Morphology is often a useless guide to species identity, and genetics might be the only way. Only recently have large-scale sequencing projects started to clarify the number of distinct types of bacteria present in environmental samples and to allow extrapolations to global diversity. The resulting numbers are very large (Bell, this volume). Explaining this diversity and its distribution in terms of ecological processes, such as colonization and local adaptation, remains a major challenge (Curtis, this volume). Aside from possible differences in the criteria used to define 'species' in the two groups, protist diversity is much lower than bacterial diversity and this cannot be readily explained by differences in body size or total number of individuals (Bell, this volume). Besides being relatively new on the face of the earth (about 2 billion years), compared to prokaryotes (>3.5 billion years), protists typically have sexual reproduction, at least occasionally. Perhaps the high frequency of sexual reproduction and recombination in protists might oppose niche specialization and speciation. In prokaryotes, gene exchange does occur by non-sexual means but recombination rates are much lower, which may allow greater diversity to evolve. Objectively definable units of diversity exist in at least some asexual taxa (Barraclough, this volume), but the units may not be equivalent to sexual species.

Speciation is a process, and this too can make it difficult to apply a species criterion. It can be very rapid, as in the origin of hybrid and polyploid species, but it may take hundreds of thousands or even millions of years from the initial restriction of gene flow to its complete elimination. During this transition time, boundaries between species are blurred and it is difficult to find objective criteria for species enumeration (Hey, this volume). For divergent, allopatric, closely related populations, there is no adequate criterion, as Mayr himself recognized (Mayr 1942), and this makes it difficult to measure beta- and gamma-diversity in consistent and meaningful ways.

## Speciation mechanisms and biodiversity

Speciation in sexual taxa is the evolution of reproductive isolation – genetically based barriers to gene flow between populations. Coyne and Orr (2004, p. 57) argue that the central problem of speciation is to understand the evolutionary forces that create the *initial* reduction in gene exchange between populations. The emphasis is on the initial barriers because genetic changes that affect hybrid production or hybrid fitness continue to accumulate long after speciation is over, yet have little to do with the process of speciation itself. For this reason, most research on speciation focuses on incompletely isolated taxa or on young, very recently formed species.

Speciation can, and probably does, occur by a variety of mechanisms (reviewed in Coyne & Orr 2004). It can occur in a single step, such as by polyploidy, or it may occur by a gradual series of allelic substitutions at multiple loci. It can happen as a by-product of adaptation to contrasting niches or environments, or it can happen without any adaptation at all. Speciation is inevitable when populations are separated for a long time by a geographic barrier, but it might also happen in the face of considerable gene flow between populations inhabiting the same small geographic region. Many speciation mechanisms have been shown to be plausible, but there remains considerable uncertainty over which of these scenarios are responsible for most of the species on earth.

For our perspective in this book, the question is whether the mechanisms of speciation – the processes that underlie the evolution of reproductive isolation – have played a role in establishing the uneven distribution of global biodiversity. One way to attempt an answer to the question is to explore whether common mechanisms of speciation affect the probability that the resulting species coexist. If so, then the number of species in a given community (alpha diversity) will increase. If not, then the derived species replace one another geographically and only beta or gamma diversity will increase. To this end, we address what speciation yields from an ecological standpoint. In particular, we compare mechanisms of speciation whose products – species – are ecologically different. Ecological divergence facilitates long-term coexistence and increases the numbers of species along environmental gradients.

Polyploidy is a mechanism of speciation that is relatively common in plants. Here, reproductive isolation evolves instantaneously upon the accidental doubling of chromosome number, because triploid hybrids between ancestral diploid and derived tetraploid individuals are sterile. When polyploids are formed from within a species, the derived form contains a subset of alleles already present in the ancestor and so is not much different genetically from the original. Nevertheless, physiological and life history differences associated with having twice the number of chromosomes can influence ecological tolerances and possibly affect niche use (Levin 2002). Functional diversification is even more likely when the polyploids are formed by hybridization between two species, because in this case entirely new gene combinations are produced. This can have two immediate ecological consequences. Firstly, the new gene combinations might code for unique ecological traits. Secondly, the polyploids might immediately be exposed to strong natural and sexual selection to purge the population of incompatible or unfit gene combinations, and favour specific gene combinations in the environments in which the polyploids occur. Polyploidy is actually one of the few speciation mechanisms for which it is possible to estimate its contribution to the generation of diversity: it accounts for about 7% of speciation events in ferns and 2–4% of speciation events in angiosperms (Otto & Whitton 2000). The same cannot be said for other types of chromosomal rearrangements, the importance of which remains unclear. Although many pairs of species differ in chromosome number or form, the evidence for a causal association with speciation is very limited (Coyne & Orr 2004; Butlin 2005).

Hybridization between species can have similar effects on the speciation process regardless of whether the resulting hybrid is polyploid or diploid, because in both cases new gene combinations are produced that are immediately distinct ecologically or immediately subjected to strong selection in their environments, provided hybridization is associated with a restriction of gene flow from their parental species. The best examples are again from plants (Rieseberg et al. 2007), but examples of hybrid species are also known in animals (Mallet, this volume; Mavarez et al. 2006).

Over the past decade evidence has been accumulating that many, and possibly most, speciation events involve some form of natural and sexual selection. The evidence comes from both genetics and ecology. First, the few genes, which have been discovered to date, that underlie reproductive isolation between species tend to exhibit statistical signatures of selection, such as rapid rates of nucleotide substitution in coding sequences (Noor 2003). Second, a growing number of studies find that reproductive isolation evolves most quickly between species when ecological differences also accumulate (Schluter 2001; Funk *et al.*, this volume). Third, a number of studies have found evidence of gene flow having occurred during the speciation process (Via 2001; Barluenga *et al.* 2006; Savolainen *et al.* 2006; Gavrilets and Vose, this volume). This provides indirect evidence for selection because without it gene flow would have prevented the evolution of reproductive isolation.

Natural and/or sexual selection can bring about the evolution of new species by a variety of mechanisms that have very different ecological consequences. We can group these mechanisms into two broad categories: reproductive isolation driven by divergent natural selection between environments or niches ('ecological speciation') and reproductive isolation resulting from divergent genetic responses to the same environmental selection pressures ('non-ecological speciation'). A key difference between the two categories of mechanisms is that whereas ecological speciation yields species that are ecologically different and can coexist, non-ecological speciation produces forms that are ecologically equivalent. Coexistence of ecologically equivalent taxa is possible, at least for considerable time periods (Hubbell 2001), but establishing sympatry may be slowed. Divergent selection is also likely to produce reproductive isolation more rapidly than uniform selection. Furthermore, divergent selection continues to select against hybrids if they have an intermediate phenotype (extrinsic postzygotic isolation), and this might allow species to coexist relatively soon after they formed, and to persist in sympatry in the face of gene flow. In other respects, the types of reproductive isolation that evolve are not necessarily different between the two categories of speciation mechanisms. Both mechanisms can lead to intrinsic postzygotic isolation due to genetic incompatibilities and prezygotic isolation due to divergence in mating cues. Reinforcement of prezygotic isolation can also occur under either category of mechanism.

The threespine sticklebacks, *Gasterosteus aculeatus*, inhabiting postglacial lakes and streams of the northern hemisphere provide multiple examples in which divergent natural selection has led to both extrinsic reproductive isolation and incidental assortative mating by body size (Rundle *et al.* 2000; McKinnon *et al.* 2004; Vines & Schluter 2006). There is also evidence for reinforcement of premating isolation in sympatry (Rundle & Schluter, 1998; Albert & Schluter 2004). In general, divergence driven by selection may occur more rapidly the greater the number of ecological dimensions that are involved (Nosil & Harmon, this volume).

Cases of speciation by non-ecological mechanisms are difficult to confirm. The evolution of intrinsic postzygotic isolation between *Drosophila melanogaster* and *Drosophila simulans* involving nucleotide substitutions at *Nup96* might represent an example (Presgraves & Stephan 2007). The gene interacts with an unknown factor on the X chromosome to create sterility of F1 hybrid males. It codes for a protein in the nuclear pore complex. Excessive non-synonymous substitutions in the coding sequence of the gene indicate that *Nup96* was under selection, but as it functions in the nuclear pore complex it is not easy to see how a change of environment could be the driving mechanism.

Divergent sexual selection may also drive the evolution of reproductive isolation (Lande 1981; Gavrilets 2004). Comparative evidence suggests that sexual selection indeed plays a role in generating species diversity. For example, sexual dichromatism in bird clades is positively correlated with number of species in the clade (Barraclough *et al.* 1995). Beta, rather than alpha diversity, is mainly affected (Price 1998). Direct evidence for a primary role of sexual selection in speciation is more equivocal (Panhuis *et al.* 2001; Ritchie 2007).

Mechanisms of sexual selection can be grouped under the ecological or non-ecological categories of speciation mechanism depending on the ultimate cause of genetic divergence in the mating preferences. For example, mating signal transmission may be strongly affected by environment, which can bring about the evolution of divergent signals between populations inhabiting different environments (Endler 1992; Boughman 2002). A powerful example of this interaction between natural and sexual selection is provided by Lake Victoria cichlids in the genus *Pundamilia* (Seehausen, this volume). This case is particularly significant because previous explanations for the exceptionally rapid diversification of these fish have put the central driving role on sexual selection, whereas divergent natural selection was overlooked (Turner & Burrows 1995). In general, signal-response systems are likely to diverge along with ecological divergence rather than independently (but see Irwin *et al.* 2008). On the other hand, divergent mating preferences arising from sexual conflict can happen independently of the environment. It is difficult to find clear-cut examples in nature but experimental manipulation of sexual conflict in laboratory populations of *Sepsis* dung flies showed increased isolation in response to high levels of conflict (Martin & Hosken 2003). The process of sexual conflict might still be influenced by the environment, for example through the impact of population density on remating rate.

The possibility of speciation without selection, by mutation and genetic drift alone, is hotly debated. There can be no doubt that, given sufficient time and strong enough extrinsic barriers, mutation and genetic drift will result in speciation through the fixation of incompatible alleles in separate populations. However, the time required is probably too long for this mechanism of speciation to be a major contributor to biodiversity. Much more attention has focused on the role of drift in very small populations that colonize new areas: the 'founder effect'. According to theory there are serious obstacles to founder-effect speciation (Barton & Charlesworth 1984). In addition, colonization of new areas is very likely to involve a habitat difference and hence strong selection. It is therefore difficult to separate effects of drift and selection in driving divergence and speciation following colonization of new environments. The evidence for a role of founder effects in nature is weak (Coyne & Orr 2004). For example, the radiation of Hawaiian *Drosophila* has been thought of as an example of rapid speciation by repeated founder events. Yet, genetic diversity within these *Drosophila* species is typically high, which seems inconsistent with repeated bottlenecks during speciation events (Hunt *et al.* 1989).

Ecology can play a role in driving speciation apart from its influence on selection. For example, successful colonization of new areas, a prerequisite for allopatric speciation, might be easier if diverse resources are present and few other species utilize them (Mayr 1942; Phillimore & Price, this volume). At the same time, reproductive interference might prevent species from coexisting even if they are ecologically different (Goldberg & Lande 2006). This is clear from the large number of sharp parapatric boundaries between populations at hybrid zones (Barton & Hewitt 1985). For example, incomplete assortative mating and low hybrid fitness may prevent sympatry in *Bombina* toads despite clear ecological differences (Vines *et al.* 2003). In other hybrid zones, the interacting populations have no detectable ecological differentiation and yet show significant pre- and postzygotic isolation. For example, in the meadow grasshopper *Chorthippus parallelus* there is some evidence that the colonization process promoted the evolution of incompatibilities (Butlin 1998; Tregenza *et al.* 2000, Tregenza *et al.* 2002).

Generation of distinct and persistent taxa that coexist is clearly a crucial endpoint of speciation from the point of view of diversity patterns. Speciation driven by divergent selection is likely to make a qualitatively different contribution to diversity patterns compared with speciation driven by other 'nonecological' forces, including drift, some forms of sexual selection, and divergent response to uniform natural selection. Ecological speciation generates species that have distinct ecological roles, in contrast to other forms of speciation. Perhaps the ecological mechanisms of speciation are also more likely to occur in the presence of high ecological opportunity and to be dependent on the number of species already present in the community. Therefore, it would be helpful to understand the contributions of different mechanisms to speciation rates and their variation. Unfortunately, as Coyne and Orr (2004) have emphasized in their thorough review of the literature, this is a question about which we know rather little. Until recently, empirical speciation research has concentrated on case studies and one can find well-supported examples of many different modes of speciation. The approach of Barraclough and Vogler (2000) to sympatric and allopatric speciation is a notable exception, although it is not clear if a similar approach can distinguish speciation mechanisms. Recently, comparative analysis of diversification rates has provided an alternative approach to finding general patterns and one that clearly has the potential to forge links between speciation process and macroecological pattern.

## Biodiversity and speciation rates

Speciation is the ultimate source of new species, in the same way that mutation is the ultimate source of genetic variation within species (and extinction is analogous to loss of alleles). Inequities in the rates of speciation are thus likely to contribute to large scale biodiversity patterns. It has often been proposed, in cases where some parts of the globe have excessive numbers of species, that the taxa in those regions have experienced unusually high speciation rates. Conversely, regions of the globe with fewer species are inferred to be speciation-limited.

For example, to explain the latitudinal gradient in species diversity, Dobzhansky argued that speciation rates were higher in the tropics than in the temperate zone, because of the greater opportunity for co-evolution (Schemske, this volume). However, diversity is the outcome of the *difference* between speciation and extinction, not just speciation alone. New estimates of speciation rates in birds and mammals in the recent past suggest that they are actually highest at temperate latitudes (Weir & Schluter 2007), even while net diversification appears to be highest in the tropics (Cardillo 1999; Ricklefs 2006). The implication is that lower extinction, not higher speciation, is behind the faster accumulation of species in the tropics (Weir & Schluter 2007). On the other hand, these estimated rates might be a temporary outcome of the turmoil of the Pleistocene glaciations, and speciation rates may really have been higher in the tropics over the longer term. This example indicates that the extent to which diversity patterns reflect speciation rates is an open question for research.

It is possible to model how speciation rates might impact biodiversity patterns. In his 'neutral theory' Hubbell (2001) showed that metacommunity diversity depends only on community size (the total number of individuals) and speciation rate. Extinction occurs as a consequence of stochastic variation in population sizes, which are on average lower the more species are present. In these models, quantitative features of the species-abundance relationship, and species' longevities, depend on the mode of speciation: if new species result from random fission of the ancestral species then they begin with relatively large population sizes and so have relatively long persistence times. Under the alternative 'point mutation' mode of speciation, each new species begins as a single individual, in which case they are expected to be very short-lived. In response to criticism by Ricklefs (2003) that these extreme patterns are unrealistic, Hubbell and Lake (2003) introduced a third mode of speciation into the Hubbell neutral model, called 'peripheral isolate speciation'. This allows speciation to occur in an isolate of variable population size and has two parameters: the mean and the variance of isolate size. By varying these parameters, a wide range of species-abundance relationships and persistence distributions can be obtained, thus overcoming Ricklefs' objection.

This exchange is instructive for two reasons. First, it demonstrates that, in the simplest of models, both the rate and the mode of speciation influence patterns of diversity. Second, it emphasizes our ignorance because even these models require parameters for which we have few, if any, good estimates (the rate of speciation and the mean and variance of the population size in which speciation is initiated). Without limits on these parameters, the model lacks heuristic value because it can produce too wide a range of outcomes. This is partly why Chave (2004) concluded that neutral models are unlikely to benefit from more complex models of speciation.

Neutral models are based on the neutrality assumption: all species are assumed to be ecologically equivalent, which is the expectation only if all speciation is non-ecological (and species fail to diverge ecologically post-speciation). But if speciation is commonly ecological (or species diverge ecologically post-speciation) then species persistence is affected by niche differences, and the stage is set for a much more complex set of interactions between the mode and rate of speciation on the one hand and patterns of diversity on the other. The first steps in unravelling these interactions must be to document speciation rates and, where they vary, test for ecological correlates of that variation.

Speciation rates do seem to be highly variable across taxa (Fig. 1.1). One can imagine many conditions that might favour local adaptation and so increase the

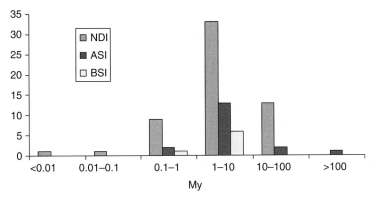

Figure 1.1 Variation in speciation intervals among taxa. NDI is 'net diversification interval', which is the average time between the origin of a lineage and the next branching event on that lineage assuming a pure birth model (i.e. without extinction); ASI is 'apparent speciation interval', which includes a correction for extinction, estimated either from a phylogeny or from fossil data; and BSI is 'biological speciation interval', estimated from extant taxa by determining divergence times between sister taxa or extrapolating the observed increase in reproductive isolation with time to the point of complete isolation. Data from Coyne and Orr (2004), Table 12.1. See their discussion, pp. 416–425, for further details.

rate of ecological speciation, such as new ecological opportunities, selection along multiple dimensions (Nosil & Harmon, this volume), high productivity, wide dispersal, or fragmented habitats (especially host associations). Conversely, there are clearly factors that limit the ability of populations to respond to selection, especially where the environment changes too quickly in time or space (Bridle *et al.* this volume). Diversity itself is one potential driver for speciation, although one can make arguments in either direction: high diversity creates more ways of making a living and so favours speciation, but it also tends to decrease population sizes, working against speciation and in favour of extinction. For example, evidence that speciation rate is positively correlated with diversity on islands (Emerson & Kolm 2005) is controversial (Cadena *et al.* 2005). A different set of factors might be expected to enhance, or depress, rates of non-ecological speciation. Large ranges, limited dispersal and environmental change all increase the opportunities for extrinsic isolation and so for accumulation of genetic incompatibilities, for example, while frequent remating increases sexual conflict and so the opportunity for conflict to drive speciation (Arnqvist *et al.* 2000).

However, speciation rates are very difficult to estimate and this makes it hard to test hypotheses about the causes of rate variation. Some of the wide variation in estimated speciation intervals is, undoubtedly, due to methodological limitations. The most problematic of these is disentangling speciation from extinction. This problem exists for interpretation both of the fossil record (Alroy, this volume) and of molecular phylogenies (Phillimore & Price, this volume; Ricklefs, this volume). Estimating speciation and extinction rates requires finding appropriate models for diversification if interesting questions are to be addressed.

For example, Ricklefs (this volume) and Alroy (this volume) both ask whether diversity is at a steady state, with extinction rate approximately equal to speciation rate. Alroy, analysing fossil mammals, demonstrates that alpha diversity is more stable than expected from observed rates of species turnover, suggesting that diversity is bounded. Ricklefs, analysing phylogenetic trees of birds and plants, finds a poor fit across taxa to the expectations of constant birth–death models. He suggests that a steady-state model in which speciation roughly matches extinction fits the data best. These results are inconsistent with previous phylogenetic studies showing that speciation always outpaces extinction, leading to a rise in diversity. Ricklefs suggests that the phylogenetic methods are biased. We tend to work on taxa with a reasonable number of species, which dooms us to find that speciation has outpaced extinction. Declines in diversity are also undetectable in phylogenetic trees. Similar methodological problems plague tests of apparent changes in speciation and extinction rate, such as a decrease in diversification rate towards the present (Phillimore & Price, this volume). These problems need to be overcome before important ecological questions can be addressed.

Comparative methods allow tests for ecological correlates with net diversification rate and many such analyses have been conducted. An early example was the demonstration that the more species rich of sister-pairs of bird families are more likely to be sexually dichromatic, suggesting that sexual selection promotes speciation (Barraclough *et al.* 1995). More recent analyses seek to make simultaneous tests of multiple predictors of diversification rate and to use the information in the phylogeny more efficiently. A powerful recent study by Phillimore *et al.* (2006) demonstrated that birds with more generalist feeding niches and greater dispersal tend to diversify more rapidly. Their models explain more than 50% of the variation in diversification rate among clades and so make a major contribution to understanding patterns of bird diversity. However, attempts to make mechanistic interpretations of these relationships again run up against the problem of separating the contributions of extinction and speciation. Do generalist feeders have high speciation rates because they are exposed to a wide range of habitats or do generalist feeding habits buffer populations against extinction due to resource fluctuations?

## Prospects

This book addresses whether and how mechanisms and rates of speciation affect global biodiversity patterns. In the present chapter, we have introduced some possible connections between diversity and speciation, and raised some of the methodological issues that need to be solved if further progress is to be made. The rest of the book examines these issues in more detail.

Speciation research has made great strides over the past decade in identifying the mechanisms of speciation. But, except perhaps for polyploidy, we still do not have a good idea as to which mechanisms account for most species in nature. Yet, there are reasons to suspect that mechanism would matter a great deal to biodiversity. At the very least, the outcome of speciation should be quite different when it occurs by divergent natural selection between ecological niches than when it results via divergent genetic responses to identical niches and environments. The former mechanism yields ecologically distinct forms that can readily coexist, all else being equal, whereas the other does not. Coexistence should also be affected by the types of reproductive isolation that result. The lack of premating isolation, for example, might prevent coexistence even if postmating isolation is complete. This may explain why even ecologically distinct forms sometimes fail to achieve range overlap, but instead meet at hybrid zones. Such reproductive interference can have dramatic consequences for species distribution and abundance, no less than the more familiar ecological interactions such as competition and predation. Concrete connections between speciation and diversity are still few, but should grow as our understanding of the mechanisms of speciation improves.

The link between biodiversity patterns and speciation rate is straightforward to envision, but perhaps even more difficult to establish. As we have seen, a major impediment is obtaining accurate estimates of speciation rates. Phylogenetic trees will continue to be the main source of data for such estimates because they are more readily available than fossils for the majority of taxonomic groups. Yet it is difficult to remove the effects of extinction on diversification when all we have to work with are the survivors of the process. Further work is needed to reduce the bias in resulting estimates of speciation rate.

## References

Albert, A. Y. K. and Schluter, D. (2004) Reproductive character displacement of male stickleback mate preference: reinforcement or direct selection? *Evolution* **58**, 1099–1107.

Alroy, J. (1998) Cope's rule and the dynamics of body mass evolution in North American fossil mammals. *Science* **280**, 731–734.

Arnqvist, G., Edvardsson, M., Friberg, U. and Nilsson, T. (2000) Sexual conflict promotes speciation in insects. *Proceedings of the National Academy of Sciences of the United States of America* **97**, 10460–10464.

Barluenga, M., Stolting, K. N., Salzburger, W., Muschick, M. and Meyer, A. (2006) Sympatric speciation in Nicaraguan crater lake cichlid fish. *Nature* **439**, 719–723.

Barraclough, T. G., Harvey, P. H. and Nee, S. (1995) Sexual selection and taxonomic diversity in passerine birds. *Proceedings of the Royal Society of London Series B-Biological Sciences* **259**, 211–215.

Barraclough, T. G. and Vogler, A. P. (2000) Detecting the geographical pattern of speciation from species-level phylogenies. *American Naturalist* **155**, 419–434.

Barton, N. H. and Charlesworth, B. (1984) Genetic revolutions, founder effects, and speciation. *Annual Review of Ecology and Systematics* **15**, 133–164.

Barton, N. H. and Hewitt, G. M. (1985) Analysis of hybrid zones. *Annual Review of Ecology and Systematics* **16**, 113–148.

Blackburn, T. M. and Gaston, K. J. (2003) *Macroecology: Concepts and Consequences.* Cambridge University Press, Cambridge.

Boughman, J. W. (2002) How sensory drive can promote speciation. *Trends in Ecology & Evolution* **17**, 571–577.

Butlin, R. (1998) What do hybrid zones in general, and the *Chorthippus parallelus* zone in particular, tell us about speciation? In: *Endless Forms: Species and Speciation* (ed. D. J. Howard and C. H. Berlocher), pp. 367–378. Oxford University Press, New York.

Butlin, R. K. (2005) Recombination and speciation. *Molecular Ecology* **14**, 2621–2635.

Cadena, C. D., Ricklefs, R. E., Jimenez, I. and Bermingham, E. (2005) Ecology – is speciation driven by species diversity? *Nature* **438**, E1–E2.

Cardillo, M. (1999) Latitude and rates of diversification in birds and butterflies. *Proceedings of the Royal Society of London Series B-Biological Sciences* **266**, 1221–1225.

Chave, J. (2004) Neutral theory and community ecology. *Ecology Letters* **7**, 241–253.

Coyne, J. A. and Orr, H. A. (2004) *Speciation.* Sinauer Associates, Sunderland, MA.

Emerson, B. C. and Kolm, N. (2005) Species diversity can drive speciation. *Nature* **434**, 1015–1017.

Endler, J. A. (1992) Signals, signal conditions, and the direction of evolution. *American Naturalist* **139**, S125–S153.

Fisher, R. A., Corbet, A. S. and Williams, C. B. (1943) The relation between the number of species and the number of individuals in a random sample of an animal population. *Journal of Animal Ecology* **12**, 42–58.

Gaston, K. J. and Blackburn, T. M. (2000) *Pattern and Process in Macroecology*. Blackwell Science, Oxford.

Gavrilets, S. (2004) *Fitness Landscapes and the Origin of Species*. Princeton University Press, Princeton, NJ.

Goldberg, E. E. and Lande, R. (2006) Ecological and reproductive character displacement on an environmental gradient. *Evolution* **60**, 1344–1357.

Hubbell, S. P. (2001) *The Unified Neutral Theory of Biodiversity and Biogeography*. Princeton University Press, Princeton, NJ.

Hubbell, S. P. and Lake, J. (2003) The neutral theory of biogeography and biodiversity: and beyond. In: *Macroecology: Concepts and Consequences* (ed. T. M. Blackburn and K. J. Gaston), pp. 45–63. Cambridge University Press, Cambridge.

Hunt, J. A., Houtchens, K. A., Brezinsky, L., Shadravan, F. and Bishop, J. G. (1989) Genomic DNA variation within and between closely related species of Hawaiian *Drosophila*. In: *Genetics, Speciation and the Founder Prinicple* (ed. L. V. Giddings, K. Y. Kaneshiro and W. W. Anderson), pp. 167–180. Oxford University Press, New York.

Irwin, D. E., Thimgan, M. P. and Irwin, J. H. (2008) Call divergence is correlated with geographic and genetic distance in greenish warblers (*Phylloscopus trochiloides*): a strong role for stochasticity in signal evolution? *Journal of Evolutionary Biology* **21**, 435–448.

Lande, R. (1981) Models of speciation by sexual selection on polygenic traits. *Proceedings of the National Academy of Sciences of the United States of America-Biological Sciences* **78**, 3721–3725.

Levin, D. A. (2002) *The Role of Chromosomal Change in Plant Evolution*. Oxford University Press, Oxford.

Mace, G. M., Gittleman, J. L. and Purvis, A. (2003) Preserving the tree of life. *Science* **300**, 1707–1709.

Martin, O. Y. and Hosken, D. J. (2003) The evolution of reproductive isolation through sexual conflict. *Nature* **423**, 979–982.

Mavarez, J., Salazar, C. A., Bermingham, E., *et al.* (2006) Speciation by hybridization in Heliconius butterflies. *Nature* **441**, 868–871.

Mayhew, P. J. (2002) Shifts in hexapod diversification and what Haldane could have said. *Proceedings of the Royal Society of London Series B-Biological Sciences* **269**, 969–974.

Mayr, E. (1942) *Systematics and the Origin of Species from the Viewpoint of a Zoologist*. Harvard University Press, Cambridge, MA.

McKinnon, J. S., Mori, S., Blackman, B. K., *et al.* (2004) Evidence for ecology's role in speciation. *Nature* **429**, 294–298.

Noor, M. A. F. (2003) Evolutionary biology – genes to make new species. *Nature* **423**, 699–700.

Otto, S. P. and Whitton, J. (2000) Polyploid incidence and evolution. *Annual Review of Genetics* **34**, 401–437.

Panhuis, T. M., Butlin, R., Zuk, M. and Tregenza, T. (2001) Sexual selection and speciation. *Trends in Ecology & Evolution* **16**, 364–371.

Phillimore, A. B., Freckleton, R. P., Orme, C. D. L. and Owens, I. P. F. (2006) Ecology predicts large-scale patterns of phylogenetic diversification in birds. *American Naturalist* **168**, 220–229.

Presgraves, D. C. and Stephan, W. (2007) Pervasive adaptive evolution among interactors of the Drosophila hybrid inviability gene, Nup96. *Molecular Biology and Evolution* **24**, 306–314.

Price, T. (1998) Sexual selection and natural selection in bird speciation. *Philosophical Transactions of the Royal Society of London Series B-Biological Sciences* **353**, 251–260.

Purvis, A., Orme, C. D. L. and Dolphin, K. (2003) Why are most species small-bodied? A phylogenetic view. In: *Macroecology: Concepts and Consequences* (ed. T. M. Blackburn and K. J. Gaston), pp. 155–173. Cambridge University Press, Cambridge.

Ricklefs, R. E. (2003) Genetics, evolution, and ecological communities. *Ecology* **84**, 588–591.

Ricklefs, R. E. (2006) Global variation in the diversification rate of passerine birds. *Ecology* **87**, 2468–2478.

Rieseberg, L. H., Kim, S. C., Randell, R. A., *et al.* (2007) Hybridization and the colonization of novel habitats by annual sunflowers. *Genetica* **129**, 149–165.

Ritchie, M. G. (2007) Sexual selection and speciation. *Annual Review of Ecology, Evolution, and Systematics* **38**, 79–102.

Rundle, H. D., Nagel, L., Boughman, J. W. and Schluter, D. (2000) Natural selection and parallel speciation in sympatric sticklebacks. *Science* **287**, 306–308.

Rundle, H. D. and Schluter, D. (1998) Reinforcement of stickleback mate preferences: sympatry breeds contempt. *Evolution* **52**, 200–208.

Savolainen, V., Anstett, M. C., Lexer, C., *et al.* (2006) Sympatric speciation in palms on an oceanic island. *Nature* **441**, 210–213.

Schluter, D. (2001) Ecology and the origin of species. *Trends in Ecology & Evolution* **16**, 372–380.

Tregenza, T. and Butlin, R. K. (1999) Genetic diversity: do marker genes tell us the whole story?. In: *Evolution of Biological Diversity* (ed. A. E. Magurran and R. M. May), pp. 37–55. Oxford University Press, Oxford.

Tregenza, T., Pritchard, V. L. and Butlin, R. K. (2000) The origins of premating reproductive isolation: testing hypotheses in the grasshopper *Chorthippus parallelus*. *Evolution* **54**, 1687–1698.

Tregenza, T., Pritchard, V. L. and Butlin R. K. (2002) The origins of postmating reproductive isolation: testing hypotheses in the grasshopper *Chorthippus parallelus*. *Population Ecology* **44**, 137–144.

Turner, G. F. and Burrows, M. T. (1995) A model of sympatric speciation by sexual selection. *Proceedings of the Royal Society of London Series B-Biological Sciences* **260**, 287–292.

Via, S. (2001) Sympatric speciation in animals: the ugly duckling grows up. *Trends in Ecology & Evolution* **16**, 381–390.

Vines, T. H., Kohler, S. C., Thiel, A., *et al.* (2003) The maintenance of reproductive isolation in a mosaic hybrid zone between the fire-bellied toads *Bombina bombina* and *B. variegata*. *Evolution* **57**, 1876–1888.

Vines, T. H. and Schluter, D. (2006) Strong assortative mating between allopatric sticklebacks as a by-product of adaptation to different environments. *Proceedings of the Royal Society B-Biological Sciences* **273**, 911–916.

Weir, J. T. and Schluter, D. (2007) The latitudinal gradient in recent speciation and extinction rates of birds and mammals. *Science* **315**, 1574–1576.

CHAPTER TWO

# On the arbitrary identification of real species

JODY HEY

## Introduction

The detection of a new species is the result of a decision-making process, one that has traditionally and primarily been based upon the discovery of distinguishing characters (Cronquist 1978; Mayr 1982; Winston 1999; Sites & Marshall 2004). This process, called diagnosis, is acutely important as it necessarily lies at the crux of the discovery of biodiversity, including the identification of conservation units. With the rise of quantitative phylogenetic methods, and the increasing accessibility of molecular data, numerous methods for diagnosis have been proposed in recent years (Sites & Marshall 2004).

For many situations the identification of a new species does not particularly require quantitative methodology. These are the cases where the organisms of the putative new species are conspicuously divergent from all known species and, thus, where the new species is identified as a sister taxon to previously identified groups. But increasingly, as more new species are described and as more species are the subject of additional investigation, the questions of diagnosis arise within previously described species, wherein patterns of differentiation among populations of the same species must be interpreted in taxonomic terms. In short: how do we decide when a closer look at one taxonomic species actually reveals the presence of more than one species?

A similar question arises in conservation contexts: do the data from one species (or one conservation unit) actually reveal the presence of multiple units, each of which merit recognition and possibly protection? This question may not be cast in terms of the taxonomic rank of species, and so the criteria used for diagnosis of a new species may differ from those used for diagnosing a population in terms of meriting conservation status. For example, in the United States the language of the amended Endangered Species Act (16 USC §§1531–1544) refers to a 'distinct population segment of any species of vertebrate fish or wildlife which interbreeds when mature'. In effect this conservation policy has identified a taxonomic category of 'distinct population segment' or 'DPS' as it is called in discussions on biodiversity conservation, one for which criteria and protocols for diagnosis have been much discussed (Waples 1991a; Moritz 1994;

*Speciation and Patterns of Diversity*, ed. Roger K. Butlin, Jon R. Bridle and Dolph Schluter. Published by Cambridge University Press. © British Ecological Society 2009.

Vogler & DeSalle 1994; Waples 1995; Pennock & Dimmick 1997; Dimmick *et al.* 1999; Goldstein *et al.* 2000).

This chapter addresses basic questions about quantitative approaches to the diagnosis of closely related taxa, such as those that might be applied when a single variable species is studied, and where investigators are faced with the question of whether or not different populations of one species each merit some taxonomic status. It is shown how most quantitative methods for determining the taxonomic status of populations base the taxonomic decision on a finding of statistically detectable divergence, and thus base the decision inherently upon the sample size of the study.

## On the nature of natural populations

Given that most described species consist of multiple partly differentiated populations (Hughes *et al.* 1997), how might we consider the reality or the objectivity of local populations? Consider as a starting point a single species taxon which possibly includes multiple populations that merit their own taxonomic status. One possible observation is that the different populations are found to be clearly differentiated from one another on the basis of multiple readily observable characters. In such cases, the diagnosis of additional taxa is likely to be supported by any diagnostic method. At the other extreme it is possible that no geographic, morphological or genetic differentiation is observed. Such a case would suggest the presence of a single species and would not meet the diagnostic criteria for additional taxa under any method. It is in between these two extremes, of strong differentiation on the one hand and of random mating with an absence of differentiation on the other hand, where lies the wilderness of taxon diagnosis. This difficult territory includes most of the millions of species that occur as structured populations. In this context 'population structure' is a catch-all that includes a wide array of phenomena, including: hierarchical patterns of population structure, multiple separate populations, isolation by distance and metapopulation dynamics.

Importantly for conspecific populations, all of the processes that give rise to population structure are expected to vary continuously – they are not 'all or none' phenomena. Even locally fixed alleles, which are restricted to one or a subset of populations, will occur effectively as a matter of degree in many situations. Such fixations can occur by chance, or by selection, and they can occur anywhere in the genome. For populations that have been separated for some time with limited gene flow, private alleles are likely to be present to some degree somewhere in the genome. In short, population differentiation is expected to vary continuously in complex ways.

The key corollary of these points is that we do not have a basic expectation that evolution will generate a tangible threshold or trigger point of differentiation, beyond which new taxa are unambiguously diagnosable. Finally, it follows necessarily that there can be no objective criterion for diagnosis over a broad range of levels and patterns of population differentiation.

These points are commonplace observations and are really just a roundabout way of affirming Darwin's point of 'how entirely vague and arbitrary is the distinction between species and varieties' (Darwin 1859, p. 48). If we are going to give some taxonomic status to varieties, whether they are species or subspecies or even some smaller conservation-based taxonomic rank, then this continuum of differentiation will be a dominating fact of the matter for investigators involved in diagnosis. The general and inherent ambiguity is well summarized by Sites and Marshall (2004):

> there is no objective criterion for how much morphological divergence is enough to delimit a species, what threshold frequency of intermediates is needed to delimit species by genotypic clusters (Mallet 1995), what proportion of unlinked loci are needed to delimit coalescent species (Hudson & Coyne 2002), or what frequency cut-off most appropriately indicates that no significant gene flow is occurring between populations.
>
> (Wiens & Servedio 2000)

### The challenges of using hypothesis tests for species diagnosis

It is common to propose that species be diagnosed in a hypothesis-testing framework (Mallet 1995; Sites & Crandall 1997; Wiens & Servedio 2000; Templeton 2001; DeSalle *et al.* 2005). Of the many methods proposed for diagnosing species in recent years, nearly all are cast in either the general language of hypothesis testing or explicitly as a mathematical and statistical method of hypothesis testing (Sites & Marshall 2004). Consider the following conventional hypotheses:

Null model = one single species = no significant differentiation
Alternative model = two (or more) species = a statistical finding of differentiation between populations

Take note that the alternative hypothesis explicitly equates a statistically significant finding of differentiation with the presence of more than one species. Now recall the reality that many species actually do consist of multiple partly differentiated populations. Suppose that these two hypotheses are to be considered in light of typical real world data from two related but partly differentiated populations. It is possible that small samples drawn from multiple populations of the same species might not reveal differentiation. However, as the sample size grows, some level of differentiation between the samples will be found to be greater than what could be expected by chance. In other words, we might expect to reject the null hypothesis for virtually any situation given enough data. In effect, the decision (one species or more than one species) is expected to be a function of the sample size. Figure 2.1 demonstrates this general relationship between sample size and a conclusion of multiple species.

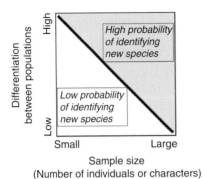

Figure 2.1 The likelihood of species diagnosis is described as a function of both sample size and divergence between populations. Studies with small sample sizes are unlikely to reveal a statistically significant finding, regardless of the degree of divergence, whereas large studies will find divergence even when it is slight. Areas above and to the right of the line will be associated with high rates of species diagnosis, whereas areas below and to the left will have low rates of species diagnosis.

The source of this difficulty lies in the specific choice of the alternative hypothesis which states that additional species are recognized with a statistical finding of *any* differentiation, no matter how slight it is. Since virtually all local populations will be somewhat differentiated from others, this particular hypothesis framework is a recipe for finding as many species as one might have resources for sampling.

Sites and Marshall (2004) reviewed a dozen methods, most of which exhibit the property described here. For example, Good and Wake (1992) proposed a test for assessing whether divergence is due to more than isolation by distance, using a regression of genetic distances among populations against geographic distances. If a significant non-zero intercept is observed, then isolation by distance is rejected. Clearly the more data there is the more likely a finding of a non-zero intercept. In this context 'more data' means more pairs of populations.

Under Population Aggregation Analysis (PAA) (Davis & Nixon 1992), populations are grouped together if they share character states. Successive aggregation of populations is done until only those populations that are separated by fixed character state differences remain. Clearly the more characters or genes that are studied, the greater the chance that one of them will show a fixed difference distinguishing more populations (Wiens 1999; Yoder *et al.* 2000).

Templeton proposed a variation of Nested Clade Analysis (NCA) (Templeton 2001) for identifying species. Like other applications of NCA (Templeton *et al.* 1995; Templeton 1998) the method begins with a statistical test of association of geographic distribution with phylogenetic patterns. Once again, larger sample sizes will mean that a significant association is more likely to be observed.

### Methods having a bias in the reverse direction

Another set of methods also exhibit a direct dependency on sample sizes, but exhibit that effect in the opposite direction such that smaller samples are more likely to support the alternative hypothesis and lead to a rejection of the null hypothesis of a single species. These methods employ the same general null and alternative hypotheses as described above; however, they treat the variation that is observed in a sample with what is observed in an entire population. It is

| Population 1 individuals | Characters |
|---|---|
| | A B C D E F G H |
| #1 | 1 0 1 0 1 1 0 1 |
| #2 | 1 1 1 0 0 0 1 0 |
| #3 | 0 1 1 0 0 1 0 1 |
| #4 | 1 1 1 0 1 0 1 0 |
| #5 | 1 0 1 0 1 0 0 1 |
| #6 | 1 0 1 0 1 0 1 1 |
| #7 | 0 0 1 0 1 0 1 1 |
| #8 | 1 0 1 0 1 0 1 1 |
| Population 2 individuals | |
| #1 | 1 0 1 0 0 0 1 1 |
| #2 | 1 0 1 0 0 1 1 0 |
| #3 | 1 0 1 0 0 1 1 0 |
| #4 | 1 0 1 0 0 1 1 1 |
| #5 | 1 0 1 0 0 1 0 0 |
| #6 | 1 1 1 0 0 0 1 1 |
| #7 | 1 0 1 0 0 1 1 1 |
| #8 | 1 1 1 0 0 1 0 1 |

Figure 2.2 An example of the effect of sample size on the discovery of fixed differences between populations. Eight characters are shown for samples from two populations, with no character showing a fixed difference in the full sample. However, smaller samples (grey areas) reveal fixed differences in characters E and F.

possible then to observe a pattern of variation in a sample – such as a fixed difference between populations – that would not be observed in a larger sample. Paradoxically then, such methods are more likely to lead to conclusions of multiple species when sample sizes are small.

The method of PAA (Davis & Nixon 1992) suffers from this problem precisely because it implicitly assumes that character values, which are observed as fixed within samples, are also fixed in species (Wiens 1999; DeSalle & Amato 2004). The problem is demonstrated in Fig. 2.2 which shows a hypothetical data set of eight binary characters for samples of size eight for each of two populations. As constructed the sample from two populations now show fixed differences and would be combined in a PAA analysis. But also highlighted in Fig. 2.2 are smaller samples which do reveal fixed differences (at characters E and F).

This kind of sample size effect will also occur under the method sketched out by Baum and Shaw (1995) to identify genealogically exclusive groups. This approach is based on gene tree estimates obtained for multiple loci, in which multiple sequences have been obtained from each population under consideration and the trees are combined into a strict consensus tree. Any clusters of individuals that are present on the multi-locus strict consensus tree are identified as exclusive groups and meet the species criterion. But clearly the more individuals sampled per population, and the more loci studied, the less likely that a strict consensus tree will reveal exclusive groups.

## Estimator bias and consistency

By considering the decision process in a slightly different light we can perceive the dependency of species diagnosis on sample size as a kind of estimator bias. By definition, estimator bias is the difference between the true value of a parameter and the expected value of an estimate of that parameter. If in the present case the parameter is the true number of species that actually exist (under some criterion), then we would hope that the expected value of an estimate of that value would be close to the true value. But if the expected value changes with the sample size, then clearly there must be a bias, at least over some ranges of sample size.

The property of being biased is not necessarily a large problem for an estimator, particularly if the bias is small. Certainly many estimators in many different contexts have some degree of bias. Also, it is common for estimators who are biased to be *consistent*, meaning that the bias becomes less as the sample size grows, so that with larger sample sizes the expected value of the estimate converges on the true value of the parameter. However, the bias that is described here arises because the estimation procedure is directly equated with hypothesis testing and thus with statistical power. This necessarily creates a strong and direct link between the finding of additional species and the sample size. For those species that consist of populations with complex and varying degrees of differentiation, we might expect that the number of detected species will continue to increase with sample size almost indefinitely as larger and larger samples reveal finer and finer, but still detectable, patterns of differentiation.

Finally, it bears noting that this kind of bias is not one that is associated with any particular species concept, except insofar that some species concepts do not, or have not yet, lent themselves to being cast in terms of a hypothesis-testing framework. Any species diagnostic protocol in which the presence of additional species is equated simply with a $p$ value (i.e. probability of a Type 1 error), or the equivalent thereof, can be expected to suffer this difficulty.

## Overcoming sample size effects using cut-off criteria

One way to partly overcome a direct dependence of species diagnosis on sample size is to require that the degree of differentiation between populations (by whatever method is being used) pass some particular threshold, or cut-off, value. Methods like this can also be, and have been, cast in explicit hypothesis-testing frameworks. In such cases the general hypotheses are

Null model = one single species = differentiation does not significantly exceed the previously specified threshold value

Alternative model = two (or more) species = a finding of differentiation between populations that does significantly exceed the threshold value

Cut-off values have been proposed in several different ways. For example, Porter (1990) proposed that an estimated population migration rate ($Nm$, the product of the effective population size and the migration rate per generation) which is less than 0.5 was likely to be an indicator of populations that are in fact diverging from one another and that may merit some taxonomic status. Similarly Highton proposed that a genetic distance value greater than 0.15, for Nei's $D$ measure of allozyme distance (Nei 1972), was a likely indication of species status, based on the observed distribution of $D$ for known species (Highton *et al.* 1989; Highton 1990). Wiens and Servedio (2000) suggested that rather than requiring species to have fixed differences, which is difficult to assess without complete sampling, a cut-off frequency of 0.05 be used as part of a statistical test of species status. Hebert *et al.* (2004) suggested that a difference between DNA sequences of the *Cytochrome Oxidase 1* mitochondrial gene, which is greater than ten times that found within species, would be a useful indication of species status.

However, cut-off values raise a new set of issues. One is that the index or indices of differentiation that are used, whatever they may be, must be well motivated by our understanding of the process of divergence. Such a motivation could possibly come from any of several sources, including evolutionary models of divergence, population genetic theory or particular species concepts. Another general concern that arises for any particular method based on cut-off values is that it may not be suitable for different taxonomic groups that necessarily vary in the ways that populations tend to be structured and in the ways that speciation occur (DeSalle *et al.* 2005). It would be better if the cut-off methods developed and used actually make sense for a wide range of taxonomic groups.

Another possible concern is that cut-off values must be partly arbitrary. If a protocol is designed with a divergence criterion $x$, then it follows that more new species are going to be identified than would be the case if $2x$ were the criterion. Given the inherent continuous nature of divergence among populations within species, this is unavoidable. However this component of arbitrariness may be difficult to accept in contexts where species are thought of as inherently fundamental and unitary.

## The insufficiency of overall summaries of differentiation

In the continuum of degrees of differentiation among populations, divergence is a complex process that is not likely to be well encapsulated by any single index or parameter. Consider the issues that arise in the use of $F_{ST}$, Wright's (1951) fixation index. $F_{ST}$ is readily estimated for most kinds of genetic variation and it is a ubiquitous feature of studies on population structure. Under a model in which population structure is at an equilibrium of genetic drift and gene flow, $F_{ST}$ can be used to estimate a population gene flow level. However, $F_{ST}$ is calculated just as easily for any pair of populations, regardless of the role of gene flow in their divergence and regardless of whether divergence is at equilibrium (Whitlock & McCauley 1999).

It is useful to consider $F_{ST}$ with respect to two complementary population genetic models. If two populations are exchanging genes at some level for a long period of time, then they will approach an equilibrium level of $F_{ST}$ and a corresponding equilibrium level of divergence. Under such situations, $F_{ST}$ can be used to estimate the population migration rate $Nm$ (Wright 1951; Slatkin & Voelm 1991). But now consider a radically different model in which two populations separate out of a single population and exchange zero migrants thereafter. At any point in time after the split there will be some level of divergence, and $F_{ST}$ can be calculated and will reflect that divergence. Similarly it is possible to estimate the time of splitting using $F_{ST}$, assuming zero gene flow (Takahata & Nei 1985). From these two examples we see that $F_{ST}$ can be used to estimate entirely different quantities, migration rate and splitting time, under two radically different models – one of which predicts that divergence will increase and the other which has divergence at an unchanging equilibrium. Clearly $F_{ST}$ by itself cannot be a suitable measure of differentiation upon which to base a cut-off value for species diagnosis, nor is the problem limited to $F_{ST}$. Because of the very different nature of population splitting time and gene flow, it is likely that no single measure of divergence can capture both of these key aspects of the divergence process.

### Considering model-based approaches

Notwithstanding the complexity of the divergence process, it is possible to capture many of the dynamics of divergence in a quantitative model. Figure 2.3 shows the Isolation with Migration model (Wakeley & Hey 1998; Nielsen & Wakeley 2001; Hey & Machado 2003), which is intended to represent the divergence history of a pair of sister populations or species. The model includes six parameters, including: the time when the populations separated from one another; the effective population sizes N, for the two populations; the effective population size for the ancestral population (before the splitting time); and two unidirectional migration rates. Together these parameters capture many of the demographic components of the divergence process. For example, divergence will be reduced if genetic drift is slow due to large effective population sizes; if the time of population splitting was recent; or if migration rates are high.

Clearly, the Isolation with Migration model cannot capture many things. In particular it does not contain any parameters that correspond to population-specific adaptation. However, the model can indirectly inform on adaptation in

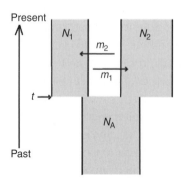

Figure 2.3 The Isolation with Migration model.

some circumstances. For example, if two populations show evidence of gene flow, then this elevates the relevance of any additional information on differential adaptation. Since only a small amount of gene flow is needed to retard divergence, the presence of phenotypic differentiation *together* with evidence of gene flow can indicate that the phenotypic differences are indeed under divergent selection (Machado *et al.* 2002; Bull *et al.* 2006).

## Suggestions for criteria for species diagnosis

Any discussion of numerical cut-off values for taxon diagnosis requires a motivating argument that justifies some particular index of differentiation and some particular cut-off value. We can use as a motivating claim the idea that putative new taxa should be evolutionarily independent to a sufficient degree for them to be expected to continue to diverge. This is a claim similar to that for 'independent evolutionary trajectories', which was proposed for a well-used (and much discussed) concept of an 'evolutionary significant unit' or ESU (Waples 1991b), but one which is also consistent with or implicit in many species concepts (Mayden 1997).

Given the roles that population size, population splitting time and migration all play in the divergence process, it seems unlikely that a single numerical criterion could be used as an indication of whether or not divergence is likely to continue. However, some basic population genetic theory does tell us that we can expect gene flow to be low for diverging populations, because a population migration rate of $Nm$ greater than 1 is sufficient to strongly limit divergence (Wright 1931). Therefore, one choice of a cut-off value would be that $Nm$ be less than 1. However, estimating gene flow also requires that confounding factors, such as the time of population splitting, be accounted for (Whitlock & McCauley 1999).

An estimate of the time since two populations split can also be a useful indicator of their evolutionary independence, as the greater the time the more opportunity there has been for fixation of alleles, including selected alleles, in individual populations. Also populations that split many generations ago have necessarily stayed at least partly separate throughout whatever other demographic vagaries have occurred in the history of the populations since that time. The trouble with splitting time, as an indicator of divergence, is that it cannot be used directly if it is expressed in units of years. The genetics of the divergence process plays out on a timescale of generations, so that it would make sense for a splitting time cut-off to be expressed in units of generations. A related point is that divergence time, even when cast in terms of generations, will matter more for small populations than for large populations. This is because small populations experience rapid genetic drift and will fix alleles more rapidly than will large populations. Thus it would be useful to express a splitting time criterion on a scale of generations, and in a way that reflects effective population size. In fact this kind of timescale is regularly used in population genetic models of divergence, where the time parameter is literally given on a scale of 'effective population size generations'. If we let $\tau$ be the time since separation, scaled by effective population size and generation, then a scaled time of 1 (i.e. $\tau = 1$) would mean that $N$ generations had passed. For example in a *Drosophila* species with 10 generations per year and an effective size of 1 million, a value of $\tau = 1$ corresponds to $0.1 \times 1\,000\,000 = 100\,000$ years.

### An example using specific migration and splitting time cut-offs

From the arguments given above we can consider a species diagnostic protocol with two criteria:

1. The time since splitting, in units of $N_e$ generations, $\tau$, should be significantly greater than 1 for each of the populations.
2. The population migration rate in each direction with the most closely related population should be significantly less than 1.

Together these criteria could be the basis of a statistical approach to species diagnosis. Both criteria 1 and 2 are required because it is not sufficient to simply reject a null hypothesis of a single species on the basis of either divergence time or a low level of gene flow.

Statistical cut-off criteria like these will still necessarily retain a dependency upon sample size. In general small samples are not expected to meet either criterion 1 or 2, regardless of the true values of $\tau$ and $Nm$. However, unlike methods that equate *any* significant non-zero differentiation with species diagnosis, the bias that arises with the use of joint cut-off values should diminish as sample size grows. This diminishing effect of sample size means that the protocol should thereby be statistically consistent. This is because as the sample size grows the estimates of $\tau$ and $Nm$ should become more and more precise, so that it will become clearer whether or not criteria 1 and 2 are met.

For example, consider the case of two subspecies of the common chimpanzee *Pan troglodytes troglodytes* (the central African chimpanzee) and *Pan troglodytes verus* (the western African chimpanzee), the divergence of which were studied using a data set of 48 loci (Won & Hey 2005). The estimated effective sizes were 27 900 for *P. t. troglodytes*, 7 600 for *P. t. verus* and 5 300 for their ancestral population; their splitting time was estimated to be 422 000 years; and a clear signal of gene flow was observed from *P. t. verus* into *P. t. troglodytes*, but not in the reverse direction. Figure 2.4 shows posterior probability density estimates for $\tau$ and $Nm$ in both directions. By calculating the area under the curves with respect to the two criteria listed above, we find that the probability that $Nm \geq 1$ for gene flow into *P. t. troglodytes* is 0.004, while the probability that $Nm \geq 1$ for gene flow into *P. t. verus* is 0.0. Clearly the gene flow criterion is met in this case. Turning to $\tau$, we find that the density curve for *P. t. verus* is far to the right and clearly exceeds 1. However, in

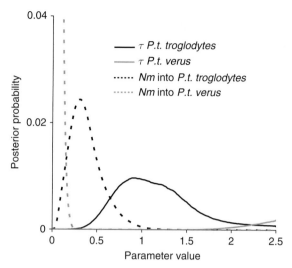

**Figure 2.4** Posterior probability density estimates of $\tau$ and $Nm$. The IM program (Hey & Nielsen 2004) was run using the same data and protocol as in Won and Hey (2005). In that program, the parameters were as follows: $tu$, the product of number of generations since splitting and the mutation rate; $4Nu$, the product of four times the effective population size and the mutation rate; and $m/u$, the ratio of the migration rate per generation and the mutation rate. Since $\tau = 4 \times tu/4\,Nu$ and $Nm = 4\,Nu \times m/u/4$, it was possible to record the necessary quantities over the course of the analysis. For $Nm$ into *P. t. verus* the peak of the curve is at 0, with a probability value that is above the axis limit shown. In the case of $\tau$ for *P. t. verus* the majority of the probability density was far to the right of the portion of parameter value axis that is shown.

the case of $\tau$ for *P. t. troglodytes* we find that the probability that $\tau \leq 1$ is 0.3878. In this case, we cannot reject the null hypothesis of a single species on the basis of criterion 1. Figure 2.4 also shows clearly how the decision-making process can depend strongly upon the selected cut-off values. If instead we use a value of 0.5 for criterion 1, then we find that for *P. t. troglodytes* the probability that $\tau \leq 0.5$ is only 0.0168, which would lead to the rejection of the null hypothesis of a single species and a conclusion that the two subspecies be elevated to species status.

## Conclusion
If the diagnosis of new species is to be based on hypothesis testing then it will be difficult if not impossible to completely remove the statistical bias, such that large samples reveal more new species than do small samples. However, if non-zero cut-off values for well-motivated indicators of divergence are used, then tests using large samples should converge on a robust result of either acceptance or rejection of the null hypothesis.

The indicators of divergence that are presented here, $\tau$ and $Nm$, were selected based on a population genetic understanding of the divergence process. However, they are not the only possible indicators, and some points should be emphasized when indicators are considered:

- No single measure of divergence can be expected to capture all of the key dynamics of the divergence process, including: separation time, gene flow and adaptation.
- Even a large set of divergence measures or estimated parameters is unlikely to fully capture the divergence history in any particular case. Whatever method is used, there will necessarily be a trade-off between the accessibility and simplicity of the diagnostic method and the completeness of the divergence estimates.
- Divergence measures should be selected to be applicable to as wide an array of kinds of organisms as possible. It would be nice if a finding of a new species within one group of organisms conveyed a similar degree of distinction as a finding of a new species in a distantly related group of organisms.

Finally, the use of specific cut-off values does highlight an apparent arbitrariness to species diagnosis. This is simply because the cut-off values must be selected by human investigators and because shifting them will lead to different rates of species diagnosis. This element of arbitrariness is inherent to the diagnosis process whenever closely related populations are being considered, and no quantitative protocol can remove it. However, this does not mean that divergence is not real and objective, and it does not mean that species that are identified do not reflect real divergence and are not, in a similar sense, real and objective.

## Acknowledgements

Thanks to Jack Sites, John Bridle, Jim Mallet and Roger Butlin for comments on the manuscript.

## References

Baum, D. A. and Shaw, K. L. (1995) Genealogical perspectives on the species problem. In: *Experimental and Molecular Approaches to Plant Biosystematics* (ed. P. C. Hock and A. G. Stevenson), pp. 289–303. Missouri Botanical Garden, St. Louis.

Bull, V., Beltran, M., Jiggins, C. D., *et al.* (2006) Polyphyly and gene flow between non-sibling Heliconius species. *BMC Biology* **4**, 11.

Cronquist, A. (1978) Once again, what is a species? In: *Biosystematics in Agriculture* (ed. J. A. Romberger), pp. 3–20. Allanheld & Osmun, Montclair, NJ.

Darwin, C. (1859) *On the Origin of Species by Means of Natural Selection*. Murray, London.

Davis, J. I. and Nixon, K. C. (1992) Populations, genetic variation, and the delimitation of phylogenetic species. *Systematic Biology* **41**, 421–435.

DeSalle, R. and Amato, G. (2004) The expansion of conservation genetics. *Nature Reviews Genetics* **5**, 702–712.

DeSalle, R., Egan, M. and Siddall, M. (2005) The unholy trinity: taxonomy, species delimitation and DNA barcoding. *Philosophical Transactions of the Royal Society B: Biological Sciences* **360**, 1905–1916.

Dimmick, W. W., Ghedotti, M. J., Grose, M. J., et al. (1999) The importance of systematic biology in defining units of conservation units. *Conservation Biology* **13**, 653–660.

Goldstein, P. Z., Desalle, R., Amato, G. and Vogler, A. P. (2000) Conservation genetics at the species boundary. *Conservation Biology* **14**, 120–131.

Good, D. A. and Wake, D. B. (1992) Geographic variation and speciation in the torrent salamanders of the genus Rhyacotriton (Caudata: Rhyacotritonidae). *University of California Publications in Zoology* **126**, 1–91.

Hebert, P. D. N., Stoeckle, M. Y., Zemlak, T. S. and Francis, C. M. (2004) Identification of birds through DNA barcodes. *PLoS Biology* **2**, e312.

Hey, J. and Machado, C. A. (2003) The study of structured populations – new hope for a difficult and divided science. *Nature Reviews Genetics* **4**, 535–543.

Hey, J. and Nielsen, R. (2004) Multilocus methods for estimating population sizes, migration rates and divergence time, with applications to the divergence of *Drosophila pseudoobscura* and *D. persimilis*. *Genetics* **167**, 747–760.

Highton, R. (1990) Taxonomic treatment of genetically differentiated populations. *Herpetologica* **46**, 114–121.

Highton, R., Maha, G. C. and Maxson L. R. (1989) Biochemical evolution in the slimy salamanders of the *Plethodon glutinosus* complex in the eastern United States. *Illinois Biological Monographs* **57**, 1–78.

Hudson, R. R. and Coyne, J. A. (2002) Mathematical consequences of the genealogical species concept. *Evolution* **56**, 1557–1565.

Hughes, J. B., Daily, G. C. and Ehrlich, P. R. (1997) Population diversity: its extent and extinction. *Science* **278**, 689–692.

Machado, C. A., Kliman, R. M., Markert, J. M. and Hey, J. (2002) Inferring the history of speciation from multilocus DNA sequence data: the case of *Drosophila pseudoobscura* and its close relatives. *Molecular Biology and Evolution* **19**, 472–488.

Mallet, J. (1995) A species definition for the modern synthesis. *Trends in Ecology and Evolution* **10**, 294–299.

Mayden, R. L. (1997) A hierarchy of species concepts: the denouement in the saga of the species problem. In: *Species: The Units of Biodiversity* (ed. M. F. Claridge, H. A. Dawah and M. R. Wilson), pp. 381–424. Chapman and Hall, London.

Mayr, E. (1982) *The Growth of Biological Thought.* Harvard University Press, Cambridge, MA.

Moritz, C. (1994) Defining 'evolutionary significant units' for conservation. *Trends in Ecology and Evolution* **9**, 373–375.

Nei, M. (1972) Genetic distance between populations. *American Naturalist* **106**, 283–292.

Nielsen, R. and Wakeley, J. (2001) Distinguishing migration from isolation: a Markov chain Monte Carlo approach. *Genetics* **158**, 885–896.

Pennock, D. S. and Dimmick, W. W. (1997) Critique of the evolutionary significant unit as a defnition for 'distinct population segments' under the U.S. Endangered Species Act. *Conservation Biology* **11**, 611–619.

Porter, A. H. (1990) Testing nominal species boundaries using gene flow statistics: the taxonomy of two hybridizing admiral butterflies (Limenitis: Nymphalidae). *Systematic Zoology* **39**, 131–147.

Sites, J. W. and Crandall, K. A. (1997) Testing species boundaries in biodiversity studies. *Conservation Biology* **11**, 1289–1297.

Sites, J. W. and Marshall, J. C. (2004) Operational criteria for delimiting species. *Annual Review of Ecology Evolution and Systematics* **35**, 199–227.

Slatkin, M. and Voelm, L. (1991) F$_{ST}$ in a hierarchical island model. *Genetics* **127**, 627–629.

Takahata, N. and Nei, M. (1985) Gene genealogy and variance of interpopulational nucleotide differences. *Genetics* **110**, 325–344.

Templeton, A. R. (1998) Nested clade analyses of phylogeographic data: testing hypotheses about gene flow and population history. *Molecular Ecology* **7**, 381–397.

Templeton, A. R. (2001) Using phylogeographic analyses of gene trees to test species status and processes. *Molecular Ecology* **10**, 779–791.

Templeton, A. R., Routman, E. and Phillips, C. A. (1995) Separating population structure from population history: a cladistic analysis of the geographical distribution of mitochondrial DNA haplotypes in the tiger salamander, Ambystoma tigrinum. *Genetics* **140**, 767–782.

Vogler, A. P. and DeSalle, R. (1994) Diagnosing units of conservation management. *Conservation Biology* **8**, 354–363.

Wakeley, J. and Hey, J. (1998) Testing speciation models with DNA sequence data. In: *Molecular Approaches to Ecology and Evolution* (ed. R. DeSalle and B. Schierwater), pp. 157–175. Birkhäuser Verlag, Basel.

Waples, R. S. (1991a) *Definition of 'Species' Under the Endangered Species Act: Application to Pacific Salmon*, p. 29. National Marine Fisheries Service, Seattle, WA.

Waples, R. S. (1991b) Pacific salmon, Oncorhynchus spp., and the definition of 'species' under the Endangered Species Act. *Marine Fisheries Review* **53**, 11–22.

Waples, R. S. (1995) Evolutionarily significant units and the conservation of biological diversity under the Endangered Species Act. In: *Evolution and the Aquatic Ecosystem: Defining Unique Units in Population Conservation* (ed. J. L. Nielsen), pp. 8–27. American Fisheries Society, Bethesda, MD.

Whitlock, M. C. and McCauley, D. E. (1999) Indirect measures of gene flow and migration: FST not equal to 1/(4 Nm + 1). *Heredity* **82** (Pt 2), 117–125.

Wiens, J. J. (1999) Polymorphism in systematics and comparative biology. *Annual Review of Ecology and Systematics* **30**, 327–362.

Wiens, J. J. and Servedio, M. R. (2000) Species delimitation in systematics: inferring diagnostic differences between species. *Proceedings of the Royal Society of London Series B-Biological Sciences* **267**, 631–636.

Winston, J. (1999) *Describing Species*. Columbia University Press, New York, NY.

Won, Y. J. and Hey, J. (2005) Divergence population genetics of chimpanzees. *Molecular Biology and Evolution* **22**, 297–307.

Wright, S. (1931) Evolution in Mendelian populations. *Genetics* **16**, 97–159.

Wright, S. (1951) The genetical structure of populations. *Annals of Eugenics* **15**, 323–354.

Yoder, A. D., Irwin, J. A., Goodman, S. M. and Rakotoarisoa, S. V. (2000) Genetic tests of the taxonomic status of the ring-tailed lemur (Lemur catta) from the high mountain zone of the Andringitra Massif, Madagascar. *Journal of Zoology* **252**, 1–9.

CHAPTER THREE

# The evolutionary nature of diversification in sexuals and asexuals

TIMOTHY G. BARRACLOUGH, DIEGO FONTANETO,
ELISABETH A. HERNIOU AND CLAUDIA RICCI

Species are fundamental units of biology, but there remains uncertainty on both the pattern and processes of species existence. Are species real evolutionary entities or not? If they exist, what are the main processes causing independent evolution and character divergence to occur? This chapter describes how systematic analyses of combined DNA and morphological data can be used to shed light on the evolutionary nature and origin of species. One widely debated test-case has been the question of whether asexual organisms diversify into species; but empirical studies are rare and discussions have often been hampered by allegiance to restrictive species concepts. We present an alternative approach, testing a set of hypotheses for what evolutionary entities might be present in a classic asexual clade, the bdelloid rotifers. Combined analyses of genetic and morphological data reveal the existence of distinct entities conforming to the predicted effects of independent evolution and isolation between sub-lineages. Interestingly, different components of what is meant by 'species' do not strictly coincide. We discuss the applicability of related methods to sexual lineages and to the question of whether higher taxa are real.

## Introduction

Diversity is a fundamental property of the living world. However, despite long interest in the causes of diversification, we still have an incomplete understanding of patterns and processes behind the evolution of biodiversity. Most evolutionary theory has focused around the concept of species – diversity appears to be packaged in distinct units and a wealth of theory recognizes independent evolution as the cause for this phenomenon (Coyne & Orr 2004; Gavrilets 2004). However, in by far most clades, even well-known ones, species are still defined by methods familiar at the time of Linnaeus, without any use of formal quantitative techniques (Sites & Marshall 2003; Rieseberg *et al.* 2006). In addition, independent evolution can take a number of different forms but there has been a tendency to emphasize particular aspects in the quest for a single species concept rather than to break down the 'species question' into its components and test different aspects (Hey *et al.* 2003). There has been growing interest in

*Speciation and Patterns of Diversity*, ed. Roger K. Butlin, Jon R. Bridle and Dolph Schluter. Published by Cambridge University Press. © British Ecological Society 2009.

quantitative methods for species delimitation, but the majority assumes that a simple entity called species exists, encompassing various aspects of diversification, and then focus on a particular signature to delimit those entities (Sites & Marshall 2003).

Several recent accounts have argued for a shift in how biologists think about species (Hey *et al.* 2003; de Queiroz 2005). Rather than searching for a single, unifying species concept, a quest that has proved fruitless to date, they argue that biologists should recognize the different facets of diversity that the term 'species' is used to embrace. For example, Hey (2001) argued that much of the confusion over species concepts concerned failure to distinguish two fundamentally different goals, namely delimiting units of diversity for utilitarian purposes and uncovering the evolutionary processes behind the origin of diverse units. Moreover, it was argued that speciation is a dynamic process in which it is to be expected to observe entities at all stages of the process from populations through to unambiguously separate species. Different features will tend to evolve at different rates to one another and from one speciation event to the next. For example, reproductive isolation (RI) may evolve faster or concomitant with morphological or ecological divergence. Most authors would still assert that species are real biological entities, albeit of an often fuzzy and dynamic nature. However, very few studies have considered the null alternative that significantly discrete and independently evolving entities are not present in a given clade of interest (Rieseberg *et al.* 2006). Moreover, most studies strive to identify a single unit, species, rather than testing whether different aspects of diversity do indeed coincide on a single basic unit.

This chapter shows how these ideas can be put into practice in evolutionary studies of diversification within clades, yielding far greater insights into the patterns and processes of diversification than possible with a classical approach that assumes simple species units. The type of study considered is one in which large samples of individuals are collected for a monophyletic clade. Either species limits are unclear or the pattern of diversity is to be tested. For example, it could comprise a survey of a marine bacterial clade from several sampling points or collections for a genus of plants living in tropical South America. The assumption is that multiple individuals are sampled for a reasonable number of any evolutionarily discrete units that are present.

The approach is to adopt a general evolutionary species concept, namely that species are independently evolving and distinct entities, but then to break the species problem down into a series of testable hypotheses derived from population genetic predictions (Hey *et al.* 2003). We use the word 'entity' to refer to a set of individuals comprising a unit of diversity according to a given criterion or test: the question of whether to call those entities 'species' will be returned to below. Alternative scenarios for the true history of diversification in a clade, including the null scenario of no diversification, can then be tested using combined DNA and morphological data. The approach is illustrated with a case study on bdelloid rotifers, exploring the question of whether asexuals diversify into 'species' or not. Some rather surprising ideas emerge about diversification in sexual organisms and the reality of higher taxa.

## Diversification in asexuals: a brief review

The question of whether asexuals should diversify in similar ways to sexual organisms has occupied a central position in debates over the nature of species (Coyne & Orr 2004). One prevalent view has been that species in sexual clades arise mainly because interbreeding maintains cohesion within species whereas RI causes divergence between species (Maynard Smith & Szathmary 1995). If so, asexuals might not diversify into distinct entities equivalent to sexual species, because there is no interbreeding to maintain cohesive units above the level of the individual. However, other authors have shown that processes causing diversification in sexual organisms, namely geographical isolation and divergent selection (also called ecological speciation), have qualitatively similar effects on asexual populations, i.e. causing divergence into discrete genetic and morphological clusters of individuals separated by gaps from other such clusters (Fisher 1930; Templeton 1989; Barraclough *et al.* 2003). The magnitude of diversification in sexuals versus asexuals remained uncertain. On the one hand, sexual reproduction can hinder divergence by breaking up favourable combinations of genes (Felsenstein 1981; Bell, this volume). Asexuals do not need to evolve special mechanisms of RI to permit divergence; therefore, some modes of speciation, such as sympatric speciation, might occur more readily. On the other hand, asexuals are expected to have slow rates of adaptation relative to sexuals; therefore, if the rate of adaptation to new conditions tended to be the limiting step in speciation, rather than the origin of RI, then asexuals might have impaired abilities to diversify than sexuals (Barraclough & Herniou 2003; Barraclough *et al.* 2003).

These ideas have stimulated lots of debate, but empirical evidence has been rare. Most asexual animals and plants are of recent origin (Burt 2000; Barraclough *et al.* 2003). The diffuse patterns of genetic and phenotypic variation typically found in such taxa (Sepp & Paal 1998) could simply reflect that they do not survive long enough for speciation to occur or the effects of ongoing gene flow from their sexual ancestors (Burt 2000; Barraclough *et al.* 2003). Other studies that are discussed in relation to asexual speciation are those looking for evidence of genetic and phenotypic clusters in bacteria (Barrett & Sneath 1994; Roberts & Cohan 1995). However, all the bacteria considered to date engage in rare or frequent recombination as well as clonal reproduction (Roberts & Cohan 1995; Hanage *et al.* 2005, 2006). Although horizontal gene transfer can occur between distantly related bacteria, homologous recombination occurs only at appreciable frequency between closely related strains (Ochman *et al.* 2005). Therefore, clusters in bacteria could arise from similar processes to interbreeding and RI in sexual eukaryotes (Ochman *et al.* 2005), even though the mechanisms of recombination do differ from those in eukaryotes (Bell, this volume). For example, plasmids and integrons may have much wider host ranges than those delimiting homologous recombination of core genes. Aside from issues of sexuality, previous studies of bacterial clusters have been mostly descriptive, relying on visual interpretation of plots of genetic or phenotypic variation rather than formal tests of predictions under null and alternative evolutionary scenarios (Coyne & Orr 2004).

### Bdelloid rotifers: an ancient asexual scandal

Several authors have argued that rare cases of ancient asexual clades might provide suitable systems for exploring the nature of diversification in asexuals. While there are several candidates (Hijri & Sanders 2005; Pawlowska 2005; Heethoff *et al.* 2007), the most celebrated by evolutionary biologists have been the bdelloid rotifers, dubbed by John Maynard Smith as an 'evolutionary scandal' (Maynard Smith 1986). Bdelloids are abundant microscopic animals in wet or occasionally damp terrestrial habitats, such as ponds, lakes, and moss, and represent one of the best-supported clades of ancient asexuals (Ricci 1987; Butlin 2002; Birky 2004). They reproduce solely via parthenogenetic eggs and no males or traces of meiosis have ever been observed. Molecular evidence that bdelloid genomes contain only divergent copies of nuclear genes present as two similar copies (alleles) in diploid sexual organisms rules out anything but extremely rare recombination (Mark Welch & Meselson 2000; Mark Welch *et al.* 2004a,b). Other unusual features of their genomes offer further support for their obligate asexual status (Arkhipova & Meselson 2005). Yet, bdelloids have survived for more than 100 million years and comprise around 460 described species taxa (Segers 2007), making them the largest clade of obligate asexual plants or animals (Butlin 2002). Their diversity challenges the idea that sex is essential for long-term survival and diversification. However, taxonomy does not constitute strong evidence for evolutionarily independent and distinct species: the taxonomic species could simply be arbitrary labels of morphological variation among a swarm of clones (Maynard Smith & Szathmary 1995).

### Are bdelloid species real?

Previous work has attempted to use bdelloid rotifers to explore questions of asexual diversification. An early example was the paper by Holman (1987) that compared how consistently bdelloid species were recognized compared to the monogonont rotifers, a related clade of facultative sexuals. He found that bdelloid species were recognized more consistently between successive taxonomic treatments than monogonont species were, perhaps indicating that bdelloid species are more distinct and recognizable than monogononts. However, Holman's result was strongly affected by the possibility of taxonomic bias. Bdelloids have received far less taxonomic attention than monogononts and the smaller community of bdelloid taxonomists may have been less inclined to revise previous work than the wider community of monogonont specialists (Fontaneto *et al.* 2007b). Moreover, bdelloids display fewer unambiguous, discrete characters identifiable by light microscopy than monogononts, perhaps limiting the scope for alternative treatments.

More recent work has begun quantitative investigations into the pattern of diversity in bdelloids. For example, Birky *et al.* (2005) assembled DNA sequence data for the mitochondrial cox1 gene from a taxonomically broad sample of bdelloid individuals sampled mainly from the south-west of the USA. Analyses revealed the existence of discrete genetic clusters separated by other such clusters by longer internal branches. Neighbouring clusters were identified using a simple criterion derived from neutral coalescent theory, namely that clusters in which divergence between clusters is more than four times greater than the maximum variation within clusters are significantly more distinct than expected assuming neutral coalescence in a single population (Rosenberg 2003; Birky *et al.* 2005). Different clusters could be found in the same sample and significant differences in food preferences and growth rate at different temperatures were observed in two cases of related clusters. Therefore, the results supported the existence of genetically distinct forms capable of co-existing in sympatry and with apparent ecological differences in those cases that were studied.

Another recent study took an alternative approach, namely to quantify the pattern of morphological variation within a clade of bdelloids, the genus *Rotaria*, and thereby to test the morphological distinctiveness of bdelloid species (Fontaneto *et al.* 2007a,b). Scanning electron microscopy was used to measure the shape and size of the only hard parts of bdelloids, their masticatory apparatus named trophi, from over 1000 individuals belonging to 48 populations. Bdelloid taxonomy had previously not used trophi as characters, rather gross body morphology (Fig. 3.1). Geometric morphometrics confirmed the distinctness of traditional species taxa, while also revealing substantial previously undescribed diversity within some traditional species. The authors argued that the distinctiveness and recognizability of bdelloid species were not qualitatively different from those observed in other groups of microscopic animals, notably in the facultatively sexual monogonont rotifers, where cryptic species are also common.

While together these studies support the hypothesis that bdelloids have diversified in a way at least broadly equivalent to sexual species, the exact nature of bdelloid diversity remains unclear. For example, genetic clusters are consistent with independent evolution between different clades of bdelloids, but such clusters may arise through strict or even partial geographic isolation. Culture experiments can show apparent ecological differences but is it possible

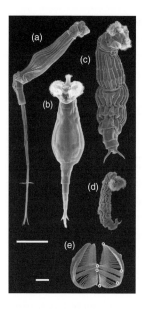

**Figure 3.1** Scanning electron micrographs of some species of the genus *Rotaria* of bdelloid rotifers. (a) *R. neptunia*, (b) *R. macruca*, (c) *R. tardigrada*, (d) *R. sordida* (e) Trophi of *R. tardigrada*. Scale bars: 100 μm for animals, 10 μm for trophi. Reprinted from Fontaneto *et al.* (2007a).

to perform a more systematic test to distinguish neutral divergence from divergence driven by divergent selection? It also remains unclear whether ecology and morphology strictly coincide with genetic clusters to define simple 'species' units or whether more complex patterns of diversification exist. Combined analyses can help distinguish the role of neutral and adaptive processes in driving diversification and test whether simple units exist.

## Testing the evolutionary nature of diversity in bdelloid rotifers

We consider several broad alternative scenarios for the diversification of bdelloids as follows (this section reviews work published in Fontaneto *et al.* (2007a) to which the reader is referred for a full account). First, the entire clade might represent a single species, i.e. a swarm of clones with no diversification into independently evolving subsets of individuals. Second, the clade may have diversified into a series of independently evolving entities. In asexuals, the term 'independently evolving' means that the evolutionary processes of selection and drift operate separately in different entities (Fisher 1930; Barraclough *et al.* 2003), such that genotypes can spread only within a single entity. Possible causes of independence include geographical isolation or adaptation to different ecological niches (Cohan 2001, 2006). In the case of geographical isolation alone, drift will eventually cause all individuals within each isolated population to descend from a more recent common ancestor than the ancestor of individuals across all populations (Barraclough *et al.* 2003). In the case of adaptation to different niches, the community will comprise distinct entities each of which comprises all of the surviving descendents of the individual in which a beneficial mutation adapting to a particular niche first occurred.

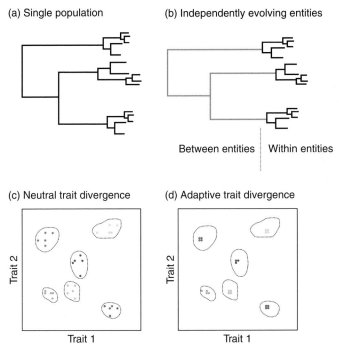

(a) Single population

(b) Independently evolving entities

Between entities | Within entities

(c) Neutral trait divergence

(d) Adaptive trait divergence

**Figure 3.2** Scheme showing the predicted patterns of genetic and morphological variation under alternative scenarios of diversification. (a) Hypothetical trees showing expected genetic relationships among a sample of individuals under the null model that the sample is drawn from a single asexual population. (b) Hypothetical gene tree under the alternative model that the clade has diversified into a set of independently evolving entities. (c) Expected variation in two eco-morphological traits (dots) evolving neutrally relative to expected variation based on variation in genetic marker (dashed lines). (d) Expected variation for eco-morphological traits under divergent selection between the six clusters. Redrawn from figure 2 of Fontaneto *et al.* (2007a).

Subsequent beneficial mutations specific to a niche will further strengthen the distinctiveness between groups of individuals in each niche (Cohan 2001).

Population genetic theory can be used to test for the presence of independently evolving entities. Under the null scenario of no diversification, genetic relationships should conform to those expected for a sample of individuals from a single asexual population (Fig. 3.2a). Under the alternative scenario that independently evolving entities are present, we expect to observe distinct clusters of closely related individuals separated by long branches from other such clusters (Fig. 3.2b, Barraclough *et al.* 2003). Coalescent models can be used to distinguish the two scenarios (Birky *et al.* 2005; Pons *et al.* 2006). Failure to reject the null model would indicate a lack of evidence for the existence of independently evolving entities.

The role of adaptation to different niches in generating and maintaining diversity can be tested directly by extending classical methods from population genetics (McDonald & Kreitman 1991) to test directly for adaptive divergence of eco-morphological traits. If trait diversity evolves solely by neutral divergence in geographic isolation, we expect morphological variation within and between entities to be proportional to levels of neutral genetic variation (Fig. 3.2c). If instead, different entities experience divergent selection on their morphology, we expect greater morphological variation between clusters than within them, relative to neutral expectations (Fig. 3.2d). Note that not all clusters need have diverged in trait values: there could be uniform purifying selection operating across some related clusters.

These tests were applied to an intensive sample of bdelloid rotifers from the genus *Rotaria*, collected from the UK, Italy and worldwide (Fontaneto *et al.* 2007a). DNA sequence data were collected for *cox1* mtDNA and *28S* ribosomal nuclear DNA. Morphology of the trophi was measured by scanning electron microscopy as in the study of Fontaneto *et al.* (2007b). Although the details of how the trophi process food remain unknown, both the size and shape of trophi are likely to reflect different types or sizes of particulate food consumed. Both DNA and morphological variation supported the monophyly of traditionally recognized species within the genus, with the exception of one species, *Rotaria rotatoria*, long suspected to comprise disparate forms. However, in common with molecular studies of sexual organisms, especially small and less-studied animals like bdelloids, both molecular and morphological data indicated the existence of additional unrecognized taxa within certain species.

Congruence between molecules and morphology confirms that most traditional *Rotaria* species are monophyletic clades, but does not rule out the possibility that taxa reflect variation within a single asexual species or swarm of evolutionarily interacting clones. Some discreteness is expected even in a single asexual population, because of the tree-like nature of inheritance in strict asexuals (Higgs & Derrida 1992; Barraclough *et al.* 2003). Therefore, coalescent models were applied to the gene trees to test for significant evolutionary independence (Pons *et al.* 2006). A coalescent model with separate subclades evolving independently of one another was significantly preferred to the null model that the entire sample derived from a single population of interacting clones. As already indicated, the clusters optimized by the coalescent model occurred at a finer resolution than traditional taxonomic species, i.e. represent cryptic taxa within those clades.

Finally, the morphological and genetic data were analysed jointly to explore the possibility of adaptive divergence between bdelloid lineages. Genetic clusters might simply represent geographically isolated, or even partially geographically isolated, populations evolving neutrally (Hubbell 2001; Wakeley & Aliacar 2001). Alternatively, the clade might have diversified into ecologically distinct entities experiencing divergent selection pressures. To resolve these alternatives, the authors tested directly for divergent selection on trophi size and shape between different lineages (McDonald & Kreitman 1991). If rotifers have experienced divergent selection on trophi morphology between entities, for example adapting to changes in habitat or resource use, there should be low variation within entities and high variation between entities, relative to the same ratio for neutral changes.

Maximum likelihood models for morphological trait evolution confirmed that different lineages of *Rotaria* have experienced divergent selection on feeding morphology. However, interestingly, repeating the test both between traditionally recognized species and between genetic clusters showed that divergent selection on trophi acts at the level of traditional species, not between genetic clusters. Separate genetic clusters within a traditional species tend to have similar morphology, even when genetic divergence between them is relatively high. Together with informal evidence of ecological differences among the species, this result shows that divergent selection has driven bdelloid diversification. Genetic clusters with similar morphology could represent geographically isolated but ecologically equivalent entities or perhaps have diverged in other adaptive traits that we were unable to measure, such as gross body morphology or behaviour.

## A species by any other name?

To conclude the discussion of bdelloids, we return to the thorny issue of species. The above studies clearly show that bdelloids display a similar pattern of genetic and morphological variation to sexual clades. There may be quantitative differences but overall the findings would be unsurprising for a study of any such group of animals. Beyond this, the analyses confirm theoretical results that the same processes behind diversification in sexuals, namely geographic isolation and divergent selection, can cause diversification in an asexual clade. However, it remains unclear whether asexuals diversify more or less than equivalent sexual clades – few sexual clades have been interrogated in the same way and similar empirical studies of a related or co-occurring sexual clade would be extremely useful. But should we call the identified units 'species'?

One argument would be that the pattern of diversity looks the same, and the underlying causes are also broadly the same; therefore, the same term should be used in both sexuals and asexuals to summarize the outcome of diversification. Interbreeding and RI is a major difference between sexuals and asexuals, with RI a critical requirement that sexuals must meet to be able to diversify (Felsenstein 1981). However, if the causes and outcome are the same, one could argue that RI is a special requirement of sexual diversification but that diversification is not so different as to require different names for its units in sexuals versus asexuals.

A second argument would be that the same problems of defining species exist in both sexuals and asexuals. In our rotifer study, we could not unambiguously point to a single level defining 'species' units. The genetic clusters provide statistical evidence for independent evolution and, therefore, are a useful starting point for evolutionary studies, for example into how bdelloids might adapt to changing environments. On the other hand, traditional species coincide with units of divergent selection and may be more relevant for ecological studies of functional diversity. However, the important point is that these issues apply to sexual organisms as well. Genetic surveys often reveal cryptic species within morphologically coherent sexual species and elicit the same arguments over their interpretation (Avise & Walker 1999; Hey *et al.* 2003; Agapow *et al.* 2004).

Many decades of disagreements show that there is no easy resolution of these arguments. However, we believe that keeping 'species' as a theoretical concept and then testing separate components of this concept offers considerable benefits to evolutionary studies: complex patterns of diversification can be treated explicitly without the constraint of assuming *a priori* that all aspects of diversification coincide on a single unit.

## Sexual diversification: neutral versus non-neutral species

Leaving behind asexuals, we consider alternative scenarios for diversification in sexuals. Many authors have focused on RI as the critical requirement for independent evolution to arise in sexual clades, and rightfully so because without it diversification across multiple loci would be lost within a generation or two. However, as described below, RI is not the only criterion for determining the degree of independent evolution in sexuals or for predicting the resulting pattern of diversification. We distinguish three phenomena.

1. *RI*. In a sexual clade, the evolution of RI between different sets of individuals is essential for diversification affecting multiple loci. Mutations arising in different reproductively isolated sets cannot be recombined together. Such isolation can be caused by geographic separation or, in sympatry, by isolating barriers such as different mate preferences or post-zygotic isolation (Turelli *et al.* 2001; Gavrilets 2004). Without reproductive isolation, balanced polymorphism at a single locus or linked loci can arise or perhaps adaptive divergence between the sexes, but not distinct forms differing in quantitative traits determined by several loci.

2. *Independent limitation*. This has been identified as a key requirement for strongly distinct clusters to arise in asexual clades (Fisher 1930; Templeton 1989; Cohan 2001; Barraclough *et al.* 2003), but has been neglected as a separate issue in sexual clades. Independent limitation means that, for two or more sets of individuals, the chance of an individual contributing genes to subsequent generations depends only on the contribution of individuals within its own set, not on the contribution of individuals in the other set. It could be caused by geographical isolation or by ecological

(a)

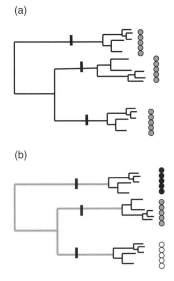

(b)

**Figure 3.3** Contrasting scenarios of diversification in a sexual clade. (a) Three reproductively isolated entities are present, indicated by bars, but they are ecologically equivalent species. The gene tree for an arbitrary marker resembles the expected gene tree if just a single population were present, and ecological traits (circles next to tips) are under uniform purifying selection across the entire clade. (b) Same three reproductively isolated entities present, but now they are ecologically distinct, hence governed by separate coalescent processes and their eco-morphological traits experience divergent selection (into 'white', 'grey' and 'black' trait values).

differences in sympatry. The result is that drift and natural selection operate separately in each population. The term 'demographic exchangeability' introduced by Templeton (1989) has also been widely used for this phenomenon. The important point is that independent limitation need not coincide strictly with reproductive isolation. For example, imagine two reproductively isolated and co-occurring groups of individuals A and B, in which survival and recruitment occurs irrespective of whether individuals belong to A or B (Fig. 3.3). This situation is broadly equivalent to the reproductively isolated but ecologically neutral species proposed by Hubbell (2001), who also proposed a mechanism for coexistence of such species. In such a scenario, both drift affecting neutral loci and natural selection on loci involved in survival and recruitment rather than sexual traits occur within an arena defined by the entire clade, and not separately within A and B. Gene combinations can be recombined only within A or within B, but gene combinations in A can out compete gene combinations in B and vice versa. Without independent limitation, the process of coalescence, i.e. tendency of individuals in the system to be descended from a common ancestor at a particular time, will act at the level of the meta-community (Hubbell 2001), not separately in each species. As a result, patterns of variation in neutral and non-sexually related traits would be similar to a single species case, even if multiple reproductively isolated entities were present. Conversely, strong patterns of discreteness in non-reproductive genes or traits will only be apparent if there is independent limitation. For example, the observation that well-defined clusters tend to be apparent for surveys of DNA barcoding markers, such as cytochrome oxidase I (*cox1*) mtDNA or plastid sequences (Hebert *et al*. 2003; Pons *et al*. 2006), says nothing about RI among groups (how could a single, maternally inherited marker tell us about RI?) but rather that each group is independently limited either by geographic isolation or ecological differences in sympatry from its sister groups.

3.  *Divergent selection*. Divergent selection can play an important role promoting the establishment of both RI and independent limitation, but has an additional effect in driving genetic and phenotypic divergence. For example, a clade with geographically isolated sets of individuals has automatic reproductive isolation and independent limitation, yet clusters will be more distinct if divergent selection acts on the genes or traits being measured rather than just drift or uniform selection pressures. Once again, divergent selection need not act at the same level as reproductive isolation or independent limitation: selection on a given trait or set of traits could be uniform across a clade of several reproductively isolated entities or geographically isolated populations, being divergent only between higher clades.

Clearly in some circumstances, all three conditions will coincide; for example populations isolated in different areas are both reproductively isolated and independently limited, but also likely to experience different environmental selection pressures (Schluter 2001). Similarly, divergent selection on non-reproductive traits implies some form of independent limitation, whether in sympatry or through geographic isolation. The primacy of reproductive isolation is also clear, since independent limitation and divergent selection cannot cause divergence unless reproductively isolated entities are present or emerge (e.g. driven by divergent selection). However, independent limitation and divergent selection could operate at broader taxonomic levels than reproductive isolation, and this would be expected to influence patterns of diversification.

A hypothetical example serves to illustrate these distinctions further. Consider the distinction between neutral reproductively isolated species sensu Hubbell (2001), and a clade diverging into ecologically distinct and independently limited species (Fig. 3.3). In the former case, coalescence occurs at the level of the entire clade, i.e. individuals take part in a single zero-sum game. The genealogy of a single arbitrary marker would resemble that for a single population, for example a neutral coalescent, with no appearance of distinct genetic clusters. Non-sexual traits would either evolve by drift or, in the case of traits determining the survival of individuals in the shared environment, by uniform purifying selection across the entire clade (assuming a single optimum phenotype for the shared environment). Reproductively isolated entities could be identified by crossing experiments, evidence of divergent selection on reproductive traits or by multi-locus tests such as concordance analyses.

This scenario contrasts with the one in which reproductively isolated entities are also ecologically distinct. Now, the genealogy of an arbitrary marker is expected to display distinct clusters, because each entity takes part in its own separate zero-sum game, i.e. separate coalescent process. This pattern would be detectable by coalescent analyses such as those described in the bdelloid example above. Moreover, adaptation to different ecological niches should be apparent as divergent selection on genes or traits involved in ecological performance, such as resource acquisition or environmental tolerances. Modifications to the McDonald–Kreitman approach outlined above could test this scenario against one of neutral divergence between species. To conclude, these two fundamentally different scenarios for the nature of diversification in a sexual clade are testable using methods akin to those applied in the bdelloids. Other approaches been have proposed to test neutral community models from phylogenetic analyses of ecological data, which can also address some of these questions (Bell *et al.* 2006; Johnson & Stinchcombe 2007). By extending these approaches to the within-species level, it is possible to gain additional insights into the evolutionary mechanisms generating different patterns of diversity.

## Do real higher taxa exist?

Prior to the 1960s, species and higher taxa such as genera or families were often treated in fairly equivalent ways as units of biodiversity. Species had special status as the arenas of selection and drift, but higher taxa were used as equally valid units of diversity (Simpson 1953; Stanley 1979). This shifted with the rise of cladistics. While still used for convenience, higher taxa have been dismissed as arbitrary units of diversity. Taxa of the same rank may differ substantially in their age and hence cannot be compared as equivalent units. The growth of quantitative cladistics and phylogenetics led to an alternative paradigm in which the only reality of higher groupings is whether they comprise a mono-phyletic clade or not. Clades are real in the sense of being all living descendants of a common ancestor but no level in the hierarchy is more significant than any other (Coyne & Orr 2004). If higher taxa such as genera appear distinct from one another, it is just because of chance gaps in the branching of the tree of life.

Extension of the ideas in the previous section provides a scenario for the evolutionary reality of higher taxa. As argued above, independent evolution through separate limitation is predicted to have a major effect on the degree of distinctiveness between different lineages within a clade. A neutral scenario leads to a situation in which coalescence acts within a higher clade, rather than separately in different reproductively isolated entities. If the sister of that higher clade comprised individuals inhabiting either a different region or specialized on a different ecological niche, then this could generate distinct higher taxa. Separate coalescence in the two clades A and B could lead to longer internal branches separating two more clustered sets of species (Fig. 3.4). Drift alone

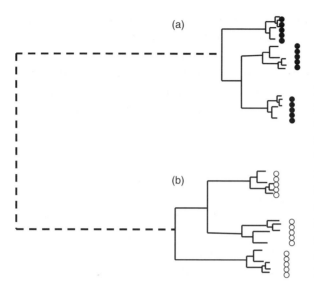

**Figure 3.4** Hypothetical scenario with two clades, (a) and (b). Both clades comprise ecologically equivalent, reproductively isolated entities, but the clades are specialized on different ecological niches. For example, clade (a) might comprise trees and clade (b) might comprise lianas. Drift and selection for ecological performance operate separately within clade (a) and clade (b), leading to long internal branches (dashed) between them.

might be rather a weak force to generate diversity patterns at this scale (Ricklefs, this volume). However, if the environment periodically changes, and because of ecological equivalence, descendents from just one species within each clade sweep to adapt to the new conditions (because offspring inheriting beneficial combinations of mutations would outcompete other RI entities as well as individuals in their own RI entity), then this would strengthen the pattern of distinctiveness between clades A and B. In other words, units of evolutionary independence and diversity might exist above the level of reproductively isolated entities: there may exist evolutionarily significant higher taxa above the level of species.

Many complexities affect this theoretical scenario; for example, the possibility of hierarchical levels of increasing independence or of evolutionary non-independence among distant relatives (the fitness of a predator influences the fitness of its prey). However, if close relatives tend to share broadly similar ecological guilds and inhabit shared geographic areas, and therefore take part in a shared zero-sum game, then this could generate distinctiveness and diversity at higher levels. Tree-based methods like those outlined above could be used to test this possibility.

## Conclusions

Rather than assuming that simple entities called species exist, we advocate an analytical approach to exploring diversity patterns. Species are vital and useful units of biodiversity for a wider range of applications and species concepts provide the theoretical framework for interpreting diversity patterns. However, evolutionary studies striving to understand the nature of diversification can benefit greatly from relaxing the single unit approach. We strongly emphasize that this does not belittle the importance of RI, which is critical in sexual organisms to permit diversification beyond single loci or sexual dimorphism. However, understanding the pattern of diversity in nature cannot be reduced solely to the study of RI: we also need to understand independent evolution in the broader sense, such as when and where divergent selection and independent limitation arise.

## Acknowledgements

We thank Giulio Melone, Bill Birky Jr, Austin Burt, Chiara Boschetti and Manuela Caprioli for their help and Roger Butlin, Graham Bell and Jody Hey for comments on the manuscript. This research was supported by Natural Environment Research Council (NERC) UK grant NER/A/S/2001/01133, an EU Marie Curie Intra-European Fellowship to DF, a Royal Society University Research Fellowship to TGB, a Royal Society Dorothy Hodgkin Fellowship to EAH and a Royal Society International Joint Project grant to TGB and CR.

## References

Agapow, P. M., Bininda-Emonds, O. R. P., Crandall, K. A., et al. (2004) The impact of species concept on biodiversity studies. Quarterly Review of Biology 79, 161–179.

Arkhipova, I. R. and Meselson, M. (2005) Diverse DNA transposons in rotifers of the class Bdelloidea. Proceedings of the National Academy of Sciences of the United States of America 102, 11781–11786.

Avise, J. C. and Walker, D. E. (1999) Species realities and numbers in sexual vertebrates: perspectives from an asexually transmitted genome. Proceedings of the National Academy of Sciences of the United States of America 96, 992–995.

Barraclough, T. G., Birky, C. W. and Burt, A. (2003) Diversification in sexual and asexual organisms. Evolution 57, 2166–2172.

Barraclough, T. G. and Herniou, E. (2003) Why do species exist? Insights from sexuals and asexuals. Zoology 106, 275–282.

Barrett, S. J. and Sneath, P. H. A. (1994) A numerical phenotypic taxonomic study of the genus Neisseria. Microbiology-UK 140, 2867–2891.

Bell, G., Lechowicz, M. J. and Waterway, M. J. (2006) The comparative evidence relating to functional and neutral interpretations of biological communities. Ecology 87, 1378–1386.

Birky, C. W. (2004) Bdelloid rotifers revisited. Proceedings of the National Academy of Sciences of the United States of America 101, 2651–2652.

Birky, C. W., Wolf, C., Maughan, H. Herbertson, L. and Henry, E. (2005) Speciation and selection without sex. Hydrobiologia 546, 29–45.

Burt, A. (2000) Perspective: sex, recombination, and the efficacy of selection – was Weismann right? Evolution 54, 337–351.

Butlin, R. (2002) The costs and benefits of sex: new insights from old asexual lineages. Nature Reviews Genetics 3, 311–317.

Cohan, F. M. (2001) Bacterial species and speciation. Systematic Biology 50, 513–524.

Cohan, F. M. (2006) Towards a conceptual and operational union of bacterial systematics, ecology, and evolution. Philosophical Transactions of the Royal Society B-Biological Sciences 361, 1985–1996.

Coyne, J. A. and Orr, H. A. (2004) Speciation. Sinauer Associates, Sunderland, MA.

de Queiroz, K. (2005) Different species problems and their resolution. Bioessays 27, 1263–1269.

Felsenstein, J. (1981) Skepticism towards Santa Rosalia, or why are there so few kinds of animals? *Evolution* **35**, 124–138.

Fisher, R. A. (1930) *The Genetical Theory of Natural Selection*. Oxford University Press, Oxford.

Fontaneto, D., Herniou, E. A., Boschetti, C., *et al.* (2007a) Independently evolving species in asexual bdelloid rotifers. *PLoS Biology* **5**, 914–921.

Fontaneto, D., Herniou, E. A., Barraclough, T. G., Ricci, C. and Melone, G. (2007b) On the reality and recognisability of asexual organisms: morphological analysis of the masticatory apparatus of bdelloid rotifers. *Zoologica Scripta* **36**, 361–370.

Gavrilets, S. (2004) *Fitness Landscapes and the Origin of Species*. Princeton University Press, Princeton, NJ.

Hanage, W. P., Fraser, C. and Spratt, B. G. (2005) Fuzzy species among recombinogenic bacteria. *BMC Biology* **3**, 6.

Hanage, W. P., Fraser, C. and Spratt, B. G. (2006) The impact of homologous recombination on the generation of diversity in bacteria. *Journal of Theoretical Biology* **239**, 210–219.

Hebert, P. D. N., Cywinska, A., Ball, S. L. and DeWaard, J. R. (2003) Biological identifications through DNA barcodes. *Proceedings of the Royal Society of London Series B-Biological Sciences* **270**, 313–321.

Heethoff, M., Domes, K., Laumann, M., *et al.* (2007) High genetic divergences indicate ancient separation of parthenogenetic lineages of the oribatid mite Platynothrus peltifer (Acari, Oribatida). *Journal of Evolutionary Biology* **20**, 392–402.

Hey, J. (2001) The mind of the species problem. *Trends in Ecology & Evolution* **16**, 326–329.

Hey, J., Waples, R. S., Arnold, M. L., Butlin, R. K. and Harrison, R. G. (2003) Understanding and confronting species uncertainty in biology and conservation. *Trends in Ecology & Evolution* **18**, 597–603.

Higgs, P. G. and Derrida, B. (1992) Genetic distance and species formation in evolving populations. *Journal of Molecular Evolution* **35**, 454–465.

Hijri, M. and Sanders, I. R. (2005) Low gene copy number shows that arbuscular mycorrhizal fungi inherit genetically different nuclei. *Nature* **433**, 160–163.

Holman, E. W. (1987) Recognizability of sexual and asexual species of rotifers. *Systematic Zoology* **36**, 381–386.

Hubbell, S. P. (2001) *The Unified Neutral Theory of Biodiversity and Biogeography*. Princeton University Press, Princeton, NJ.

Johnson, M. T. J. and Stinchcombe, J. R. (2007) An emerging synthesis between community ecology and evolutionary theory. *Trends in Ecology & Evolution*. **22**, 250–257.

Mark Welch, D. and Meselson, M. (2000) Evidence for the evolution of bdelloid rotifers without sexual reproduction or genetic exchange. *Science* **288**, 1211–1215.

Mark Welch, D. B., Cummings, M. P., Hillis, D. M. and Meselson, M. (2004a) Divergent gene copies in the asexual class Bdelloidea (Rotifera) separated before the bdelloid radiation of within bdelloid families. *Proceedings of the National Academy of Sciences of the United States of America* **101**, 1622–1625.

Mark Welch, J. L., Mark Welch, D. B. and Meselson, M. (2004b) Cytogenetic evidence for asexual evolution of bdelloid rotifers. *Proceedings of the National Academy of Sciences of the United States of America* **101**, 1618–1621.

Maynard Smith, J. (1986) Contemplating life without sex. *Nature* **324**, 300–301.

Maynard Smith, J. and Szathmary, E. (1995) *The Major Transitions in Evolution*. W. H. Freeman, Oxford.

McDonald, J. H. and Kreitman, M. (1991) Adaptive protein evolution at the Adh locus in *Drosophila*. *Nature* **351**, 652–654.

Ochman, H., Lerat, E. and Daubin, V. (2005) Examining bacterial species under the specter of gene transfer and exchange.

*Proceedings of the National Academy of Sciences of the United States of America* **102**, 6595–6599.

Pawlowska, T. E. (2005) Genetic processes in arbuscular mycorrhizal fungi. *FEMS Microbiology Letters* **251**, 185–192.

Pons, J., Barraclough, T. G., Gomez-Zurita, J., *et al.* (2006) Sequence-based species delimitation for the DNA taxonomy of undescribed insects. *Systematic Biology* **55**, 595–609.

Ricci, C. N. (1987) Ecology of bdelloids – how to be successful. *Hydrobiologia* **147**, 117–127.

Rieseberg, L. H., Wood, T. E. and Baack, E. J. (2006) The nature of plant species. *Nature* **440**, 524–527.

Roberts, M. S. and Cohan, F. M. (1995) Recombination and migration rates in natural populations of *Bacillus subtilis* and *Bacillus mojavensis*. *Evolution* **49**, 1081–1094.

Rosenberg, N. A. (2003) The shapes of neutral gene genealogies in two species: probabilities of monophyly, paraphyly, and polyphyly in a coalescent model. *Evolution* **57**, 1465–1477.

Segers, H. (2007) Annotated checklist of the rotifers (Phylum Rotifera), with notes on nomenclature, taxonomy and distribution. *Zootaxa* **1564**, 1–104.

Schluter, D. (2001) Ecology and the origin of species. *Trends in Ecology & Evolution* **16**, 372–380.

Sepp, S. and Paal, J. (1998) Taxonomic continuum of *Alchemilla* (Rosaceae) in Estonia. *Nordic Journal of Botany* **18**, 519–535.

Simpson, G. G. (1953) *The Major Features of Evolution*. Columbia University Press, New York.

Sites, J. W. and Marshall, J. C. (2003) Delimiting species: a Renaissance issue in systematic biology. *Trends in Ecology & Evolution* **18**, 462–470.

Stanley, S. M. (1979) *Macroevolution: Pattern and Process*. W. H. Freeman, San Francisco.

Templeton, A. (1989) The meaning of species and speciation: a population genetics approach. In: *Speciation and Its Consequences* (ed. D. Otte and J. Endler). Sinauer Associates, Sunderland, MA.

Turelli, M., Barton, N. H. and Coyne, J. A. (2001) Theory and speciation. *Trends in Ecology & Evolution* **16**, 330–343.

Wakeley, J. and Aliacar, N. (2001) Gene genealogies in a metapopulation. *Genetics* **159**, 893–905.

CHAPTER FOUR

# The poverty of the protists

GRAHAM BELL

## Experimental adaptive radiation in bacteria

Bacteria are readily dispersed, so a lineage will often encounter unfamiliar conditions of growth to which it must adapt or die. This process can be studied in the laboratory by supplying a microcosm with medium containing a single limiting substrate and then inoculating it with an isogenic culture of bacteria. The culture will die if the substrate is refractory, but in practice often adapts to novel conditions through the successive substitution of beneficial mutations that confer a metabolic capability it previously lacked. This constitutes periodic selection, with one mutation after another sweeping through the population. This kind of selection experiment is particularly valuable because it allows the mechanism of adaptation to be studied directly, and in the last 20 years it has enabled us to discover a great deal about how bacteria adapt to simple environments (reviewed by Mortlock 1984; Elena & Lenski 2003). Fitness typically increases rapidly at first and then slows down, because beneficial mutations of larger effect are likely to be substituted earlier, so that over time there a non-linear approach to a plateau, with a half-life of some hundreds of generations. Hence, if a bacterial culture is inoculated into a spatially structured environment (such as a set of glass vials) in which each site offers a different substrate it will undergo an adaptive radiation through divergent specialization to each of the available substrates. We inoculated a landscape containing 95 different substrates with a single genotype of the plant-associated bacterium *Pseudomonas fluorescens* and propagated this system by serial transfer for about 1000 bacterial generations (MacLean & Bell 2002). The ancestral genotype was capable of growing on only one-half of the substrates, leaving a broad range of unexploited ecological opportunities. After 1000 generations, most of these sites had been occupied by newly adapted strains that were capable of using the novel substrate but that had lost the ability to use two or three of the substrates in the ancestral niche. To this point, the experiment exemplified a classical adaptive radiation – the evolution of specialists that had gained function but paid a cost of adaptation that impaired their performance when replaced in their former environment. There was an unexpected feature of this radiation, however.

*Speciation and Patterns of Diversity*, ed. Roger K. Butlin, Jon R. Bridle and Dolph Schluter. Published by Cambridge University Press. © British Ecological Society 2009.

When substrates not used by the ancestor were included, the evolved lines could actually use more substrates, not fewer: that is, the metabolic niche was larger, not smaller, in the specialists than in their ancestor. Consequently, although some lineages failed to adapt to the substrate they were cultured on, other lines were able to use that substrate, even though they had been cultured on a different one. With low rates of dispersal between sites, the landscape would soon have become fully occupied. The occupant of each site would not be a narrow specialist unable to grow elsewhere, however, but would rather be an imperfect generalist, able to grow in some sites and not in others.

## Bacterial adaptation to complex environments

Although natural environments vary in space, the conditions of growth at a site are seldom as simple as a microcosm with a single limiting resource. In most cases, natural environments will supply a complex mixture of many potential substrates, none of them limiting because all are at concentrations too low to support growth on their own. In these circumstances, the classical account of periodic selection breaks down. Selection lines will become adapted to rich complex medium in the laboratory, but selective sweeps are seldom or never seen. The reason seems to be that strains that are adapted to some fraction of the resources present spread through the population but cannot eliminate strains adapted to other resources. Consequently, the population comes to consist of a set of imperfect generalists held in equilibrium by frequency-dependent selection (MacLean et al. 2005). This conclusion can be extended by using chemically defined complex environments in which each constituent is present at a concentration high enough to support growth. We could document adaptive radiation by estimating the genetic variance of fitness at 100-generation intervals. This showed that the genetic variance supplied by mutation increases three times as fast in complex environments as in simple environments (Barrett & Bell 2006). The outcome is again a mixture of imperfect generalists, rather than a single universal generalist or a set of narrow specialists (Barrett et al. 2005). Evolution in complex environments, then, seems to resemble evolution in structured environments. This should not be very surprising, as the conditions for maintaining diversity in simple models are the same for complex and structured environments. In either case, however, the outcome seems to be unlike the predictions of simple models: the emergence of overlapping imperfect generalists.

## Themes and variations

Adaptation to simple environments may follow a stereotyped course, with the same mutations being fixed in the same sequence in every replicate experiment. Our experimental adaptive radiation did not follow this rule. We tested the phenotypes of replicate lines adapted to the same novel substrate and found they often differed markedly (MacLean & Bell 2003). Consequently, replicate lines may converge in fitness but differ in the genetic basis of adaptation. This result has been reported from a range of other systems involving bacteria, virus and artificial life. Although there may be several possible outcomes when the same selection pressure is applied to the same genotype, however, it does not seem likely that there are very many major possibilities. In many cases it is possible to identify a few main themes that consistently emerge as a response to selection and that usually correspond with the range of loci at which beneficial mutations can occur. There are usually many minor variations on these themes, however, which correspond to the range of beneficial mutations that may occur at a given locus. The historical diversity created by the themes and variations that are the outcome of replicate episodes of adaptation will supplement the functional diversity caused by different agents of selection.

## Bacterial diversity in nature

The diversity of bacteria in natural communities is difficult to estimate because most cells do not grow in culture and those that do lack easily distinguished phenotypes. Using classical methods, about 6000 Linnean names have been assigned. More recently, gene sequences have been used to estimate the abundance of genotypes in a sample, using some arbitrary level of similarity to demarcate distinct types ('phylotypes'). Kemp and Aller (2004) reviewed the field and concluded that most surveys led to estimates of fewer than 200 phylotypes in the source environment. This implies that bacteria are not very diverse, that most local diversity is captured in rather small samples of the community, and that classical taxonomy gives a roughly correct estimate of overall diversity. Curtis et al. (2002) took the underlying distribution of abundance among types to be lognormal, an empirical assumption that will be a good approximation in many cases, and use the ratio of total abundance $N_T$ to that of the most abundant single phylotype $N_{max}$ to estimate the number of types in the sample. The lowest ratios of about 1.5 are found in extreme environments such as sewage sludge, and predict about 70 types $ml^{-1}$. Marine samples have $N_T/N_{max}$ of about 4, and hence 100–200 types $ml^{-1}$. This leads to much higher estimates of diversity in many communities than Kemp and Aller's methods; the community of an anaerobic digester, for example, had $N_T/N_{max} \approx 20$ and hence about 9000 phylotypes $ml^{-1}$.

Survey methods that count clones are inevitably dependent on sample size, and are likely to be unreliable when most types are rare, as they are in bacterial surveys, where most clones are singletons. DNA reassociation of bulk DNA provides a quantitative estimate of diversity without these limitations. If DNA is sheared to produce fragments of roughly equal size and then melted, homologous strands will reassociate on cooling at a rate that depends on their concentration. If the original DNA was very complex, the concentration of any given homologous sequence will be correspondingly low, and reassociation will take a long time. For any given concentration, the time taken for 50% of strands to reassociate will therefore provide an estimate of the complexity of the original DNA, relative to some standard. Torsvik *et al.* (1990) extracted bacterial DNA from 30 g of forest soil and used reassociation kinetics relative to an *Escherichia coli* standard to estimate the complexity of the whole-community genome as about $3 \times 10^{10}$ bp. Given an average genome size (for bacteria isolated from the same sample) of about $7 \times 10^6$ bp, this implies that the sample contained about 4000 different genomes.

Dykhuizen (1998) argued that the biologically meaningful diversity is much greater than this. In the first place, lineages can be functionally different even if they share a considerable part of their genomes. He proposed 70% similarity as a conservative criterion to distinguish types that were likely to be as different, or more different, than closely related eukaryote species. In this case only 30% of the DNA belonging to different types will not reassociate, and much of the DNA that does reassociate will be from different types; in fact, the reassociation curve suggests that about 90% of the reassociating DNA belong to different types. Taking this into account, the real number of functionally different types is about 40 000. This assumes, however, that all types are equally frequent, which from the shape of the reassociation curve is manifestly untrue. Dykhuizen guesses that the ratio of rare to common types is about 25:1, given that common types contribute about half the total number of individuals. The total number of distinct types in the sample is then about 500 000, an impressive number for a cupful of soil.

Although a number of wild guesses are involved in this estimate, they are probably close to the mark. Gans *et al.* (2005) found that the reassociation kinetics of their samples were fitted well by a Zipf distribution of species abundance with parameter −2, which corresponds to a Sheldon size spectrum of equal biomass in equal log increments of cell size (Sheldon *et al.* 1972). This led to an estimate of $8 \times 10^6$ distinct types in 10 g of soil, an order of magnitude greater than Dykhuizen's estimate.

Given the Sheldon size spectrum, the distribution of overall abundance will be Pareto, i.e. $\text{Prob}(W > x) = k_1{}^{\psi} W^{-\psi}$, where $W$ is size and $\psi$ is an empirically determined parameter (Vidondo *et al.* 1997). The underlying probability density distribution is $\text{Prob}(W = x) = W^{-2}$, so the power law for total abundance $J = k_2 W^{-z}$ has $z = 1 + \psi = 2$. The number of distinct types $S$ will increase with $J$ in a manner determined by the distribution of the abundance of types. Suppose that this is described empirically by the power law $S = J^y$. Then mean abundance will be $J/S = k_3 W^{-z + zy}$. It is well established that both mean abundance and maximum abundance are proportional to $W^{-1}$ (Peters & Wassenberg 1983; Duarte *et al.* 1987); hence $-z + zy = -1$, $z = 2$, and therefore $y = 0.5$. Siemann *et al.* (1996) found that insect diversity estimated from very large samples fitted the power law

$S = 1.05 J^{0.51}$ and I shall take $S = J^{0.5}$ as a good empirical generalization. It may be an overestimate, as the mean for 28 eukaryote taxa reviewed by Hillebrand *et al.* (2001) was $y = 0.36$. The soil sample analysed by Torsvik *et al.* (1990) using fluorescence microscopy contained $1.5 \times 10^{10}$ cells $g^{-1}$, and thus a total of $J = 2\text{-}10^{11}$ cells. These should comprise $J^{0.5} = 630\,000$ distinct types, which is not far from Dykhuizen's estimate of 500 000. The sample analysed by Gans suggests that this estimate is conservative. The total number of soil bacteria worldwide is estimated to be $2.6 \times 10^{29}$ (Whitman *et al.* 1998). The number of distinct types of soil bacteria worldwide is thus about $5 \times 10^{14}$.

If this astonishing number is correct, even to within a few orders of magnitude, we are as yet ignorant of the bulk of bacterial diversity. Moreover, even more diversity will exist in the oceans and in subterranean habitats. How can it be reconciled with the much more modest estimates of diversity from survey data? In the first place, most bacterial surveys consist largely of rare species, so that the rarefaction curve (phylotype number as a function of total clones) is more or less linear, with little sign of an asymptote, and estimates of community diversity are then fragile. The plot of phylotype number $S$ on library size $J$ for the studies reviewed by Kemp and Aller (2004) is fitted (rather poorly) by $S = 1.23 J^{0.7}$, which is steeper than the relation I have used and consequently predicts even more diversity in large communities. Secondly, Dykhuizen's analysis is equivalent to assuming that $N_T/N_{max}$ is very large for soil, which may be reasonable. A survey of unimproved pasture (McCaig *et al.* 1999) gave $N_T/N_{max} = 275/9 = 30$, which would point to a global total of about $10^{10}$ phylotypes. I conclude that the balance of the evidence currently available supports the notion of very high bacterial diversity, much of which is so far undocumented.

## The eukaryotic life cycle

Eukaryotes have characteristic life cycles in which growth and reproduction are periodically interrupted by sexual episodes. Sex is the dual process of gamete fusion and meiotic restitution that switches the lineage from the haploid to the diploid state and then back again. Bacteria have mechanisms for uptake and recombination of DNA but lack sexuality. Sex is often confused with reproduction because they are obligately associated in many large multicellular organisms, but in microbes the distinction is clear. A haploid eukaryotic microbe will reproduce vegetatively for a period, then form gametes which fuse and thereafter continue to reproduce vegetatively as diploids, until the cycle is closed by meiosis to reconstitute the haploid state. The two complementary concepts that follow naturally from the sexual life cycle are species and gender: successful mating takes place between individuals of like species and unlike gender. Consequently, 'species' and by extension 'speciation' refer exclusively to sexual eukaryotes, and using the same labels for asexual organisms such as bacteria leads only to confusion. Bacterial diversity is defined in terms of arbitrary limits on genetic or functional similarity, whereas eukaryotic diversity is defined in terms of species. More sophisticated accounts of diversification in asexual groups involve the concept of independently evolving clades within whose confines the spread of beneficial mutations is restricted (see Barraclough *et al.*, this volume). This provides a more satisfactory basis for comparison with sexual species, but cannot yet be used to estimate bacterial diversity.

## Diversity of eukaryotic microbes

Most major groups of eukaryotic microbes have no more than a few hundred species: only diatoms, foraminiferans, 'radiolarians', ciliates and dinoflagellates have diversified into a few thousand species, about the same as much narrower groups such as grasses, orchids or sponges. There are only about 3000 species of ciliates, for example, and as they are well-studied and morphologically distinctive there may be few more remaining to be discovered. Other groups that may be very abundant have very few species: prasinophytes about 140, cryptophytes 200, haptophytes 500, ellobiopsids 25, and dictyostelids 25. Other major clades are even more depauperate: actinophryids 6, labyrinthulids 8, jakobids 4 and so forth. These numbers are very rough and will no doubt increase as knowledge accumulates, but it seems questionable that they will increase by many orders of magnitude. The common ancestor of eukaryotes was undoubtedly unicellular, and large complex organisms appear only late in the fossil record. Modern protists are extremely abundant, occupy every environment and follow every way of life available to eukaryotes. Nevertheless, despite their ancient origins and current ubiquity only about 50 000 species have been described, about the same as the number of snails. This is unlikely to be attributable to their lack of easily recognizable morphological characters, because many groups (such as coccolithophores, testate amoebas, or all the groups mentioned above) have very distinctive and variable appearances under the microscope and have been closely studied for two centuries. It may be attributable in part to a shortage of taxonomic work, but the same is likely to be true for fungi, insects or other more diverse groups. The most obvious interpretation of the unexpected poverty of the protists is that there really are relatively few species, most of which can be distinguished morphologically and many of which have very broad geographical distributions (Finlay 2002; Finlay *et al.* 1998).

   The alternative view is that the uniformity of cosmopolitan microbes is more apparent than real. In the first place, it is now clear that extremely small picoeukaryotes, little larger than bacteria, are extremely abundant and since they often lack features such as flagella and scales their diversity, like that of bacteria, may be greatly underestimated by morphological surveys. Secondly, large numbers of undescribed taxa have been identified by molecular surveys. Most sequences fall close to known species, but some do not, and new clades of unknown or poorly known marine heterotrophs and deep-sea picoeukaryotes have been detected (Moreira & Lopez-Garcia 2002; Behnke *et al.* 2006); even in freshwater a large fraction of sequences could not be attributed to named species (Slapeta *et al.* 2005). One obvious reason is that relatively few protist sequences are found in databases, and the survey data do not suggest extremely high levels of cryptic diversity. Slapeta *et al.* (2005) pooled 50 ml sediment samples and 1000 ml water samples from two freshwater ponds to construct a SSU rDNA library that comprised 377 sequences. These fell into 83 groups ('phylotypes') whose members were no more than 3% divergent (a much more restrictive criterion than the 30% similarity used for bacteria); the expected number, from the diversity-library size plot for bacterial studies, is 78. Behnke *et al.* (2006) identified 92 phylotypes by a criterion of 2% divergence from a library of 753 protist 18S rRNA sequences obtained by sampling an anoxic fjord; the expected number is 127. Thus, surveys of protist genes do not seem to give a distinctively different result from comparable bacterial surveys, bearing in mind that these surveys are held to demonstrate low local levels of diversity in bacteria. No DNA reassociation experiments have yet been described, unfortunately, so no quantitative assessment of diversity can be attempted. Finally, morphological uniformity may conceal genetic diversity and geographical structure. Foraminiferans are classified by their intricate coiled shells, but species defined on morphological grounds may encompass several distantly related clusters of genotypes representing cryptic species (de Vargas *et al.* 1999; Kucera & Darling 2002; Darling *et al.* 2004). The small prasinophyte *Micromonas pusilla* occurs in all the oceans of the world and genetic surveys likewise show that it consists of a series of sibling species (Slapeta *et al.* 2006).

Evidently, it is not yet possible to estimate the diversity of protists with much conviction. The real diversity is clearly greater than species lists drawn up on classical criteria, but how much greater it really is remains debatable for the time being. The situation would best be clarified by measuring the reassociation kinetics of bacterial and protistan DNA from the same sample, but as far as I know this has not yet been reported. For the present chapter, I shall accept that the superficial poverty of the protists is real, and that the real extent of protist diversity, although it may be greater than is currently recognized, nevertheless falls far short of bacterial diversity. Should this prove to be wrong, then the rest of the chapter can be consigned to the dustbin of unjustified speculations. If it holds up, then it would be interesting to know why.

## Ecological phylogenetics of bacteria

The first possibility is that there are relatively few species of protists because there are relatively few individuals. A simple neutral model will account for the difference, with most diversity being historical rather than functional. There are two reasons to doubt this hypothesis. The first is that bacteria are only about 100 times as abundant as protists, whereas the power law relating diversity $S$ to abundance $J$ certainly has exponent $y < 1$. The world population of organisms between $10^{-10}$ and $10^{-4}$ g in mass, most of which are protists, is about $4 \times 10^{27}$ individuals, given the allometric rules that I have described. This implies that there should be about $6 \times 10^{13}$ species of protists worldwide, so those currently described are only a minute fraction (about one in a billion) of the real biota. The diversity–abundance relation might be shallower for protists than for multi-cellular organisms, however. Hillebrand *et al.* (2001) report average values from a range of studies of 0.44 for multicellular organisms and 0.24 for protists. The difference is not formally significant, but taking 0.25 rather than 0.5 as the protist value yields a prediction of $8 \times 10^6$ species of protists. This is a much more modest total, but still implies that after a century of microscopy less than 1% of protist species have been described. It is very difficult to believe that this is true.

A second reason for doubting a strictly neutral interpretation of low protist diversity is that there is considerable circumstantial evidence that protists are often ecologically different. The most powerful comparative test of neutrality is that related species should be ecologically more similar or less similar than random species. They should be more similar if the ecological innovation was possessed by their most recent common ancestor, and less similar if it evolved thereafter. The application of ecological phylogenetics to bacterial communities has been reviewed by Martin (2002). Horner-Devine and Bohannan (2006) ana-lysed four large datasets and found that ecologically similar species tended to be phylogenetically clustered, both among bacteria as a whole and within partic-ular groups such as a- or b-proteobacteria. At least some part of bacterial diver-sity is therefore functional rather than merely neutral. Conversely, Fenchel (2005) argues that the cryptic diversity of protists revealed by gene sequencing may represent mostly neutral variation.

## Microbial biogeography

A second possibility is that low diversity is a consequence of high rates of dispersal. Specific adaptation to particular sites in a structured environment will be impeded by dispersal, because the effect of local selection will tend to be negated by the arrival of poorly adapted immigrants. The balance between selection and immigration that permits the maintenance of diversity depends on the productivity of sites and the variation in fitness, but in a very simple model of two alleles in two equally productive kinds of sites with symmetrical selection $s$ and fraction of immigrants $m$, both alleles will be protected if $m/s < 1$ (see Bulmer 1972). Small organisms such as protists are readily dispersed by currents of air or water, and consequently many microbes are cosmopolitan and can be found at suitable sites anywhere in the world. A litre of fresh water taken anywhere in the world, for example, is likely to contain the ciliate *Tetrahymena pyriformis*; a litre of seawater from the photic zone of any ocean will probably contain the coccolithophore *Emiliana huxleyi*. The chrysomonad *Paraphysomonas* occurs in all parts of the world, and yet 32 of the 41 named species were found in 25 uL of sediment from a single Cumbrian pond (Finlay & Clarke 1999). In effect, ecological diversification is obstructed by high levels of immigration, so that most eukaryotic microbes lack a biogeography, their global diversity being not much greater than their local diversity.

The converse position is, of course, that the lack of geographical structure is more apparent than real. The wild sister species of baker's yeast, *Saccharomyces paradoxus*, has a worldwide distribution on the surface of hardwood trees. On a single tree, however, neighbours are likely to belong to the same clone; on a larger scale, European and Asian isolates are more different than local isolates; and a deeper divide separates Eurasian and American isolates, which are now named as different species (Koufopanou *et al.* 2006). Phylogenetic analyses have shown geographic structure in other microbial fungi (Taylor *et al.* 2006). Microbes do have biogeography, therefore, but whether they have as much biogeography is doubtful. The simplest quantitative expression of geographical structure is the exponent $z$ of the species–area power law $S = kA^z$, where $A$ is the area sampled. The average value of $z$ for large multicellular organisms is about 0.3 and that for protists about 0.15 (Hillebrand *et al.* 2001). A very detailed survey of soil ascomycetes by Green *et al.* (2004) using a DNA fingerprinting method showed that community similarity decayed with distance and yielded an estimate of $z = 0.07$. Other estimates have been published for marine ciliates ($z = 0.08$), diatoms ($z = 0.07$) (Azovsky 2002) and freshwater ciliates ($z = 0.04$) (Finlay *et al.* 1998). There is a great deal of variation among taxa, but the data seem to indicate a compromise position: on average, diversity increases more slowly with area in protists, but it increases nonetheless. The most plausible reason for lower relative global diversity is high rates of dispersal.

Bacteria are even more readily dispersed, however. Soil bacteria, for example, have weak geographic structure and tend to occur where conditions are suitable rather than being restricted to a particular region (Fierer & Jackson 2006). The diversity of tree-hole bacteria estimated by fingerprinting increases with habitat volume at a rate greater than that found in studies of protists, $z \approx 0.26$ (Bell *et al.* 2005). Clone libraries have given lower values of $z \approx 0.02 – 0.04$ (Horner-Devine *et al.* 2004). Hence, dispersal alone cannot explain low protist diversity.

The predictions of worldwide diversity from applying the species–area rule to fingerprint or sequence-based surveys are much too low: fewer than 200 kinds of either bacteria (Horner-Devine *et al.* 2004) or ascomycetes (Green *et al.* 2004). The reason is that with exponents as low as 0.04 the number of kinds of bacteria in the whole world exceeds the number in a single gram of soil by a factor of less than 10. There is a dramatic difference between these unrealistically low figures and the surprisingly high estimates of bacterial diversity from applying the species–individual rule (with an exponent of about 0.5) to estimates from DNA reassociation studies. I do not understand the reason for this, but the characters used in molecular surveys (such as length of ITS region) clearly do not correspond with membership of single species, and it is possible that they saturate (fill up with increasing numbers of kinds) at modest levels of diversity. Woodcock *et al.* (2006) argue that the large disparity between sample size and community size is responsible for the very low exponents often found in microbial surveys.

## Sex and diversification

Suppose that two lineages are divergently specialized for growth in different kinds of site. If the genetic basis of the difference between them amounts to possessing different alleles at a single locus, then there is no important difference between sexual and asexual cases. If two or more loci are involved and adaptation requires particular combinations of alleles, then an asexual population will split into two well-adapted lineages provided that the relevant genotypes appear at an appreciable rate. Bacterial populations are often so large that all possible single mutants arise in every generation, and so will readily undergo adaptive radiation into a variety of specialized types, as seen in the laboratory. In sexual populations, divergent adaptation will be hindered by outcrossing, because well-adapted genotypes that are brought together by random gamete fusion are liable to be disrupted by recombination (Felsenstein 1981). The low diversity of protists might then be attributable to their fundamentally sexual life cycles.

One objection to this theory is that many eukaryotic microbes have no known sexual phase and may be perennially asexual. This is quite difficult to demonstrate conclusively, however. The Felsenstein effect implies that outcrossing populations will be close to linkage equilibrium, so that strong disequilibrium is evidence for asexuality; alternatively, different genes may give rise to different phylogenetic trees in sexual populations, whereas all loci will give concordant results for asexual populations. Such methods can identify widespread recombination even in organisms where no sexual stage is known (Burt *et al.* 1996), but they require extensive sampling and sequencing, and have not yet been applied to many protists. Moreover, quite low rates of recombination will effectively block ecological divergence, unless selection is very strong. Second-order selection for modifiers of mating or recombination may affect this conclusion, but this possibility lies outside the scope of this chapter.

Secondly, genetic recombination occurs by several non-sexual processes in bacteria. Recombination rates per locus are typically of the same order of magnitude as mutation rates, or lower, and recombination is much less frequent than in eukaryotes, unless a very long period elapses between sexual episodes. Even these low rates of recombination may be sufficient to obstruct diversification in neutral models (Fraser *et al.* 2007), although whether it has much effect in opposing selection for divergent specialization is doubtful (see Cohan 2002).

A third objection is that the theory applies with even greater force to multicellular eukaryotes, many of which are obligately sexual and outcrossing. There are very few really small multicellular organisms comprising fewer than about a thousand cells, however, and consequently multicellular eukaryotes are much less abundant than protists. The very smallest metazoans often have cosmopolitan distributions like protists, and have modest or low levels of species diversity. Major groups of large multicellular organisms, on the other hand, are often very diverse; even leaving arthropods and nematodes on one side, there are about 60 000 species of deuterostomes, 40 000 ascomycetes, 250 000 land plants, 6000 rhodophytes, 1500 phaeophytes and so forth. There is still a discrepancy between these numbers and those expected from species–individual curves. With a world population of about $4 \times 10^{17}$ individuals of 1 g in weight or more there should be about $6 \times 10^8$ species, whereas there are actually about three orders of magnitude fewer. The discrepancy is not as great as for protists, however. This might be because large organisms are often poorly dispersed, so that distant populations seldom outcross. In this case, bacteria are highly diverse because they lack sex, large organisms are fairly diverse because the effect of sex is weakened by geographical isolation, and the low diversity of protists is the result of sex and dispersal combined.

This admittedly speculative argument has an interesting consequence for speciation in large organisms. If sex is the main agent responsible for the poverty of the protists, then sexual isolation may be the main precipitating factor in speciation. Ecological diversification occurs very readily in bacteria but is obstructed by recombination in sexual microbes; hence, sexual isolation is often a necessary prerequisite for speciation in eukaryotes and will normally precede the evolution of ecological specialization. It will not be easy to confirm or refute this position, but, to return to the initial topic of this chapter, simple microcosm experiments with protists might be a good place to start.

## Acknowledgements

The bacterial work was funded by the Natural Science and Engineering Research Council of Canada.

## References

Azovsky, A. I. (2002) Size-dependent species–area relationships in benthos: is the world more diverse for microbes? *Ecography* **25**, 273–282.

Barrett, R. D. H. and Bell, G. (2006) The dynamics of diversification in evolving Pseudomonas populations. *Evolution* **60**, 484–490.

Barrett, R. D. H., MacLean, R. C. and Bell, G. (2005) Experimental evolution of *Pseudomonas fluorescens* in simple and complex environments. *The American Naturalist* **166**, 470–480.

Behnke, A., Bunge, J., Barger, K., et al. (2006) Microeukaryote community patterns along an $O_2/H_2S$ gradient in a supersulfidic anoxic fjord (Framvaren, Norway). *Applied and Environmental Microbiology* **72**, 3626–3636.

Bell, T., Ager, D., Song, J.-I., *et al.* (2005) Larger islands house more bacterial taxa. *Science* **308**, 1884.

Bulmer, M. G. (1972) Multiple niche polymorphism. *The American Naturalist* **106**, 254–257.

Burt, A., Carter, D. A., Koenig, G. J., White, T. J. and Taylor, J. W. (1996) Molecular markers reveal cryptic sex in the human pathogen Coccidioides immitis. *Proceedings of the National Academy of Sciences of the United States of America* **93**, 770–773.

Cohan, F. M. (2002) Sexual isolation and speciation in bacteria. *Genetica* **116**, 359–370.

Curtis, T. P., Sloan, W. T. and Scannell, J. W. (2002) Estimating prokaryotic diversity and its limits. *Proceedings of the National Academy of Sciences of the United States of America* **99**, 10494–10499.

Darling, K. F., Kucera, M., Pudsey, C. J. and Wade, C. M. (2004) Molecular evidence links cryptic diversification in polar planktonic protists to Quaternary climate dynamics. *Proceedings of the National Academy of Sciences of the United States of America* **101**, 7657–7662.

Duarte, C. M., Agusti, S. and Peters, R. H. (1987) An upper limit to the abundance of aquatic organisms. *Oecologia* **74**, 272–276.

Dykhuizen, D. E. (1998) Santa Rosalia revisited: why are there so many species of bacteria? *Antonie van Leeuwenhoek* **73**, 25–33.

Elena, S. F. and Lenski, R. E. (2003) Evolution experiments with microorganisms: the dynamics and genetic bases of adaptation. *Nature Reviews Genetics* **4**, 457–460.

Felsenstein, J. (1981) Skepticism towards Santa Rosalia, or why are there so few kinds of animals? *Evolution* **35**, 124–138.

Fenchel, T. (2005) Cosmopolitan microbes and their 'cryptic' species. *Aquatic Microbial Ecology* **41**, 49–54.

Fierer, N. and Jackson, R. B. (2006) The diversity and biogeography of soil bacterial communities. *Proceedings of the National Academy of Sciences of the United States of America* **103**, 626–631.

Finlay, B. J. (2002) Global dispersal of free-living microbial eukaryote species. *Science* **296**, 1061–1063.

Finlay, B. J. and Clarke, K. J. (1999) Apparent global ubiquity of species in the protist genus Paraphysomonas. *Protist* **150**, 419–430.

Finlay, B. J., Esteban, G. V. and Fenchel, T. (1998) Protozoan diversity: converging estimates of the global number of free-living ciliate species. *Protist* **149**, 29–37.

Fraser, C., Hanage, W. P. and Spratt, B. G. (2007) Recombination and the nature of bacterial speciation. *Science* **467**, 476–480.

Gans, J., Wolinsky, M. and Dunbar, J. (2005) Computational improvements reveal great bacterial diversity and high metal toxicity in soil. *Science* **309**, 1387–1390.

Green, J. L., Holmes, A. J., Westoby, M., *et al.* (2004) Spatial scaling of microbial eukaryote diversity. *Nature* **432**; 747–750.

Hillebrand, H., Watermann, F., Karez, R. and Berninger, U.-G. (2001) Differences in species richness patterns between unicellular and multicellular organisms. *Oecologia* **126**; 114–124.

Horner-Devine, M. C. and Bohannan, B. J. M. (2006) Phylogenetic clustering and overdispersion in bacterial communities. *Ecology* **87**, S100–S108.

Horner-Devine, M. C., Lage, M., Hughes, J. B. and Bohannan, B. J. M. (2004) A taxa–area relationship for bacteria. *Nature* **432**, 750–753.

Kemp, P. F. and Aller, J. Y. (2004) Bacterial diversity in aquatic and other environments: what 16S rDNA libraries can tell us. *FEMS Microbiology Ecology* **47**, 161–177.

Koufopanou, V., Hughes, J., Bell, G. and Burt, A. (2006) The spatial scale of genetic differentiation in a model organism: the wild yeast Saccharomyces paradoxus. *Philosophical Transactions of the Royal Society of London B* **361**, 1941–1946.

Kucera, M. and Darling, K. F. (2002) Cryptic species of oceanic foraminifera: their effect on palaeocenographic reconstructions. *Philosophical Transactions of the Royal Society* **360**, 695–718.

MacLean, R. C. and Bell, G. (2002) Experimental adaptive radiation in Pseudomonas. *The American Naturalist* **160**, 569–581.

MacLean, R. C. and Bell, G. (2003) Divergent evolution during an experimental adaptive radiation. *Proceedings of the Royal Society B* **270**, 1645–1650.

MacLean, R. C., Dickson, A. and Bell, G. (2005) Resource competition and adaptive radiation in a microbial microcosm. *Ecology Letters* **8**, 38–46.

Martin, A. P. (2002) Phylogenetic approaches for describing and comparing the diversity of microbial communities. *Applied and Environmental Microbiology* **68**, 3673–3682.

McCaig, A. E., Glover, L. A. and Prosser, J. I. (1999) Molecular analysis of bacterial community structure and diversity in unimproved and improved upland grass pastures. *Applied and Environmental Microbiology* **65**, 1721–1730.

Moreira, D. and Lopez-Garcia, P. (2002) The molecular ecology of microbial eukaryotes unveils a hidden world. *Trends in Microbiology* **10**, 31–38.

Mortlock, R. P. (ed.) (1984) *Microorganisms as Model Systems for Studying Evolution*. Plenum Press, New York and London.

Peters, R. H. and Wassenberg, K. (1983) The effect of body size on animal abundance. *Oecologia* **60**, 89–96.

Sheldon, R. W., Prakash, A. and Sutcliffe, W. H. (1972) The size distribution of particles in the ocean. *Limnology and Oceanography* **17**, 327–340.

Siemann, E., Tilman, D. and Haarstad, J. (1996) Insect species diversity, abundance and body size relationships. *Nature* **380**, 704–706.

Slapeta, J., Lopez-Garcia, P. and Moreira, D. (2006) Global dispersal and ancient cryptic species in the smallest marine eukaryotes. *Molecular Biology and Evolution* **23**, 23–29.

Slapeta, J., Moreira, D. and Lopez-Garcia, P. (2005) The extent of protist diversity: insights from molecular ecology of freshwater eukaryotes. *Proceedings of the Royal Society B* **272**, 2073–2081.

Taylor, J. W., Turner, E., Townsend, J. P., Dettman, J. R. and Jacobson, D. (2006) Eukaryotic microbes, species recognition and the geographic limits of species: examples from the kingdom Fungi. *Philosophical Transactions of the Royal Society* **361**, 1947–1963.

Torsvik, V., Goksoyr, J. and Daae, F. L. (1990) High diversity in DNA of soil bacteria. *Applied and Environmental Microbiology* **56**, 782–787.

de Vargas, C., Norris., R., Zaninetti, L., Gibb, S. W. and Pawlowski, J. (1999) Molecular evidence of cryptic speciation in planktonic foraminifers and their relation to oceanic provinces. *Proceedings of the National Academy of sciences of the United States of America* **96**, 2864–2868.

Vidondo, B., Priarie, Y., Blanco, J. M. and Duarte, C. M. (1997) Some aspects of the analysis of size spectra in aquatic ecology. *Limnology and Oceanography* **42**, 184–192.

Whitman, W. B., Coleman, D. C. and Wiebe, W. J. (1998) Prokaryotes: the unseen majority. *Proceedings of the National Academy of sciences of the United States of America* **95**, 6578–6583.

Woodcock, S., Curtis, T. P., Head, I. M., Lunn, M. and Sloan, W. T. (2006) Taxa–area relationships for microbes: the unsampled and the unseen. *Ecology Letters* **9**, 805–812.

# CHAPTER FIVE

# Theory, community assembly, diversity and evolution in the microbial world

THOMAS P. CURTIS, NIGEL C. WALLBRIDGE
AND WILLIAM T. SLOAN

## Theory in the microbial world

The microbial world is vast and important domain of apparently unfathomable complexity. The latest swathe of sequencing technology (Sogin *et al.* 2006; Huber *et al.* 2007) has confirmed what many had already predicted: there is an awful lot of different kinds of bacteria in the world. The number is unknown even in ostensibly well-studied environments and this is preventing us from understanding one of the most important and remarkable things about the microbial world: the way in which communities form and reform, and change.

For all our molecular sophistication, our analysis and understanding of the diversity and community assembly is still very primitive. Microbial ecology is perhaps in a situation analogous to that of general ecology before McArthur's first contributions; a situation described by Cody and Diamond (1975) who wrote:

in the 1950s, ecology was still mainly descriptive. It consisted of qualitative, situation-bound statements that had low predictive value, plus empirical facts that often seem to defy generalization.  (Cody & Diamond 1975)

What McArthur brought was theory, and theory is what microbial ecologists need now. Parameterized mathematical descriptions of community assembly will help us to make coherent quantitative predictions about the microbial world. These predictions can guide the exploration and manipulation of this domain.

In the search for theory, theoretical microbial ecologists have naturally looked to classical ecology for insight and inspiration (Horner-Devine *et al.* 2007; Prosser *et al.* 2007). This may be unwise. For much contemporary theoretical ecology is not really up to the job of predicting characteristics of the microbial world or indeed the non-microbial world. The literature tends to offer mechanistic explanations for a world that can be readily observed. Model parameter are frequently 'invented', perhaps selected at random (Mouquet & Loreau 2003; Tilman 2004), or specifically chosen using special searching algorithms (Huisman & Weissing 1999) to give a particular answer. Such models may

*Speciation and Patterns of Diversity*, ed. Roger K. Butlin, Jon R. Bridle and Dolph Schluter. Published by Cambridge University Press. © British Ecological Society 2009.

have a role in exploring the mechanisms underlying the formation of the world that is readily perceived. If a model with randomly selected parameters can give a plausible representation of the real world, the mechanisms in the model may be considered, if not proven, plausible.

In microbial ecology such an approach is at best risky and at worst dangerous. For we are not sure what the microbial world looks like (Curtis *et al*. 2006), the microbial parameter space is probably not random. Moreover, the ability of a model of a particular mechanism to reproduce a pattern observed in nature does not prove that that mechanism is at work. We are consequently poorly placed to spot models which are completely wrong and we could be profoundly deceived by un-calibrated models.

Our criticism of the application of theory in the microbial world is in many ways a mere subset of the more authoritative critique of ecology in general offered by Peters (1991) nearly two decades ago. This monograph cautions that the pervasive nebulosity of much ecology threatens the link between the discipline and reality. He warned that the weakness of the link threatened the status of ecology as a science in general and an applicable science in particular. Science must be built on hypothesis testing. Ideas which cannot be rigorously and quantitatively tied to a reality are difficult to test and difficult to put into practice. Peters had a point and his critique has all too sadly stood the test of time. His remedy was predictive ecology, for prediction would permit ideas to be tested and policy to be evaluated. But the Achilles heel of the monograph is that prediction was equated with patterns and correlations. This is an approach that has limitations.

## Patterns are perilous

The conceptual basis of a theory is also extremely important. In particular, it is essential to distinguish between mechanisms and patterns or correlations. For example, the sincere but misguided belief that taxa–area curves are one of the few cast iron rules in ecology (Pounds & Puschendorf 2004) is misleading. Taxa–area curves are patterns and patterns are a function of some underlying mechanism that may be contingent on scale, time or some other variable. The provisional nature of species–area curves can be seen most clearly in the microbial ecology literature. It can be observed easily (Bell *et al*. 2005) or with difficulty (Horner-Devine *et al*. 2004) or not at all (Baptista *et al*. 2008) depending on the nature of the system and the method used (Woodcock *et al*. 2006). In one of the better known papers (Horner-Devine *et al*. 2004), the proportion of the variation explained by random variation (about 87%) appears to vastly exceed the proportion explained by distance or area (about 13%).

Patterns are important because they can provide clues to the rules or mechanisms underlying the world around us. If at all possible, we should use these rules and mechanisms to predict. This is not a new insight. MacArthur who in drawing attention to the limitations of curve fitting commented 'A far more fruitful approach seems to be ... to predict on the basis of simple biological hypotheses' (MacArthur 1957). That was good advice then and it is good advice now.

## What makes a good theory?

Wilson (1998) has argued that theory should be evaluated on the basis of its parsimony, generality, consilience and predictiveness. Consilience is often overlooked; it means being related to the rest of knowledge. At the broadest scales, one would like to be able to relate ecology to thermodynamics or geology. In a more narrow sense, one would like theories that dealt with community assembly to link to theories about resource use and for all theoretical ecology to be grounded in evolution. We are still far from this ideal. The most important single test is predictiveness. For a model to be predictive it must be calibrated. Trying to calibrate models makes one value parsimony (of course in an ideal and consilient model one might be able to predict some parameters from first principles). Parameters calibrated independently are superior to those which are co-determined by fitting, for example. It is worthwhile noting that a model is not predictive if it merely describes the data used to calibrate it.

## Back to basics

Clearly, predictive mechanistic models are a challenge. However, to be a success, a model need not predict or explain perfectly. MacArthur and Wilson (1967) suggested that 'if a theory can explain 70% of the observed phenomena it will have served its purpose well'. Thus if one commences with the simplest possible explanation, further complexities are required to master the last 30% not to re-explain the first 70%. This may sound modest, but actually MacArthur and Wilson have set the bar rather high; recent studies in microbial ecology have succeeded in explaining less than 13% of the observed variation (Horner-Devine *et al.* 2004; Ramette & Tiedje 2007).

## Stochastic models

Thus, in looking for theories to describe the microbial world, we need to find theories that are based on real biological rules (i.e. not merely patterns), are sufficiently simple to be parameterized and at least plausibly testable. From this perspective, a niche-based perspective looks problematic. There is simply no hope of parameterizing a niche-based model at least in the short term. For example, imagine using the stochastic niche approach of Tilman (2004); even if there were only two resources of consequence and a mere 100 species, we would require 400 parameters just to describe the system at a fixed temperature and pH (Tilman 2004). The coding of such a model would be a trivial exercise in comparison to its calibration. Consequently, the central intellectual challenge of a niche-based approach is to find some simple way to determine or predict such properties in a realistic manner or to decide which of the several tens of taxa are worth characterizing. Unless and until this can be done the validity of the perspective in the microbial world is moot.

Perhaps, the simplest possible models are those that simply invoke births (i.e. growth), deaths and immigration to explain patterns in microbial communities. Such simple models are called neutral (neutral community models – NCM) by analogy with neutral models in population biology as they assume equivalence within a functional group (Bell 2001; Hubbell 2001). This notion and perhaps the self-confident tone of Hubbell's excellent monograph has meant that such models have been greeted with suspicion. But the term neutral is really a misnomer for they are birth, death and immigration models. Since most communities are subject to birth, death and immigration, the question is: Can I determine the parameters in such a model and if so what part of the real world can I explain and predict on this basis? The corollary being, if you cannot determine these simple parameters and establish this baseline what hope have you got with a more complex model?

Neutral models have their intellectual roots in statistical mechanics. Though not the first attempt to take such an approach to ecology (Maynard-Smith 1974; Caswell 1976), the current suite of neutral theory is undoubtedly the most encompassing (Alonso *et al.* 2006). Moreover the advantages of such an approach were noted decades ago by Maynard-Smith (1974) who commented that Lotka Volterra 'cannot claim to have as close correspondence with reality as Newton's Laws or as Mendel's Laws'. By contrast, births, deaths and immigration are a fundamental reality.

## Applying neutral community models in the microbial world

The original formulations of Hubbell and Bell conceived of a local community of $N_T$ individuals in which a death is replaced from a source community of diversity $\theta$ with a probability $m$ (and thus from the local community with a probability m − 1).

Typically, such models have been assessed by comparison with species abundance curves once the parameters for migration and the source community ($m$ and $\theta$) have been fitted with greater (Etienne & Alonso 2005) or lesser rigour.

Unfortunately, there is no species abundance curve, for any microbial community anywhere and such distributions are, at best, a distant prospect. One consequence of this lacuna in our knowledge is that there is no agreement, even to within a couple of orders of magnitude of the source or local diversity of most microbial communities. In short, we have little or no idea of the value of $\theta$ (Curtis *et al.* 2006). Moreover, the fitted $\theta$ affects the fitted migration parameter $m$. Therefore the two parameters should ideally be estimated independently, otherwise any uncertainty about the former will cause uncertainty about the latter.

A further and important consideration for a microbial ecologist's simple birth–death model is that the original formulations of Bell and Hubbell were mathematically unsuitable for microbial studies. The discrete formulations are computationally intractable with all but the most modestly sized communities. It is therefore necessary to use a continuous form of the model.

A number of continuous forms of NCM are available (Houchmandzadeh & Vallade 2003; Vallade & Houchmandzadeh 2003; Volkov *et al.* 2003; McKane *et al.* 2004) but the simplest is that of Sloan (Sloan *et al.* 2006, 2007). The conceptual basis of the model is identical to that of Hubbell and is predicated on the conception that, over a very small period of time, the number of individuals in a community can either: increase by one organism, decrease by one organism or not change. The probability of each of these possibilities can be expressed in terms of $N_T$, $m$ and the proportional abundance of the species in the source community ($p_i$). Based on these probabilities, it is possible to derive an equation that describes the rate of change of the probability that the species will have a particular relative abundance, $x_i$. The steady-state solution of the equation gives an expression for the probability density function for the relative abundance of $i^{th}$ species, $x_i$ is beta distributed,

$$x_i = \beta(N_T m p_i, N_T m, (1 - p_i)) \tag{5.1}$$

This relatively simple formulation describes the probability distribution of a single taxon of mean proportional abundance $p_i$ in the source community that avoids, at this stage, using the fundamental biodiversity number $\theta$.

To employ this equation, we need to determine the parameters. Fortunately, $p_i$ and $N_T$ are measurable. $N_T m$ can be inferred and in certain colonial organisms, $m$ can be directly measured (Curtis *et al.* 2006; Baptista *et al.* 2008). $N_T m$, which is almost exactly the same thing as 'the universal immigration parameter' of the more complex literature (Etienne & Alonso 2005), profoundly affects the probability distribution of a given taxon (Fig. 5.1).

An obvious qualitative prediction of NCM is that the abundance and frequency of observation are linked: i.e. an organism that is abundant at a given site should be frequently observed at high abundances. This is also an empirical observation that predates the species–area curve by a good margin, having first appeared in the *Origin of Species*. Stochastic explanations for frequency abundance patterns are simple (Sloan *et al.* 2006, 2007) and relate the precise nature of the distribution to $N_T m$ (dispersal limitation is indicated by values of $m < 1$) (Fig. 5.2). Niche-based explanations of this phenomena are tortuous and waffly (Brown 2000).

Given that we now have a value for $N_T$, $m$ and how $m$ scales, we can estimate the size of the source community that would be required for the model to reproduce the patterns seen in the real world. The term $\theta$ is used to represent this source diversity. Hubbell, but not Bell (Bell 2001), ascribes the source term with a particular biological meaning and mathematical structure that may or may not reflect the reality (Sloan *et al.* 2007).

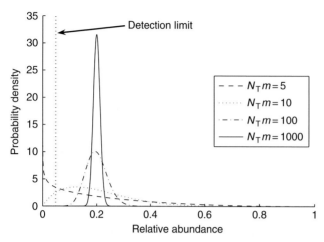

Figure 5.1 Neutral models (Sloan *et al.* 2006) predict the effect of immigration and the number of individuals on the distribution, and thus abundance and frequency of observation of a particular taxon. For a given mean source community abundance, the frequency with which an organism is observed is related to $N_Tm$, the total number of individuals multiplied by the immigration parameter (also 'the universal immigration number' (Etienne & Alonso 2005). At high $N_Tm$ values, the local distribution is tightly clustered around the mean metacommunity distribution (in this case 0.1). As the $N_Tm$ values drop, the distribution widens and eventually the mode of the curve falls below the detection limit and the organism is typically no longer observed.

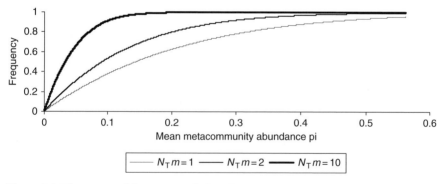

Figure 5.2 The expected frequency-relative abundance relationships observed in a local community for differing $N_Tm$ values. Thus for a given data set, the $N_Tm$ values can be found by fitting a line to these data, where the value of $N_T$ is known and $m$ can be easily inferred (Sloan *et al.* 2006).

It appears that $\theta$ has to vary a great deal to accommodate the patterns we observe. For example, ammonia-oxidizing bacteria in activated sludge plants are typically represented by a handful of species in a geometric series. This implies a very low value of theta (~2) and correspondingly low source diversity. The

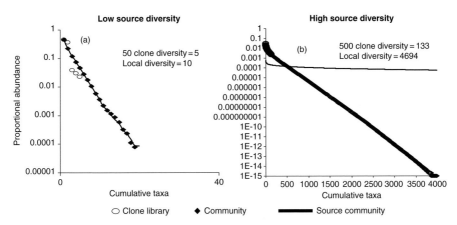

**Figure 5.3** A sample of 50 ammonia oxidizing bacteria clones and 500 bacterial (putatively heterotrophs) clones in a wastewater treatment plant with diversities of 7 and 202 respectively. Here examine the ability of the Sloan *et al.* model to qualitatively reproduce that finding using immigration parameters ($10^{-12}$ and $1.5 \times 10^{-16}$) estimated using the extrapolation procedure described in Curtis *et al.* (2006), known $N_T$ values and hypothetical source diversities of 200 (low source diversity) and $>10^5$ (high source diversity) respectively. (a) Represents the ammonia oxidizing bacteria; the diversity is similar to that found in the samples. (b) Represents the high source diversity group. The clone and local diversity are distinct. Slightly less diversity was found in the model than observed in practice, suggesting some error in the source term or the immigration parameter. This is neither proof nor disproof of the model.

heterotrophs by comparison are represented by 60 or 70 species in a 200 clone 16S rRNA gene clone library, which implies a $\theta$ value $>10^4$ and a source diversity of $>10^6$ (Fig. 5.3). Treehole communities by comparison have a $\theta$ value of about 15. These $\theta$ values are able to not only reproduce the number of detectable taxa but also the presence (Woodcock *et al.* 2006, 2007) (Fig. 5.4) or absence (Baptista *et al.* 2008) of species–area curves. It is worth noting that these simple models are incredibly plastic. With just three parameters and the most basic assumptions a wide range of phenomena can be reproduced. The parameters really matter.

## The biology is in the parameters

If biologists can conceptualize the biology implicit in the parameters of NCM and thus how change in the environment will lead to changes in the community, it might help them accept that neutral models are more than mathematical trickery. $N_T$ is simple enough being the number of individuals in a community, though determining the number of individuals in a functional group can be surprisingly tricky and we are not always sure what constitutes a microbial community. The migration parameter is the probability that a death is replaced

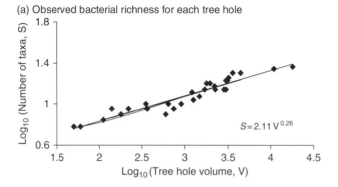

(a) Observed bacterial richness for each tree hole

$S = 2.11\,V^{0.26}$

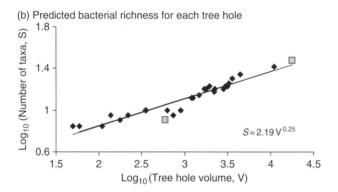

(b) Predicted bacterial richness for each tree hole

$S = 2.19\,V^{0.25}$

**Figure 5.4** (a) The observed bacterial richness in the 29 beech tree holes described by Bell *et al.* (2005). The solid line represents the power–law relationship, $S = 2.11\,V^{0.26}$, fitted using linear regression. (b) The bacterial richness predicted by the neutral model with $\theta = 15$ and $m = 10^{-6}$ calibrated using the taxa-abundance distribution of the smallest tree hole. The solid line represents the power–law relationship, $S = 2.19\,V^{0.25}$, again fitted using linear regression (Woodcock *et al.* 2007).

by an immigrant (as opposed to a birth) and thus the ratio of immigration and local growth. It is consequently a function of the rate of arrival of new cells, a physical phenomenon, and the rate of growth of new cells, a function of biology and local environmental conditions. At a large scale one might imagine that *m* is geographically variable, for as one gets nearer the poles, the lower growth rates would increase the importance of migration and lead to higher diversities. There is evidence that microbial diversity at high latitudes is surprisingly high (Neufeld & Mohn 2005). It follows that the notion of a universal immigration parameter ($N_T m$) (Etienne & Alonso 2005) is perhaps unhelpful.

### Evolution as the master variable

What about $\theta$? The size of the source community is, in essence, a measure of the net rate of evolution for a given functional group. It is the most interesting, and perhaps the most important, parameter of all. A sensitivity analysis, albeit a simple one, suggests that the diversity of the source community, not immigration, dictates local community diversity (Fig. 5.5). This parameter represents some global or local pool of diversity for a given functional group. The size of these pools is one of the more problematic unknowns in modern microbial ecology (Curtis & Sloan 2005). The doubt about the extent of microbial diversity

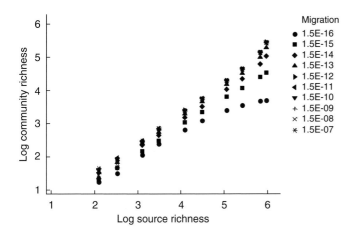

Migration
● 1.5E-16
■ 1.5E-15
◆ 1.5E-14
▲ 1.5E-13
▶ 1.5E-12
◀ 1.5E-11
▼ 1.5E-10
+ 1.5E-09
× 1.5E-08
✳ 1.5E-07

**Figure 5.5** A sensitivity analysis of the affect of the source diversity and the immigration parameter in a community of $10^{16}$ individuals using the neutral community model of Sloan *et al.* (2007). The immigration parameter varies over 10 orders of magnitude and the source diversity by 6 orders of magnitude. Clearly, the latter has a greater effect on the local diversity predicted by the model.

has its roots in the scale of the microbial world and the inadequacy of even the most sophisticated molecular methods to adequately sample it (Curtis *et al.* 2002, 2006).

The modelling above can shed some light on this conundrum and on the nature of $\theta$. There is a great deal of data on the diversity found in relatively small samples of different functional groups. If local diversity is dictated by source diversity, then the huge variation in local diversity observed in the samples very probably reflects a very wide variation in the diversity in the source community. Hubbell assumed a point mutation process and formally defined $\theta$ as $2J_m\nu$ where $J_m$ is all the individuals in the metacommunity and $\nu$ the rate of speciation (Hubbell 2001). Though not necessarily an accurate, or even inaccurate, description of evolution in bacteria, this relationship highlights a simple fact, the diversity of a given bacterial group will be some function of the number of individuals and time. In short, evolution is the master variable in microbial community assembly. Thus, even the most practically minded microbial ecologist or practitioner may find a remarkably direct link between evolution and the challenges and phenomena they face on a daily basis.

Thus, our recent estimates of $\theta$ suggest that there may be over $10^6$ species in the metacommunity of a humble sewage works. This is of course just one community and we recently used a metaanalysis of ribosomal databases to make a 'back of the envelope calculation' of a plausible upper-estimate of the total number of taxa distinguishable at the level of 97% 16S rRNA sequence identity, to be about $10^{10}$ (Curtis *et al.* 2006). Although this is supposed to be a silly upper estimate, the microbial world has had about 3.5 billion years to generate this diversity. By assuming an exponentially increasing number of species, we can see that after 3.5 billion years $10^{10}$ species would generate just 65 extra species per year (Fig. 5.6) and the mean net rate of speciation would be just $3 \times 10^{-9}$ species per year.

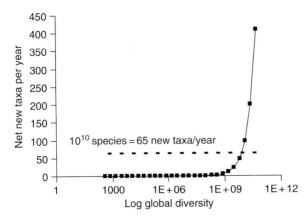

Figure 5.6 How slow is the net rate of speciation? The net rate of speciation is a function of the number of species extant today (on the $x$ axis), biological time (about 3.5 billion years) and the net rate of increase in the number of taxa per year ($y$ axis). If the number of taxa per year expanded exponentially and there were $10^{10}$ bacterial taxa extant today, the net rate of evolution would be just 65 new species a year over about $10^{30}$ individuals and presumably $>10^{30}$ births. The mean net rate over all time would be just $3 \times 10^{-9}$ new species a year.

Now $10^{10}$ is supposed to be a silly number. It seems likely that many very important and relatively abundant functional groups could well have global metacommunity diversities in the hundreds or thousands. Moreover, some of these functional groups, such as the methanogens are thought to have been around for a very long time indeed. In this context, it becomes apparent that the diversity, and thus rate of diversification, of the microbial world seems, at least superficially, improbably small and certainly incredibly variable.

Unfortunately, systematic comparisons of diversity between functional groups are not straightforward, as differences and similarities are confounded by methodological and environmental considerations. We can, however, use the plethora of studies on one class of environment, biological wastewater treatment systems, to give us some idea about the relative diversity using the ratio of the number of individuals to the most abundant taxon as a guide. This is a crude (Curtis *et al.* 2002), but not unworthy (May 1974), indicator of diversity. Such poorly constrained surveys of the diversity of these functional groups should be taken 'with a pinch of salt'. However, the rank order of diversity probably reflects the opinion of most contemporary microbiologists (Table 5.1).

### The non-paradox of the nitrifiers

Why then is there so much variation in diversity between functional groups? At first glance it is obvious. The heterotrophs are diverse because they can metabolize a wide range of substrates giving a much larger 'ecological space' into which they can evolve. However, this is not a wholly satisfactory explanation. For example, the electron acceptor (i.e. the chemical reduced by the electrons extracted from an electron donor) appears to have something to do with diversity as well. The sulphate-reducing bacteria appear to be a good deal less diverse than the denitrifying heterotrophs (nitrate reducers) (Table 5.1).

Table 5.1. *A crude comparison of the variation in an indicator of diversity ($N_T/N_{max}$; Curtis* et al. *2002) of various functional groups found in biological treatment plants and the free energy associated (under standard conditions) with that function*

| Functional group | $N_T/N_{max}$ (Curtis *et al.* 2001) | $G_r^{o'} kJ/e^{-eq}$ (Rittmann & McCarty 2001) | Reference |
|---|---|---|---|
| Aerobic heterotrophs | 16S ~ 10 | $-120.07^a$ | (Baptista *et al.* unpublished) |
| Denitrifying heterotrophs | NirK > 5 NirS > 4 16S ~ 10 | $-113.55^a$ | (Hallin *et al.* 2006) |
| Sulphate reducers | 2 | $-20.50^a$ | (Dar *et al.* 2007) |
| Ammonia oxidizers | 1.2 | $-43.61$ | (Baptista *et al.* 2006) |
| Hydrogenotrophic methanogens | 2.61 | $-21.0$ | (McHugh *et al.* 2003; Leclerc *et al.* 2004) |
| Acetoclastic methanogens | 1.3 | $-3.87$ | (McHugh *et al.* 2003; Leclerc *et al.* 2004) |

$^a$ With glucose

Moreover, why should simple resource requirements mean low diversity in the world of one set of microbial autotrophs, when it does not for others? There are many species of photoautotrophs, but not so many autotrophic ammonia oxidisers, even with the newly discovered ammonia oxidisers in the Archaea (Francis *et al.* 2005). Resource complexity is not an adequate indicator of diversity. How can we have the paradox of the plankton and the non-paradox of the nitrifiers?

Ultimately, different numbers of species in functional groups imply different net amounts of speciation in each group. This implies that evolution has to occur either over different time periods or at different rates. At least superficially, time does not appear to be a problem. The methanogens have been around for an awfully long time and yet do not seem to be very diverse. This implies that evolution must be occurring at different rates in different functional groups. Something we hypothesize, that is somehow related to the free energy available from the compounds they use for energy generation.

### Free energy and the rate of evolution

One of the most remarkable things about the microbial world is the wide range of energy sources available to microorganisms. Different functional groups use different resources that have different free energies. It is striking that the rank order of diversities is similar to the rank order of the estimated free energies (Table 5.1). It is instructive to compare the methanogens. The low energy ($\Delta G_r^{o'} - 3.87$ kJ/e$^{-eq}$) acetoclastic methanogens are much less diverse (though often more abundant) than the hydrogenotrophic methanogens which have more energy available to them ($\Delta G_r^{o'} -21$ kJ/e$^{-eq}$). Of course one must be careful to distinguish between standard conditions and conditions in nature (Finke *et al.* 2003; 2007) and be alive to the possibilities that these differences in free energy may be less extreme or even reversed in real life.

The relationship between energy and rates of evolution in larger organisms has figured prominently in the ecological literature. The work in this area has been reviewed and clarified by Gaston and colleagues (Evans & Gaston 2005; Clarke & Gaston 2006). This literature typically evaluates the diversity in terms of the energy impinging on the ecosystem. These theories may very well apply to, or at least be testable in, microbial communities. Moreover because more energy often means more individuals, at least some aspects of work on energy are implicit in neutral models and other established models in microbial ecology (Coskuner *et al.* 2005).

These established models are based on growth yield: the efficiency with which matter is transformed into individuals. Microbial physiologists and environmental engineers (McCarty 1971, 2007; Rittmann & McCarty 2001) have a good grasp of how and why yield varies between functional groups. Yield is related to the amount of energy released by a particular method of energy production and the amount of energy required to convert a particular source of carbon into biomass. In essence, each molar equivalent of electrons released by an electron donor must be divided between those used for synthesis ($f_s$) and those transferred to the electron donor to create energy ($f_e$). The true yield is the fraction $f_s$ translated into mass units.

It is possible to roughly calculate the maximum true yield for a given functional group by calculating $A$, the number of equivalents of donor electrons used in energy generation for each equivalent devoted to cell synthesis. Since the total equivalents required to generate a cell must be $1+A$:

$$f_s = \frac{1}{1+A} \tag{5.2}$$

Thus, if $\Delta G_r$ units of energy are released by an equivalent of electron donor and $\Delta G_s$ units of energy are required to make an equivalent of cells, and $\varepsilon$ represents the efficiency with which energy is converted from one form to another to maintain an energy balance, then:

$$A\varepsilon\Delta G_r + \Delta G_s = 0 \tag{5.3}$$

Rearranging:

$$A = -\frac{\Delta G_s}{\varepsilon\Delta G_r} \tag{5.4}$$

Clearly, the smaller the amount of energy in an equivalent of electrons in the electron donor, the greater the value of $A$ and the lower the yield. However, the metabolism of the cell also impinges on $\Delta G_s$, the energy needed for an equivalent of cells. $\Delta G_s$ is made of two parts: (i) the amount of energy needed to transform the source of carbon to some cellular carbon pool, $\Delta G_p$, and (ii) the amount required to transform the pool into cellular carbon $\Delta G_{pc}$.

$$\Delta G_S = \frac{\Delta G_p}{\varepsilon^n} + \frac{\Delta G_{pc}}{\varepsilon} \tag{5.5}$$

The efficiency with which energy is gained or lost is represented by $\varepsilon$. The term $n$ is 1 if energy is released and $-1$ if it is consumed. The energy is consumed to fix carbon dioxide to pyruvate and released when a heterotrophic substrate is transformed into pyruvate. On this basis, and by assuming that the cells are relatively (60%) efficient machines, maximum theoretical yields can be estimated with some success. In real life, yield may be lower than calculated if energy is diverted away from reproduction due to environmental or other considerations. There is no *a priori* lower limit on yield though the organism must be able to reproduce itself faster than it dies. Thus, lower yields may lead to fewer births and thus lower rates of mutation and evolution.

However, this is not a cast iron or wholly satisfactory explanation. Sometimes organisms with lower yields are more abundant than organisms with higher yields. For example, the acetoclastic methanogens sometimes appear to be more abundant but less diverse than their hydrogenotrophic counterparts. It is possible that the evolution of a functional group is somehow constrained by yield or energy. There could be more to this than yield. Evolution ultimately derives from random mutations in the genome. There is evidence that metabolism can affect mutation rates (Gillooly *et al.* 2005) but the theory expressly excludes diverse bacterial functional groups. Nevertheless, the idea of a link between the energy or yield and mutation rates is an attractive one. All things being equal, one might suppose that the greater the capacity of a given functional group to carry mutations, the greater its capacity to evolve (assuming a varying environment).

Population geneticists believe that there is an upper limit to the number of mutations that an organism can bear. If this limit was a function of the free energy or yield, it might create a link between the thermodynamics of a particular functional group and its capacity to evolve. There is certainly a link between mutations and fitness, a link which is well established and plays a central role in contemporary theories about the evolution of sex (Kondrashov 1988; Gillespie 2004). The fitness of *Escherichia coli* and *Salmonella typhimurium* declines as the number of mutations increases (Kibota & Lynch 1996; Elena & Lenski 1997). The literature has tended to emphasize the ability of sexual reproduction to alleviate this burden and rather tended to overlook the implications for microbial evolution. However, Maisnier-Patin (Maisnier-Patin *et al.* 2005) mentioned that the fitness reduction could have important ecological impacts.

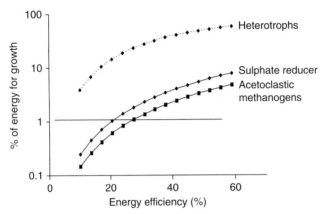

**Figure 5.7** An exploration of the proportion of electrons available for synthesis once the energetic needs of the cell are satisfied. This fraction is a function of the metabolism and efficiency of the cell. There is, we hypothesize, some lower limit on this efficiency below which the cell cannot survive. Mutations are thought to decrease the efficiency of the cell and thus bring it closer to this limit. Consequently, we reason that cells with less energy available for synthesis will be able to bear fewer mutations and thus be intrinsically less able to evolve.

There must be some lower limit of fitness below which the burden of mutation becomes intolerable and the rate of reproduction falls below the rate of death and that organism disappears. Of course the data we have is on the accessible but ecologically inconsequential 'laboratory freaks' *E. coli* and *S. typhimurium* where losses of fitness of 30% are relatively easily observed. We can attempt to interpret this finding for other organisms by considering how the loss of fitness might occur and what implications that might have for an organism.

Fitness is reflected in growth rate (Maisnier-Patin *et al.* 2005). There is evidence that, to a first approximation, growth rate is a function of the energy available for transfer, the energy required for synthesis and the efficiency of these two processes (McCarty 1971). It seems likely that deleterious mutations would decrease the efficiency of the cell. We therefore use the simple energetics-based model described above to show how a decline in efficiency would affect the fraction of energy left for growth, and thus the yield, in different functional groups all using acetate as an electron donor (Fig. 5.7). There will be some threshold below which the efficiency of an organism is too low for it to survive in that environment. It seems likely that a deleterious mutation in a lower energy group will bring an organism closer to such a threshold than in a higher energy organism. The demands of the environment, especially extreme environments, might exacerbate this finding. Thus, those organisms with the least energy and the greatest demands on the energy they have will be the worst placed to carry deleterious mutations and thus the slowest to evolve.

Unfortunately, most studies of mutation in bacteria are undertaken on tractable organisms and certainly not over the wide range of functional groups that this hypothesis addresses. However, the little we do know is supportive of this suggestion. The extremophile archaeon *Sulfolobus acidocaldarius* has lower yields (Hatzinikolaou *et al.* 2001) and inexplicably lower mutation rates (by a half) than in *E. coli* (Grogan *et al.* 2001). Interestingly, Valentine (2007) has pointed to the low rates of evolution in the Archaea and suggested that this is related to energy stress.

## Caveats and ambitions

These are tentative but testable suggestions about the mechanism by which free energy affects microbial diversity. Nevertheless, the phenomenological association between free energy and diversity is compelling and a consideration of free energy has proven invaluable and insightful in predicting (Strous *et al.* 1999) and explaining (Costa *et al.* 2006) the existence of certain functional groups. It seems likely that if, and when, we get a proper understanding of microbial diversity, we will find that the underlying mechanism, evolution, is controlled and constrained by free energy. This understanding will be an important step forward in understanding and even predicting the characteristics of microbial ecosystems from first principles. Fulfilling this aspiration would fulfil Wilson's desire for consilience, it need not stop there. For a detailed consideration of the relationship between free energy, kinetic parameters and evolution could reveal relationships between the two that might lead to a calibrated and predictive understanding of resource use in microbial ecosystems without the necessity of characterizing every single species. This in turn might open the door to usable niche-based models. This would be invaluable at every level of human endeavour from the humblest sewage works in a slum to the modelling of the whole Earth system. Both systems are hanging by a thread of a few select microbial species (Curtis 2006). Thus, Wilson's ideals (Wilson 1998) are not merely erudite and idealistic but deeply practical. This search for simple parameterized consilient models may prove to be quixotic. But given the importance of a predictive understanding of microbial ecosystems and the pace of change in climate and society and the central role of microbes to a sustainable future for humans, and indeed all life on the planet, it is a quest worth undertaking. For it is better to seek this grail and fail, and know that you have failed, than to give the illusion of progress with mathematical castles in the air.

## References

Alonso, D., Etienne, R. S. and McKane, A. J. (2006) The merits of neutral theory. *Trends in Ecology & Evolution* **21**, 451–457.

Baptista, J. C. B., Sloan, W., Head, I. M. and Curtis, T. P. (2008) The species area curve is not observed or predicted in wastewater treatment plants. In Preparation.

Bell, G. (2001) Ecology – neutral macroecology. *Science* **293**, 2413–2418.

Bell, T., Ager, D., Song, J. I., *et al.* (2005) Larger islands house more bacterial taxa. *Science* **308**, 1884.

Brown, J. H. (2000) *Macroecology.* University of Chicago Press, Chicago.

Caswell, H. (1976) Community structure – neutral model analysis. *Ecological Monographs* **46**, 327–354.

Clarke, A., and Gaston, K. J. (2006) Climate, energy and diversity. *Proceedings of the Royal Society B-Biological Sciences* **273**, 2257–2266.

Cody, M. L. and Diamond, J. M. (1975) Robert MacArthur, 1930–1972. In: *Ecology and Evolution of Communities* (ed. M. L. Cody and J. M. Diamond), pp. 1–12. Belknap Press of Harvard University Press, Cambridge, MA.

Coskuner, G., Ballinger, S. J., Davenport, R. J., *et al.* (2005) Agreement between theory and measurement in quantification of ammonia-oxidizing bacteria. *Applied and Environmental Microbiology* **71**, 6325–6334.

Costa, E., Perez, J. and Kreft, J. U. (2006) Why is metabolic labour divided in nitrification? *Trends in Microbiology* **14**, 213–219.

Curtis, T. (2006) Microbial ecologists: it's time to 'go large'. *Nature Reviews Microbiology* **4**, 488.

Curtis, T. P. and Sloan, W. T. (2005) Exploring microbial diversity – a vast below. *Science* **309**, 1331–1333.

Curtis, T. P., Sloan, W. T. and Scannell, J. W. (2002) Estimating prokaryotic diversity and its limits. *Proceedings of the National Academy of Sciences of the United States of America* **99**, 10494–10499.

Curtis, T. P., Head, I. M., Lunn, M., *et al.* (2006) What is the extent of prokaryotic diversity? *Philosophical Transactions of the Royal Society B-Biological Sciences* **361**, 2023–2037.

Dar, S. A., Yao, L., van Dongen, U., Kuenen, J. G. and Muyzer, G. (2007) Analysis of diversity and activity of sulfate-reducing bacterial communities in sulfidogenic bioreactors using 16S rRNA and dsrB genes as molecular markers. *Applied and Environmental Microbiology* **73**, 594–604.

Elena, S. F. and Lenski, R. E. (1997) Test of synergistic interactions among deleterious mutations in bacteria. *Nature* **390**, 395–398.

Etienne, R. S. and Alonso, D. (2005) A dispersal-limited sampling theory for species and alleles. *Ecology Letters* **8**, 1147–1156.

Evans, K. L. and Gaston, K. J. (2005) Can the evolutionary-rates hypothesis explain species–energy relationships? *Functional Ecology* **19**, 899–915.

Finke, N., Hoehler, T. M., and B. B. Jorgensen. (2003) Methanogenesis from methylamine and methanol at changing hydrogen concentrations. *Geochimica et Cosmochimica Acta* **67**, A97.

Finke, N., Hoehler, T. M. and Jorgensen, B. B. (2007) Hydrogen 'leakage' during methanogenesis from methanol and methylamine: implications for anaerobic carbon degradation pathways in aquatic sediments. *Environmental Microbiology* **9**, 1060–1071.

Francis, C. A., Roberts, K. J., Beman, J. M., Santoro, A. E. and Oakley, B. B. (2005) Ubiquity and diversity of ammonia-oxidizing archaea in water columns and sediments of the ocean. *Proceedings of the National Academy of Sciences of the United States of America* **102**, 14683–14688.

Gillespie, J. H. (2004) *Population Genetics: A Concise Guide.* John Hopkins University Press, Baltimore, MD.

Gillooly, J. F., Allen, A. P., West, G. B. and Brown, J. H. (2005) The rate of DNA evolution: Effects of body size and temperature on the molecular clock. *Proceedings of the National Academy of Sciences of the United States of America* **102**, 140–145.

Grogan, D. W., Carver, G. T. and Drake, J. W. (2001) Genetic fidelity under harsh conditions: analysis of spontaneous mutation in the thermoacidophilic archaeon Sulfolobus acidocaldarius.

*Proceedings of the National Academy of Sciences of the United States of America* **98**, 7928-7933.

Hallin, S., Throback, I. N., Dicksved, J. and Pell, M. (2006) Metabolic profiles and genetic diversity of denitrifying communities in activated sludge after addition of methanol or ethanol. *Applied and Environmental Microbiology* **72**, 5445-5452.

Hatzinikolaou, D. G., Kalogeris, E., Christakopoulos, P., Kekos, D. and Macris, B. J. (2001) Comparative growth studies of the extreme thermophile Sulfolobus acidocaldarius in submerged and solidified substrate cultures. *World Journal of Microbiology & Biotechnology* **17**, 229-234.

Horner-Devine, M. C., Lage, M., Hughes, J. B. and Bohannan, B. J. M. (2004) A taxa–area relationship for bacteria. *Nature* **432**, 750-753.

Horner-Devine, M. C., Silver, J. M., Leibold, M. A., *et al.* (2007) A comparison of taxon co-occurrence patterns for macro- and microorganisms. *Ecology* **88**, 1345-1353.

Houchmandzadeh, B. and Vallade, M. (2003) Clustering in neutral ecology. *Physical Review E* **68**.

Hubbell, S. P. (2001) *The Unified Neutral Theory of Biodiversity and Biogeography.* Princeton University Press, Princeton, NJ.

Huber, J. A., Welch, D. B. M., Morrison, H. G., *et al.* (2007) Microbial population structures in the deep marine biosphere. *Science* **318**, 97-100.

Huisman, J. and Weissing, F. J. (1999) Biodiversity of plankton by species oscillations and chaos. *Nature* **402**, 407-410.

Kibota, T. T. and Lynch, M. (1996) Estimate of the genomic mutation rate deleterious to overall fitness in *E-coli*. *Nature* **381**, 694-696.

Kondrashov, A. S. (1988) Deleterious mutations and the evolution of sexual reproduction. *Nature* **336**, 435-440.

Leclerc, M., Delgenes, J. P. and Godon, J. J. (2004) Diversity of the archaeal community in 44 anaerobic digesters as determined by single strand conformation polymorphism analysis and 16S rDNA sequencing. *Environmental Microbiology* **6**, 809-819.

MacArthur, R. (1957) On the relative abundance of bird species. *Proceedings of the National Academy of Science* **43**, 293-295.

MacArthur, R. and Wilson, E. (1967) *The Theory of Island Biogeography.* Princeton University Press, Princeton, NJ.

Maisnier-Patin, S., Roth, J. R., Fredriksson, A., *et al.* (2005) Genomic buffering mitigates the effects of deleterious mutations in bacteria. *Nature Genetics* **37**, 1376-1379.

May, R. M. (1974) Patterns of species abundance and diversity. In: *Ecology and Evolution of Communities* (ed. M. L. Cody and J. M. Diamond), pp. 81-120 Harvard University Press, Harvard.

Maynard-Smith, J. (1974) *Models in Ecology.* Cambridge University Press, London.

McCarty, P. L. (1971) Energetics and bacterial growth. In: *Organic Compounds in Aquatic Environments* (ed. S. D. Fraust and J. V. Hunter). Marcel Dekker Inc, New York.

McCarty, P. L. (2007) Thermodynamic electron equivalents model for bacterial yield prediction. *Modifications and Comparative Evaluations. Biotechnology and Bioengineering* **97**, 377-388.

McHugh, S., Carton, M., Mahony, T. and O'Flaherty, V. (2003) Methanogenic population structure in a variety of anaerobic bioreactors. *FEMS Microbiology Letters* **219**, 297-304.

McKane, A. J., Alonso, D. and Sole, R. V. (2004) Analytic solution of Hubbell's model of local community dynamics. *Theoretical Population Biology* **65**, 67-73.

Mouquet, N., and Loreau, M. (2003) Community patterns in source-sink metacommunities. *American Naturalist* **162**, 544-557.

Neufeld, J. D. and Mohn, W. W. (2005) Unexpectedly high bacterial diversity in arctic tundra relative to boreal forest soils, revealed by serial analysis of ribosomal sequence tags. *Applied and Environmental Microbiology* **71**, 5710–5718.

Peters, R. H. (1991) *A Critique for Ecology.* Cambridge University Press, Cambridge.

Pounds, J. A. and Puschendorf, R. (2004) Ecology – clouded futures. *Nature* **427**, 107–109.

Prosser, J. I., Bohannan, B. J. M., Curtis, T. P., *et al.* (2007) Essay – the role of ecological theory in microbial ecology. *Nature Reviews Microbiology* **5**, 384–392.

Ramette, A. and Tiedje, J. M. (2007) Multiscale responses of microbial life to spatial distance and environmental heterogeneity in a patchy ecosystem. *Proceedings of the National Academy of Sciences of the United States of America* **104**, 2761–2766.

Rittmann, B. E. and McCarty, P. L. (2001) *Environmental Biotechnology, Principles and Applications.* McGraw-Hill Inc., New York.

Sloan, W. T., Lunn, M., Woodcock, S., *et al.* (2006) Quantifying the roles of immigration and chance in shaping prokaryote community structure. *Environmental Microbiology* **8**, 732–740.

Sloan, W. T., Woodcock, S., Lunn, M., Head, I. M. and Curtis, T. P. (2007) Modeling taxa-abundance distributions in microbial communities using environmental sequence data. *Microbial Ecology* **53**, 443–455.

Sogin, M. L., Morrison, H. G., Huber, J. A., *et al.* (2006) Microbial diversity in the deep sea and the underexplored 'rare biosphere'.

*Proceedings of the National Academy of Sciences of the United States of America* **103**, 12115–12120.

Strous, M., Fuerst, J. A., Kramer, E. H. M., *et al.* (1999) Missing lithotroph identified as new planctomycete. *Nature* **400**, 446–449.

Tilman, D. (2004) Niche tradeoffs, neutrality, and community structure: a stochastic theory of resource competition, invasion, and community assembly. *Proceedings of the National Academy of Sciences of the United States of America* **101**, 10854–10861.

Valentine, D. L. (2007) Adaptations to energy stress dictate the ecology and evolution of the Archaea. *Nature Reviews Microbiology* **5**, 316–323.

Vallade, M. and Houchmandzadeh, B. (2003) Analytical solution of a neutral model of biodiversity. *Physical Review E* **68**.

Volkov, I., Banavar, J. R., Hubbell, S. P. and Maritan, A. (2003) Neutral theory and relative species abundance in ecology. *Nature* **424**, 1035–1037.

Wilson, E. O. (1998) *Consilience: The Unity of Knowledge.* Vintage, New York.

Woodcock, S., Curtis, T. P., Head, I. M., Lunn, M. and Sloan, W. T. (2006) Taxa-area relationships for microbes: the unsampled and the unseen. *Ecology Letters* **9**, 805–812.

Woodcock, S., van der Gast, C. J., Bell, T., *et al.* (2007) Neutral assembly of bacterial communities. *FEMS Microbial Ecology* **62**, 171–180.

# CHAPTER SIX

# Limits to adaptation and patterns of biodiversity

JON R. BRIDLE, JITKA POLECHOVÁ
AND TIM H. VINES

## Why do species have finite ranges in space and time?

All species have limited ecological distributions, and all species eventually become extinct. At the heart of these distributional limits is the idea of trade-offs: a single population or species cannot maximize its fitness in all environments (Woodward and Kelly 2003). Each species therefore occupies a limited range of ecological conditions, or a particular period in history, and interacts in complex ways in ecosystems consisting of many co-existing species. These interactions may in turn generate more specialization (Nosil & Harmon, this volume; Schemske, this volume). However, from an evolutionary biology perspective this explanation is incomplete. Populations clearly adapt to novel environments in some circumstances, otherwise there would be no life on land, no mammals in the ocean, and only a few species on oceanic islands such as Hawaii (Wagner & Funk 1995). What processes, therefore, act to constrain adaptation to changing environments and continually prevent the expansion of species into new habitats at the edge of their range?

Understanding the factors that limit the temporal or spatial persistence of species is of key practical importance, given ongoing changes in global climate (Root *et al.* 2003), coupled with rapid habitat loss and alteration by the introduction of exotic species of parasites, predators and competitors. Models based on species' existing ecological tolerances estimate that at least 11% of species will become extinct during this century due to climate change alone, even if populations can freely disperse to track the distribution of suitable habitat (Thomas *et al.* 2004; see Parmesan 2006). This figure will be an underestimate where dispersal is limited, or if local adaptation already exists within a species' range, meaning that ecological tolerances of single populations are actually lower than models assume (Harte *et al.* 2004). Conversely, extinction rates will be reduced if species can rapidly evolve to changing conditions, allowing the exploitation of more widespread habitats, and reducing the necessity for the large geographical range shifts (Schwartz *et al.* 2006; Bridle & Vines 2007). These issues bring together the issue of limits to adaptation in time as well as in space: why should

*Speciation and Patterns of Diversity*, ed. Roger K. Butlin, Jon R. Bridle and Dolph Schluter. Published by Cambridge University Press. © British Ecological Society 2009.

populations respond to changes in their environments over time if they are consistently unable to respond to spatially variable selection, as demonstrated by their current stable margins? Is adaptation within populations over time easier than for populations distributed across an ecological gradient? Alternatively, are ecological limits to species' distributions only stable over short evolutionary time scales?

In this chapter we review theoretical models for limits to the response of populations to temporal or spatial changes in selection. We then consider how these models are connected theoretically, and briefly review empirical data relating to the critical issues raised by these models. Finally, we ask how relevant limits to adaptation are to existing patterns of biodiversity, and highlight key areas for future research.

## Limits to adaptation in time: the cost of shifting optima

For quantitative traits determined by more than a few loci, evolutionary change can be summarized by changes in the mean and the variance of a trait, or its covariance with other traits. The rate of adaptation of this trait is determined by the additive genetic covariance with fitness. However, genetic variance in fitness comes at a cost, because it means that not all individuals match local optima. The continual loss of productivity due to genetic variance in fitness is termed the 'standing load' of a population, and is due to segregation, recombination and stabilizing selection around an optimum (Haldane 1957). Should the selective optimum change, the response of a population to directional selection generates a substitution load (the 'cost of selection'), which can be defined as the number of selective deaths required to generate a given change in trait mean (Haldane 1957; Barton & Partridge 2000). However, while the population responds to this shifting optimum, there is also a third type of load (the 'lag load' or 'evolutionary load'; Fig. 6.1), which is the fitness cost of the population remaining a given distance from the local selective optimum (Lande 1976). The magnitude of this lag, relative to the width of the stabilizing selection function (a measure of the strength of selection), determines the mean fitness of the population and therefore its growth rate, so reducing population density. Above a critical rate of environmental change, the lag load becomes so great that it matches the population's rate of growth, and extinction occurs before adaptation is possible.

Following the model presented by Lande (1976), Lynch and Lande (1993) considered the relationship between the rate of environmental change in time and extinction using a deterministic model of weak stabilizing selection towards an optimal phenotype. Lynch and Lande (1993) found that the upper bound to where the population has zero growth ($k'_c$) is given by $k'_c < V_G \sqrt{2\bar{r}_{max}/V_S}$, where $V_G$ is the genetic variance, $V_S$ is the width of the stabilizing selection function and $\bar{r}_{max} = r_{max} - V_P/2V_S$ is the maximum rate of population increase

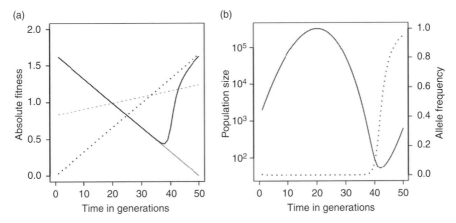

**Figure 6.1** Evolutionary load resulting from a changing optimum in time (after Roughgarden 1998). (a) The absolute mean fitness (thick solid line) and the fitness of each genotype through time. The thin solid line represents *AA*, the dashed line *Aa* and the dotted line *aa*. (b) Population size (solid line) and the frequency of the *a* allele through time. Initially, the population absolute fitness is greater than one and the population is expanding. However, as the environment changes the most common allele (*A*) becomes less fit, but the alternate allele (*a*) is still not fitter than *A* and cannot make up for the loss of fitness (Fig. 6.1a). The rate of population expansion declines, and ultimately the population shrinks (Fig. 6.1b). If the fecundity of each genotype is too low, the population goes extinct before *a* can fix. The cost of remaining a given distance from the optimum is the lag load. Even if fecundity is moderately high, the load imposed by replacing *A* with *a* can still cause a dramatic reduction in population size (Fig. 6.1b).

when the population mean trait is at the optimum. Generally speaking, these models suggest that the lag of the trait mean behind a smoothly changing optimum is the key issue determining the upper rate of ecological change that can be matched by evolution without causing the population's extinction. The magnitude of this lag is proportional to the speed of movement of the optimum times the strength of stabilizing selection, divided by the square of genetic variance. For a smooth, directional change in the environment, increasing genetic variation increases the maximum rate of evolution, thus reducing the lag load (Fig. 6.2). Similarly, the lag load decreases with the strength of selection, but the optimal genetic variance (when the population grows the fastest) increases with the strength of selection. This is because although for a given genetic variance, stronger selection increases the per generation standing load of genetic variation, this effect is more than counterbalanced by its effect in causing the population to track the shifting optimum more closely, so reducing the total size of the lag load.

Lande and Shannon (1996) explored how the pattern of environmental change affects maximum adaptive rates by contrasting smoothly changing

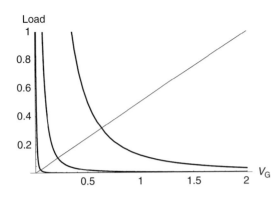

**Figure 6.2** Genetic variation ($V_G$) leads to a higher standing load, but also increases the maximum sustainable rate of adaptation for a smooth change in the environment over time ($k$). The curved lines (from left to right) represent lag load $k^2 V_S/2 V_G^2$ for the rate of change of the optimum $k = 0.01$, 0.1 and 0.5. The straight line represents standing load $V_P/2 V_S$ with $V_S$ representing the width of the selective function (a constant in this example), and assuming no environmental effects on phenotypes ($V_E = 0$, $V_G = V_P$).

environments with cyclic or randomly changing environments, and considering the effect of genetic variance on population persistence in each case. They concluded that increased genetic variance does not always enhance the probability of survival. Instead, the value of genetic variance depends on the pattern of environmental change, and its effect of reducing evolutionary load overall must be compared with the selective cost of genetic variation per generation. Where changes in the environment are predictable, but fluctuate in a cyclic or autocorrelated fashion, increased genetic variation only increases adaptability if the period of environmental change is much longer than the generation time, and if the magnitude of the oscillations is sufficiently large that the increased rate of evolution outweighs the per generation cost of genetic variation. Genetic variance is therefore a benefit where the population is often far from its optimum (and selection is sufficiently strong), but is a cost where the optimum changes too quickly for increased evolutionary rates to be an important advantage. These results highlight how the evolutionary responses of populations critically depend on the temporal pattern of environmental change. Similar conclusions were obtained by Charlesworth (1993a,b) and also by Bürger (1999) and Waxman and Peck (1999) in reference to the evolution of genetic variance in changing environments.

Bürger and Lynch (1995) extended Lynch and Lande's (1993) model to include genetic and demographic stochasticity by sampling individuals at random to found the next generation of parents, whenever the population size was greater than the population's carrying capacity given its trait mean. They estimate that demographic and genetic stochasticity alone can reduce the critical rate of environmental change to only 1% of the phenotypic standard deviation per generation (an order of magnitude less than the deterministic models described above). Stochasticity raises the probability of extinction because even though it occasionally boosts the mean phenotype towards the optimum, the negative consequences of moving the population away from, or overshooting, the optimum more than outweigh this effect. In addition, genetic variance can change substantially from generation to generation, and may take several generations to recover from periods of low variation. By contrast, high levels of genetic variance are removed by only a single generation of low population size. Importantly, Bürger and Lynch (1995) find that increasing initial population size only reduces the probability of extinction when the rate of environmental change is relatively low. This is because the amount of genetic variance rapidly levels off with increasing population size, so that with high rates of environmental change, even very large populations will rapidly become extinct.

Another class of models deals with how populations adapt to abrupt, rather than gradual, environmental change. Gomulkiewicz and Holt (1995) examine a population that is shrinking after a sudden change in the environment, and ask whether adaptation can restore a positive growth rate before the population becomes too small to avoid extinction through demographic stochasticity. They find that only large populations with high levels of additive genetic variance in fitness are likely to be rescued from extinction by adaptation, but only if the initial change in the environment is small relative to the standing phenotypic variation.

As well as generating increased lag load, a greater proportion of mutations is favourable the further the population is from the selective optimum, including those of large phenotypic effect (Fisher 1930; Poon & Otto 2000; Orr 2005; Martins & Lenormand 2006; Kopp & Hermisson 2007). This phenomenon may allow adaptation to occur at a faster rate than predicted by deterministic models. For example, individual-based simulation models developed by Boulding and Hay (2001) found that adaptation was possible even when the population mean was shifted more than two phenotypic standard deviations from the optimum. This was because, rather than assuming a fixed variance scaled according to the mean, adaptation could occur by the spread of rare, extremely beneficial genotypes provided the population's growth rate was high, and the starting population size large. However, the maladapted population must persist for sufficiently long for such highly beneficial, multilocus genotypes to arise. The relative importance of novel mutations of large effect on maximum rates of adaptation in natural populations remains an open question.

## Limits to adaptation in space

As selection changes in time within populations, it also varies across a species' geographical range. Where species show clinal divergence throughout their range, migration along the spatial environmental gradient will act to increase genetic variance in fitness above the expectations of single-population models (Felsenstein 1976; Wilson & Turrelli 1986; Lenormand 2002). In particular, for marginal populations, genetic variance will be dominated by the effects of gene flow rather than local population size. This 'spreading' effect increases the evolutionary capacity of the population to adapt towards the environmental optimum (Slatkin & Maruyama 1975; Gomulkiewicz *et al.* 1999; Alleaume-Benharira *et al.* 2006; Bolnick & Nosil 2007), as well as potentially rescuing the population from extinction by continually introducing new individuals to the margins (Holt 2003). At the same time, however, immigration from differently adapted populations displaces the population phenotypic mean away from the local optimum (a 'migration load' in this case, analogous to the lag load discussed above), reducing population growth and therefore reducing population density (a 'swamping' rather than 'spreading' effect of gene flow). Migration along environmental gradient also maintains high levels of 'standing load' because genetic variance remains high, even when the mean matches the optimum.

It has been shown by Slatkin (1973) and Nagylaki (1975) that divergence is only possible if the environment changes over a scale which is large enough relative to the ratio of dispersal over the square root of the intensity of selection per locus, $\sigma/\sqrt{s}$. If environment changes only over smaller scales, a cline in allele frequency cannot develop. Instead, allele frequencies respond to the selection averaged over the characteristic length. It follows from the more general treatment of Nagylaki (1975) that adaptation to habitats smaller than about $\sigma/\sqrt{s}$ is prevented if the difference in the ratio of selection coefficients over the characteristic length is large. In the following section, we deal with environments that are gradually changing on habitats much larger than $\sigma/\sqrt{s}$.

In the simplest models of evolution at range margins, beginning with Haldane's theory of a cline (1956), and developed by Pease *et al.* (1989) and Kirkpatrick and Barton (1997), a continuous population persists along a environmental gradient that varies smoothly in space (Fig. 6.3). Evolution is modelled by changes in the mean of a trait to match this changing optimum selective gradient. As with models for selection in time, population growth rate is a function of how closely the trait matches the environmental optimum at that point on the gradient. When the population is able to track the optimum, the species can expand along the gradient and density remains high (Fig. 6.3a). However, if the environmental gradient is too steep relative to the amount of genetic variation available, the population declines in mean fitness at a rate proportional to the distance of the trait mean from the optimum. Migration now comes mainly from the central, well-adapted parts of the range into the margins, meaning that marginal populations remain distant from the local selective optimum (Fig. 6.3b). In such models, where gene flow is biased towards the influx of genes into marginal populations, adaptation is effectively biased towards those environmental conditions where the largest number of individuals persist (Holt & Gaines 1992; Kawecki & Holt 2002).

The Kirkpatrick and Barton (1997) model focuses on the swamping consequence of gene flow, and predicts that adaptation at range margins is determined by the steepness of a environmental gradient, or the amount of gene flow along it (defined as the 'rate of change in selection'), relative to the amount of genetic variation in fitness available (Fig. 6.4). Species' margins should therefore be

(a)

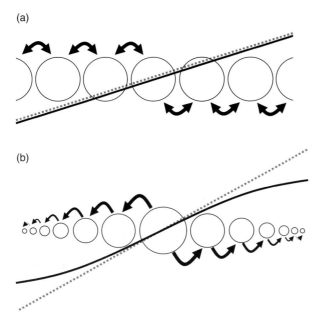

(b)

**Figure 6.3** (a) Range expansion without limit along a one-dimensional environmental gradient in space – the Kirkpatrick and Barton model (1997). Here, the trait mean (solid line) at each point along the environmental gradient matches the environmental optimum (dashed line) everywhere. Population fitness is therefore high, and population size uniformly large (indicated by the size of the circles), and the species continually expands along the gradient. The arrows depict the direction and magnitude of migration between adjacent populations. (b) Range margins generated by migration load in the Kirkpatrick and Barton model (1997) of limits to evolution in space. In this case, the well-adapted central population is also the largest and sends out many migrants to adjacent populations (solid black arrows). These immigrants prevent adjacent populations from reaching their trait optimum (the solid line is displaced from the dashed line), which reduces their fitness and hence their population size. These populations in turn send out migrants that are even less fit, further reducing the fitness and therefore the size of the more peripheral populations. Eventually, the trait mean of the peripheral populations is very far from the optimum, and fitness is so low that population growth is negative even with immigration. Reproduced, with permission from Bridle and Vines (2007).

associated with locally steep environmental gradients and/or areas where gene flow is locally increased. The restriction of gene flow along environmental gradients (by the evolution of assortative mating or habitat choice) may therefore be necessary for adaptive radiation into novel environments.

Although increasing genetic variation throughout the range increased the maximum rate of change a population could adapt to (Fig. 6.4), Kirkpatrick and Barton (1997) did not allow genetic variance within populations to be elevated by migration along the environmental gradient. Barton (2001) extended the

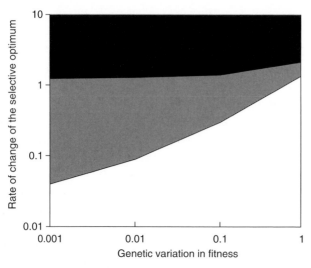

**Figure 6.4** Causes of limits to adaptation in space. Plot from Butlin *et al.* (2003), modified from Kirkpatrick and Barton (1997), to show the behaviour of an analytical model exploring adaptation in a fitness trait along a environmental gradient. Shading delimits parameter values with different outcomes: Unlimited range (adaptation everywhere); Finite range (adaptation is limited to a region of ecological space); and Extinction (the species is unable to sustain itself along any part of the environmental gradient). This model reveals two parameters to be central to predicting when and where adaptation will occur: (1) The rate of change in the selective optimum and (2) The amount of genetic variation in fitness. The rate of change in the optimum is determined by gene flow as well as local selection (see text for details) (reproduced, with permission, from Butlin *et al.* 2003).

Kirkpatrick and Barton (1997) model to allow the evolution of genetic variance, and found that, for a range of plausible genetic models, the swamping effect of gene flow was outweighed by the associated increase in genetic variance. This allowed the population to match the phenotypic optimum, even along very steep environmental gradients. However, as with models for selection in time, an absolute critical limit is reached. Although gene flow along even very steep environmental gradients provides enough variance to match the optimum perfectly, the demographic cost of these high levels of variance eventually reduces the mean fitness sufficiently to cause population extinction. Butlin *et al.* (2003) and Bridle *et al.* (in revision) considered an individual-based simulation model, the assumptions and behaviour of which closely match Barton's (2001) analytical model for weak selection on a quantitative trait. This study allows the consequences of demographic and genetic stochasticity on adaptation in space to be considered in detail. As with selection in time (Bürger & Lynch 1995), the stochasticity created by finite population size generates limits to adaptation at a wider range of parameter values than those predicted by analytical models.

As with models for selection in time, models for evolution along environ-mental gradients in space suggest that the evolutionary consequences of gene flow vary depending on how close a population is to its trait optimum. When the population is well adapted, the increased variance introduced by gene flow creates a genetic load that reduces population density. By contrast, in marginal habitats, where the trait mean is already far from the optimum, increased variance is a benefit, as some migrant individuals are close enough to the optimum to reproduce, and extend the population into novel habitat. The effect of gene flow at range margins is analogous to the evolution of recombination which is beneficial only when a population does not match the current selective optimum (see Otto & Lenormand 2002). However, differences are likely in the spatial scale at which the swamping and spreading effects of gene flow operate across a species' range. Although genetic variance (and evolutionary potential) may be significantly inflated by only small levels of gene flow, the swamping of local adaptation may require higher levels of migration, and may be important only where migration is asymmetrical (Endler 1973; Felsenstein 1976; Kawecki & Holt 2002).

## Integrating limits to adaptation in time and space

Limits to the rates of evolution in time and space are connected, in that both depend on the amount of genetic variance (either dominated by mutation/selection balance, or because of gene flow), and on the demographic cost of being displaced from the selective optimum. The distance from the optimum can change over time or vary due to gene flow along a gradient, provided selection varies in space. The key question therefore becomes: at what point does the genetic variance (the standing load), or the distance of a population's mean from the optimum (the lag or migration load), have such a high demo-graphic cost that the population goes extinct?

For adaptation in time or in space, the critical rate of environmental change above which the population becomes extinct or forms a finite spatial range is given by similar equations. Here we present the critical rates of change under joint evolution of trait mean and population density (see Lande 1976, Eq. 7; Lande & Shannon 1996, Eqs 1–3; Kirkpatrick & Barton 1997, Eqs 1 and 7), assuming that density-dependence is logistic, and that the genetic variance is fixed. Assessing adaptation in space, migration is approximated by diffusion, with $\sigma$ as the standard deviation of the migration distance. The population growth rate is then $\bar{r} = r_{max}(1 - n/K) - (\bar{z} - \theta)^2/2V_S - V_P/2V_S$, where $r_{max}$ is a constant determining maximum growth rate, $n$ is population density, $K$ carrying capacity, $\bar{Z}$ is the mean phenotypic value, $\theta$ is the optimum ($bx$ at point $x$ in space, or $kt$ at point $t$ in time), $V_S$ is the width of the stabilizing selection, and $V_P$ the phenotypic variance ($V_P = V_G + V_E$, where $V_E$ is environmental variance). For an optimum changing in time, the maximum sustainable rate of evolution is

$k_c = V_G\sqrt{2\bar{r}_{max}/V_S}$, which is the same as with no density-dependence (Lynch & Lande 1993).

Considering adaptation in space, the population can adapt on the whole range, adapt only on a limited range, or go extinct (see Kirkpatrick & Barton 1997; Barton 2001). Firstly, unless the standing load is too high, $V_P > 2\,r_{max}V_S$, the population can adapt with the trait mean matching the environmental gradient ($b$), so allowing the range to expand without limits. Secondly, as the gradient steepens above a critical value, $b_c \approx 2/\sigma\sqrt{r_{max}V_G}$, the population may fail to adapt. The gradient in trait mean is therefore shallower than the environmental gradient (creating a lag load), and the range is limited. Although both extremes are locally stable when the space is infinite, boundaries to the physical space often lead to collapse of the perfect adaptation towards the regime when gradient in trait mean is shallower than the optimum, and the range is limited. The density of a population with a limited range reflects the underlying environmental gradient as the difference between the trait mean and the local optimum increases as the gradient steepens. Such populations become extinct throughout the entire range approximately when $b > 1/\sigma\left(V_G/\sqrt{V_S} + 2\bar{r}_{max}\sqrt{V_S}\right)$ (Kirkpatrick & Barton 1997, corrected Eq. 16; we assume all genetic variance is additive. Details of the above formulae can be found in Polechová *et al.*, in preparation). When the loss of fitness due to genetic variance as described by $V_G/(V_S\bar{r}_{max})$ is large, adaptation fails abruptly at the above threshold: the parameter region, where the solution with limited range exists, tends to zero. Generally, the population is able to adapt to changing conditions if the additive genetic variance and the average growth rate at the optimum $\bar{r}_{max} = r_{max} - V_P/2V_S$ are large. The ability of the population to adapt in space (as measured by the critical gradient $b_c$) increases with weaker stabilizing selection, $1/V_S$. Similarly, when the optimum changes in time, this is true for $1/V_S > r_{max}\,/\,V_P$; the critical rate of change $k_c$ over which the population can still persist is reduced when stabilizing selection is weaker.

Adaptation in space and time are also connected in that the range of ecological conditions that a species experiences while it is adapting to changes in time is determined by the amount of clinal divergence occurring within the species. A wider range will also typically show larger local population sizes (Gaston 2000), which is consistent with the predictions of Kirkpatrick and Barton (1997). Larger populations have increased mutational input and can sustain higher demographic cost during adaptation. More widely distributed species should therefore be better at adapting to environmental change over time because their maximum sustainable rate of evolution will be higher. Such species are also likely to maintain particular genotypes within their range that are important for adaptation to temporal environmental gradients. Furthermore, populations arrayed across steep environmental gradients may be better at evolving to match shifting optima in time, because a given amount of gene flow has a greater effect on genetic variance. There have been only a few attempts to address joint adaptation in space and time: most notably, Pease *et al.* (1989) give predictions for such a scenario, which are valid when the genetic variance is much smaller than the strength of selection. Detailed models exploring adaptation in environments which vary both in space and time are currently being developed (Polechová *et al.*, in preparation).

## Predicting maximum rates of adaptation in natural populations

These theoretical models highlight three key factors that limit the ability of populations to adapt to changing conditions in time or space. For selection in time, such limits result in extinction, whereas for selection in space, they result in the distributional limits to species that generate geographical patterns of biodiversity and determine how species interact in ecological communities. These factors are: (1) The strength of stabilizing selection; (2) the rate of change of the selection optima; and (3) the amount of genetic variation in fitness. Below we consider each of these parameters, and briefly summarize the empirical data available.

### The strength of stabilizing selection

The width of the selective function ($V_S$) describes the relative fitness of phenotypes that differ from the optimum for that point on the spatial or temporal environmental gradient. A narrow stabilizing selection function increases the standing load due to the existing genetic variance (generated both by mutation and by gene flow from other populations), but for a smoothly changing optimum, strong selection reduces the magnitude of the lag load generated by a given shift in the optimum.

Endler (1986) reviewed selection in natural populations and concluded that strong selection may be quite common. Contemporary studies of adaptive divergence also suggest that substantial changes can occur over short time scales, again implying strong selection (Stockwell *et al.* 2003). However, estimates of the strength of selection using quantitative genetic data vary depending on whether selection is standardized according to the trait mean (Houle 1992), or according to the trait variance (Lande 1979; Lande & Arnold 1983). Standardizing by the trait mean estimates the increase in relative fitness for a proportional change in the trait, and generates unrealistically strong estimates of selection (Hereford *et al.* 2004). By contrast, standardizing by trait variance estimates the change in relative fitness for change of one standard deviation in a given trait. This more data-intensive method generates estimates of selection that are too weak to have been detected by the sample sizes used in most cases, implying a significant publication bias (Kingsolver *et al.* 2001). In addition, these estimates are typically based on studies of one or two traits, whereas selection will usually act on particular combinations of many traits (Blows 2007; see p. 89). The likely width of the selective function in nature, and how much it typically varies between traits, populations and species therefore remains uncertain (Hendry 2005). Estimates of selection also tend to consider a small number of generations, whereas the strength of the selection function may fluctuate substantially over time, and may be less powerful in the long term than studies of natural and experimental populations suggest. To our knowledge, temporal variation in the strength of selection on a trait is something that has yet to be considered in theoretical models.

### The rate of change of the selective optima

Another key issue in predicting adaptation is the way that the environment changes in time (Felsenstein 1976; Charlesworth 1993a,b; Lande & Shannon 1996; Bürger 1999; Waxman & Peck 1999) as well as in space (Case & Taper, 2000; Barton 2001; Bridle *et al.*, in revision). This affects both the role of genetic variance, and the effect of the strength of selection on maximum adaptation (Pease *et al.* 1989; Lande & Shannon 1996).

If defining selective optima is difficult in a single population, then describing how rapidly selective optima change over time or space is much harder. In addition, the rate of change of selection in space is a product not only of the ecological gradient, but also of the rate and distance moved by individuals along this gradient (Kirkpatrick & Barton 1997; Fig. 6.3). The difficulty of estimating spatially variable selection accurately in the field is compounded by the fact that when we measure clines in quantitative traits we measure the response to selection, not the environmental gradient itself, which will almost always be an underestimate (see Case & Taper 2000; Barton 2001). Conversely, when we measure environmental gradients alone, we are ignoring interactions among different environmental gradients, as well as biotic interactions between species, which may act to locally increase or reduce the environmental gradient (Case & Taper 2000; Case *et al.* 2005). Such biotic interactions are likely to be important determinants of range margins in many cases (Davis *et al.* 1998). The nature and magnitude of these interactions can also evolve in response to selection.

There is evidence that species' limits coincide with regions where the environment changes quickly over short distances, rather than being arrayed randomly with respect to the steepness of ecological gradients (Thomas & Kunin 1999; Gaston 2000; Parmesan *et al.* 2005). However, the causal importance of ecological or population genetic factors in determining these species' limits remains unclear (Roy *et al.* 1998; Holt 2003; Parmesan *et al.* 2005). In particular, regions of low population density observed where ecological gradients change steeply could be associated either with a lack of suitable genetic variation in fitness, or with high levels of genetic variation resulting from swamping gene flow (Barton 2001; Butlin *et al.* 2003). Addressing this issue would provide valuable insight into causes of range margins. However, accurately estimating trait means and levels of genetic variation in such situations is challenging because it requires establishing large numbers of samples from low density populations at species' margins (Bridle *et al.*, submitted).

## The amount of genetic variation in fitness

As we have seen, the amount of genetic variation in fitness is a key factor affecting maximum rates of adaptation, although its effect depends on the pattern of change in selection (Lande & Shannon 1996). The rapid responses of small laboratory populations and domesticated organisms to artificial selection suggest that even small populations typically harbour high levels of genetic variation in quantitative traits (Barton & Keightley 2002). This means that the correlation between population size and genetic variation emphasized by models for adaptation to temporal change may apply only at very low population sizes (Willi *et al.* 2006, 2007). Instead, levels of genetic variation may be dominated by gene flow along spatial environmental gradients, as emphasized by models for adaptation in space.

Estimates of genetic variance in fitness are the key to estimating maximum evolutionary rates (Fisher 1930). However, the degree to which genetic variation in single traits relates to actual fitness variation is unclear (Blows & Hoffmann 2005; Blows 2007). The critical issue is defining the trait or combination of traits that allows a population to match the local selective optimum (Blows *et al.* 2004; Jones *et al.* 2004; McGuigan 2006; Blows 2007; Hellman & Pineda-Krch 2007). The measurement of single traits is not sufficient, because they are only rarely the sole target of selection and because divergence in any one trait is constrained by pleiotropy (where each gene affects variation in more than one trait), as well as by linkage disequlibria between loci affecting different traits (Mitchell-Olds 1996; Johnson & Barton 2005). Empirical evidence for constraints resulting from such genetic correlations between traits comes from the very rapid and sustained rates of phenotypic change generated by artificial selection (where trade-offs can be offset by husbandry) compared with natural selection (Barton & Keightley 2002). Widespread pleiotropy is also suggested by the high rate of input of mutational variance to quantitative traits relative to the per genome mutational rate (Lynch and Walsh, 1998; Chapter 12). Such genetic correlations are likely to increase as organisms become more complex (Fisher 1930; Orr 2005).

Genetic correlations between traits define 'genetic lines of least resistance' (Schluter 1996, 2000), and may be summarized by the variance–covariance matrix, or G matrix (Lande 1979; Lande & Arnold 1983; Houle 1992; Price *et al.* 1993; Steppan *et al.* 2002). Evolution will be biased in the direction of largest genetic covariance ($G_{max}$) should the fitness optimum shift, regardless of its actual position relative to the population's starting point (Fig. 6.5a). Genetic covariance slows the response to selection relative to its maximal rates, as well as significantly increasing the lag load by increasing the number of generations for which the population remains distant from the optimum (Hellmann & Pineda-Krch 2007). This reduces the maximum

(a)

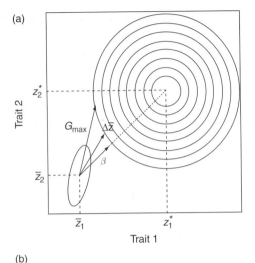

(b)

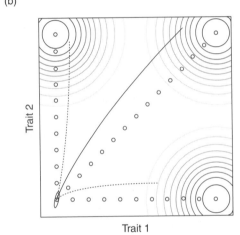

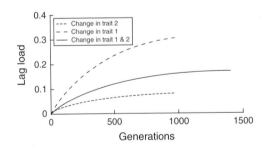

Figure 6.5 Genetic correlations and their effect on lag load. Reproduced with permission from Hellmann and Pineda-Krch (2007). (a) The effect of selection on a population some distance from a selective optimum. The variance–covariance ellipse describes the distribution of trait value for traits 1 and 2, and the direction and strength of covariation between them. The fitness peak is represented by concentric circles. The vector $\beta$ indicates the direction of movement demanded by selection, while $G_{max}$ describes the genetic line of least resistance, where genetic variation is maximized. The actual phenotypic change effected ($\Delta Z$) is biased by $G_{max}$. (b) Illustrates how the orientation of these traits affects the magnitude of the lag load for selection on each trait alone, or in combination over time. The population starts on the fitness optimum, which moves at a constant rate per generation (circles). The solid lines indicate the evolutionary response, with dots and circles representing the population mean and the optimum respectively, at 100 generation intervals. The bottom panel shows the lag, or evolutionary load, for each case, which is substantially increased, depending on the orientation of $G_{max}$ relative to $\beta$.

sustainable rate of evolution in a way proportional to the difference between the direction of the selective gradient and the direction of largest genetic covariance (Fig. 6.5b). This effect is increased if the environment fluctuates periodically, rather than changing smoothly over time (Hellmann & Pineda-Krch 2007). Migration between populations diverging in response to spatial environmental gradients may also strongly affect the G matrix, potentially constraining evolutionary change more than predicted by single-population models (Guillaume & Whitlock 2007).

Potentially high levels of covariance among traits led Blows and Hoffmann (2005) to argue that genetic variation in fitness is an important limiting factor in natural populations, despite its abundance in laboratory stocks. For example, Grant and Grant (1995) showed that covariances between traits were more successful than variance in individual traits in predicting the short-term response to natural selection of different beak characters in *Geospiza* finches. Low levels of additive variation in fitness traits also explain the lack of response to selection in *Drosophila birchii* along latitudinal gradients (Hoffmann *et al.* 2003; Kellerman *et al.* 2006). Genetic correlations may also be restricting adaptation to climate change in legumes (Etterson & Shaw 2001), as well as the longer-term evolution of fish morphology and *Drosophila* wing shape (McGuigan & Blows 2007). By contrast, McGuigan *et al.* (2005) showed that adaptive divergence within fish species was proportional to genetic variance, rather than being constrained by genetic correlations. Similarly, Caruso *et al.* (2005) suggested that evolutionary change in lobeliads was limited by a lack of genetic variation rather than by genetic correlations *per se*. Such variation between taxa and traits in the degree to which genetic correlations constrain evolutionary rates may reflect increased stability of the G matrix for some traits compared to others. This has been shown theoretically by Jones *et al.* (2003).

Long-term studies of selection in wild populations have provided valuable insight into spatial and temporal variation in selection and genetic variation in fitness in natural populations. Such studies are especially powerful when coupled with detailed pedigree information, as in Scottish island populations of Soay sheep and red deer, great tits in Oxfordshire, and collared flycatchers in Gotland (Qvarnstrom 1999; Kruuk *et al.* 2001; Merila *et al.* 2001; Sheldon *et al.* 2003; Wilson *et al.* 2006; Garant *et al.* 2007). These studies show that genetic variation in fitness in natural populations is complex, and is often masked by environmental effects on phenotypes, such as condition-dependence and environmental plasticity, particularly where the environment fluctuates substantially between generations. Selection may therefore fail to generate adaptive evolution, despite demonstrably high variation in the reproductive success of phenotypes. Although the weakening of selection on condition-dependent traits reduces the standing load of genetic variance, it is likely to increase both substitutional and lag load. For example, in red deer, climatic variation only generates evolutionary change in the small proportion of lineages that already experience favourable ecological conditions (Nussey *et al.* 2005). Substantial evolution may also occur in individual growth rate without being reflected by a change in body size or population mean fitness (Wilson *et al.* 2007). Condition-dependence may therefore reduce maximum sustainable rates of evolution below the estimates of theoretical models.

Alternatively, condition-dependent sexual selection could accelerate the rate of adaptation were females to actively select males closest to the phenotypic optimum, so increasing variation in fitness in the direction required by selection (Lorch *et al*. 2003). At least one experiment provides some support for this idea (Fricke & Arnqvist 2007). However, other studies in *Drosophila* failed to find any effect of sexual selection on adaptation (Holland 2002; Rundle *et al*. 2006), perhaps because intersexual conflict erased the positive effects of mate choice (Rundle *et al*. 2006), or because very little of the additive variance is in fitness as defined by female choice (Brooks *et al*. 2005; Van Homrigh *et al*. 2007). Also, trade-offs between sexual selection and natural selection on secondary sexual characters, such as antler and horn size in red deer and Soay sheep, reduce fitness and may be beneficial in benign years, but costly where competition for food (rather than females) is high (Kruuk *et al*. 2002; Robinson *et al*. 2006). In addition, female choice can only increase adaptation to current, rather than future, conditions. Sexual selection will therefore only promote adaptation to temporal ecological change if the optimum changes smoothly over time, rather than fluctuating substantially between generations.

The effects of phenotypic plasticity on rates of adaptation are also complex, and have recently been reviewed by Ghalambor *et al*. (2007). On one hand, plasticity may increase rates of adaptation by allowing populations to persist in novel situations so that selection has time to act. On the other hand, plasticity may slow rates of adaptation by reducing the amount of genetic variance exposed to selection in marginal environments.

## Do limits to adaptation determine species' distributions?

Evolutionary explanations for distributional limits assume that species are in a quasi-equilibrium state, with their persistence in time or space depending on their tracking environmental change at some sustainable rate. However, limits to adaptation are not the only causes of extinction. For example, extinction may occur when populations lie well within possible rates of evolution, but adapt successfully to a resource as it becomes increasingly rare or unproductive. Furthermore, some species survive well even when transferred outside their range, at least in the short term (Prince & Carter 1985). Some European tree species are also still expanding their ranges following post-glacial warming, and so have a more limited geographical distribution than their ecological tolerances suggest (Jump & Penuelas 2005).

There is a large literature concerning range expansions of organisms during historical climate change, particularly during Quaternary climate change (Hewitt 1999; Parmesan *et al*. 2005). It is likely that prehistoric range shifts, as well as the persistence *in situ* of trees and herbaceous plant species involved substantial evolutionary change on a timescale comparable with likely rates of ecological change (Davis *et al*. 2005). However, few data are available concerning what species or populations became extinct during this period, or at sufficient

(a)

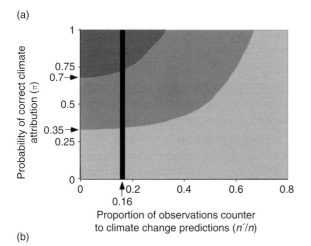

(b)

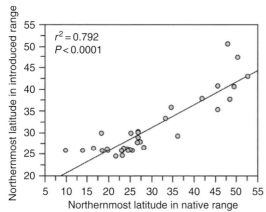

Figure 6.6 Evidence for niche conservatism. (a) Trends in shifts in species' ranges are on average well predicted by climatic envelope models. Probabilistic models for 677 plant and animal taxa show that most species show changes in distribution that closely match expectations based on recent climate change (from Parmesan and Yohe 2003). The black line shows the expected confidence level when best estimates of distributional changes were used, and $\pi$ (the probability that climate change is the principal cause of biological change) was allowed to vary freely. (Reprinted by permission from Macmillan Publishers Ltd: *Nature* **421**, 37–42. ©2003). (b) Highly significant linear regression between the northern range limits for 35 species of reptile and amphibian in their native and introduced ranges (in north America). This suggests that successful invasion depends on preadaptation (from Wiens and Graham 2005). Both of these examples suggest that limits to adaptation significantly affect species' distributions in space and time. (Reprinted, with permission, from the *Annual Review of Ecology, Evolution, and Systematics*, Volume 36 ©2005 by Annual Reviews www.annualreviews.org.)

detail to compare rates of range expansion or adaptation in extant taxa or in different ecological scenarios.

Habitat destruction and climate change in recent times provide a more direct opportunity to test the importance of evolutionary change in determining the persistence of species in response to rapidly changing ecological conditions. Parmesan and Yohe (2003) analysed data for more than 1700 animal and plant species and showed that 73% have recently shifted their ranges during the latter half of this century (Fig. 6.6a). Similar results are reported by Root *et al.* (2003). These range shifts are similar to those predicted based on the existing ecological tolerances of species, rather than expansion to occupy new habitats, suggesting that little evolutionary change has occurred over these timescales (Parmesan

2006). However, more than a quarter of species failed to show evidence for range shifts in response to climate change, particularly at equatorial margins. This observation may reflect rapid adaptation to changing conditions, or may reflect the technical difficulty of detecting range contractions using species' presence/absence rather than abundance data, or in situations where sampling occurs at large spatial scales (Hill *et al.* 2002). The latter interpretation is supported by Wilson *et al.* (2005) who demonstrated that many butterfly species whose ranges appear static at large spatial scales show rapid and predictable altitudinal range contractions when examined in more detail. However, several of the butterfly species studied still persist at low altitude sites despite significantly increased local temperatures (R. J. Wilson and D. Guttierez, pers. comm.), which may reflect rapid adaptation to changing conditions. Similarly rapid evolutionary change has been observed in several species undergoing range expansions at their poleward margins, typically involving increased dispersal (Simmons & Thomas 2004), shifts in host preference (Thomas *et al.* 2001), or the timing of bud burst or seasonal reproduction (Bradshaw & Holzapfel 2006). There is also historical and contemporary evidence that some species of small mammals compensate for temperature changes by changes in body size, without showing substantial range shifts (Smith & Betancourt 1998; Smith *et al.* 1998).

Studies of invasive species also provide an opportunity to test the relative importance of evolutionary change in determining species' distributions. Wiens and Graham (2005) reviewed data for 35 invasive species of reptiles and amphibians, and showed a highly significant relationship between their northernmost latitudinal limits in their native and introduced ranges (Fig. 6.6b). This suggests that most successful invasive species are preadapted to the habitat into which they are successfully introduced. Whether or not a species becomes invasive therefore appears to be largely determined by its existing ecological characteristics rather than by its propensity for evolutionary change.

Studies of current responses to climate change and contemporary species invasion therefore suggest that limits to adaptation do affect where species are found: most species are shifting their ranges to track the availability to currently suitable habitat, and invasions into novel habitats typically fail. As predicted by theoretical models, ecological change is occurring too rapidly in many cases to allow population persistence *in situ*, meaning that evolutionary change may have little impact on the future distributions of many species. However, which of the critical parameters described above is actually preventing evolution in these natural situations? Is there too much, or too little genetic variation? Are population sizes sufficiently large to sustain the genetic load created by shifting optima? Closely related species showing contrasting responses to ecological change provide invaluable opportunities to test the predictions of theoretical models. Conservation efforts could then focus on managing populations to maintain maximum rates of evolutionary change.

## Future prospects

Models exploring the ability of a population to match a shifting optimum in time or space have identified parameter values of particular interest, and provide a robust theoretical framework for testing their relative importance. The problem, however, is that these critical parameters interact in complex ways, and they are difficult to estimate in natural populations. Currently therefore, theoretical models are of limited practical use in predicting maximum rates and spatial scales of evolution in real populations.

If adaptive rates are limited by low levels of genetic variance in fitness due to correlations between traits (Blows & Hoffmann 2005), rapid evolutionary change may depend on breaking up genetic correlations to allow novel responses to selection. The observation that more closely-related species are more ecologically similar than predicted by chance provides some support for this (Ricklefs & Latham 1992; Peterson *et al.* 1999; Schluter 2000). Any changes to the G matrix demanded by ecological change are likely to depend most critically on the spread of new mutations rather than on recombination between existing genotypes. For example, gene duplication and sub-functionalization allow negative pleiotropic interactions to be overcome by the evolution of increased gene number (Otto & Yong 2002). The emerging discipline of evo-devo, coupled with modern genomic analysis, suggests several promising directions to investigate this in more detail (reviewed by Brakefield *et al.* 2003; Brakefield 2006; Roff 2007a,b; Reusch & Wood 2007), and to explore the distribution of mutational variation available for selection in natural populations (Orr 2005). In particular, studies of experimental evolution in microorganisms allow the tracking of novel mutations in situations where the pattern of selective change, gene flow and recombination can be manipulated and observed over thousands of generations within a single research grant. The integration of such genomic and experimental studies with population genetic and ecological approaches will allow evaluation of the factors which are most crucial in limiting evolutionary responses to ecological change. However, more detailed empirical data are required to relate theoretical models for adaptation to natural situations, and predict how the distributions of species, and interactions between them, will respond to the profound ecological changes that will be a predominant feature of coming centuries.

## Acknowledgements

We are very grateful to Dolph Schluter, Trevor Price, James Buckley, and two anonymous referees for insightful comments on this manuscript. Roger Butlin gave detailed feedback and advice on earlier drafts. We would also like to thank to Nick Barton, Konrad Lohse and Glenn Marion for their comments on an earlier draft. JP received support from the EPSRC-funded NANIA network (GRT11777).

## References

Alleaume-Benharira, M., Pen, I. R. and Ronce, O. (2006) Geographical patterns of adaptation within a species' range: interactions between drift and gene flow. *Journal of Evolutionary Biology* **19**, 203–215.

Barton, N. H. (2001) Adaptation at the edge of a species' range. In: *Integrating Ecology and Evolution in a Spatial Context* (ed. J. Silvertown and J. Antonovics), pp. 365–392. Blackwell Sciences, Oxford, UK.

Barton, N. H. and Keightley, P. D. (2002) Understanding quantitative genetic variation. *Nature Reviews Genetics* **3**, 11–21.

Barton, N. H. and Partridge, L. (2000) Limits to natural selection. *Bioessays* **22**, 1075–1084.

Blows, M. W. (2007) A tale of two matrices: multivariate approaches in evolutionary biology. *Journal of Evolutionary Biology* **20**: 1–8.

Blows, M. W. and Hoffmann, A. A. (2005) A reassessment of genetic limits to evolutionary change. *Ecology* **86**, 1371–1384.

Blows, M. W., Chenoweth, S. F. and Hine, E. (2004) Orientation of the genetic variance-covariance matrix and the fitness surface for multiple male sexually selected traits. *American Naturalist* **163**, 329–340.

Bolnick, D. I. and Nosil, P. (2007) Natural selection in populations subject to a migration load. *Evolution* **61**, 2229–2243.

Boulding, E. G. and Hay, T. (2001) Genetic and demographic parameters determining population persistence after a discrete change in the environment. *Heredity* **86**, 313–324.

Bradshaw, W. E. and Holzapfel, C. M. (2006) Evolutionary responses to rapid climate change. *Science* **312**, 1477–1478.

Brakefield, P. M. (2006) Evo-devo and constraints on selection. *Trends in Ecology & Evolution* **21**, 362–368.

Brakefield, P. M., French, V. and Zwann, B. J. (2003) Development and the genetics of evolutionary change within insect species. *Annual Review of Ecology and Systematics* **34**, 633–660.

Bridle, J. R. and Vines, T. H. (2007) Limits to evolution at range margins: when and why does adaptation fail? *Trends in Ecology & Evolution* **22**, 140–147.

Bridle, J. R., Gavaz, S. and Kennington, W. J. (submitted) Limits to adaptation along repeated ecological gradients in Australian rainforest *Drosophila*.

Bridle, J. R., Polechová, J., Kawata, M. and Butlin, R. K. (in revision) Adaptation is prevented at range margins if population size is limited.

Brooks, R, Hunt, J., Blows, M. W., *et al*. (2005) Experimental evidence for multivariate stabilizing sexual selection. *Evolution* **59**, 871–880.

Bürger, R. (1999) Evolution of genetic variability and the advantage of sex and recombination in changing environments. *Genetics* **153**, 1055–1069.

Bürger, R. and Lynch, M. (1995) Evolution and extinction in a changing environment. *Evolution* **49**, 151–163.

Butlin, R. K., Bridle, J. R. and Kawata, M. (2003) Genetics and the boundaries of species' distributions. In: *Macroecology: Concepts and Consequences* (ed. T. Blackburn and K. Gaston). Blackwell, Oxford.

Caruso, C. M., Maherali, H., Mikulyuk, A., Carlson, K. and Jackon, R. B. (2005) Genetic variance and covariance for physiological traits in *Lobelia*: are there constraints on adaptive evolution? *Evolution* **59**, 826–837.

Case, T. J. and Taper, M. L. (2000) Interspecific competition, environmental gradients, gene flow, and the coevolution of species' borders. *American Naturalist*. **155**, 583–605.

Case, T. J., Holt, R. D., McPeek, M. A. and Keitt, T. H. (2005) The community context of species' borders, ecological and evolutionary perspectives. *Oikos* **88**, 28–46

Charlesworth, B. (1993a) The evolution of sex and recombination in a varying environment. *Journal of Heredity* **84**, 345–350.

Charlesworth, B. (1993b) Directional selection and the evolution of sex and recombination. *Genetical Research* **61**, 205–224.

Davis, M. B., Shaw, R. G. and Etterson, J. R. (2005) Evolutionary responses to changing climate. *Ecology* **86**, 1704–1714.

Davis, A. J., Jenkinson, L. S., Lawton, J. H., Shorrocks, B. and Wood, S. (1998) Making mistakes when predicting shifts in species' ranges in response to global warming. *Nature* **391**, 783–786.

Endler, J. A. (1973) Gene flow and population differentiation. *Science* **179**, 243–250.

Endler, J. A. (1986) *Natural Selection in the Wild*. Princeton University Press, USA.

Etterson, J. R. and Shaw, R. G. (2001) Constraint to adaptive evolution in response to global warming. *Science* **294**, 151–154.

Felsenstein, J. (1976) The theoretical population genetics of variable selection and migration. *Annual Review of Genetics* **10**, 253–280.

Fisher, R. A. (1930). *The Genetical Theory of Natural Selection*. Reprinted, 1999. Clarendon Press, Oxford.

Fricke, C. and Arnqvist, G. (2007) Rapid adaptation to novel host in a seed beetle (*Callosobruchus maculatus*): the role of sexual selection. *Evolution* **61**, 440–454.

Garant, D., Kruuk, L. E. B., McCleery, R. H. and Sheldon, B. C. (2007) The effects of environmental heterogeneity on multivariate selection on reproductive traits in female Great Tits. *Evolution* **61**, 1546–1559.

Gaston, K. (2000) *The Structure and Dynamics of Geographical Ranges*. Oxford, Surveys in Ecology and Evolution.

Ghalambor, C. K., MacKay, J. K., Carroll, S. P. and Reznick, D. N. (2007) Adaptive versus non-adaptive phenotypic plasticity and the potential for contemporary adaptation in new environments. *Functional Ecology* **21**, 394–407.

Gomulkiewicz, R. and Holt, R. D. (1995) When does evolution by natural selection prevent extinction? *Evolution* **49**, 201–207.

Gomulkiewicz, R., Holt, R. D. and Barfield, M. (1999) The effects of density-dependence and immigration on local adaptation and niche evolution in a black-hole sink environment. *Theoretical Population Biology* **55**, 283–296.

Grant, P. R. and Grant, B. R. (1995) Predicting microevolutionary responses to directional selection on heritable variable. *Evolution* **49**, 241–251.

Guillaume, F. and Whitlock, M. C. (2007) Effects of migration on the genetic covariance matrix. *Evolution* **61**, 2398–2409.

Haldane, J. B. S. (1956) The relation between density regulation and natural selection. *Proceedings of the Royal Society B* **145**, 306–308.

Haldane, J. B. S. (1957) The cost of natural selection. *Journal of Genetics* **55**, 511–524.

Harte, J., Ostling, A., Green, J. L. and Kinzig, A. (2004) Biodiversity conservation: climate change and extinction risk. *Nature* **427**, 6995.

Hellmann, J. J. and Pineda-Krch, M. (2007). Constraints and reinforcement on adaptation under climate change: selection of genetically correlated traits. *Biological Conservation* **137**, 599–609.

Hendry, A. P. (2005) The power of natural selection. *Nature* **433**, 694–695.

Hereford, J., Hansen, T. F. and Houle, D. (2004) Comparing strengths of directional selection: how strong is strong? *Evolution* **53**, 2133–2143.

Hewitt, G. M. (1999) Postglacial colonisation of the European biota. *Biological Journal of the Linnean Society* **68**, 87–112.

Hill, J. K., Thomas, C. D., Fox, R., *et al.* (2002) Responses of butterflies to twentieth century climate warming: implications for future ranges. *Proceedings of the Royal Society of London Series B* **269**, 2163–2171.

Hoffmann, A. A., Hallas, R. J., Dean, J. A. and Schiffer, M. (2003) Low potential for climatic stress adaptation in a rainforest *Drosophila* species. *Science* **301**, 100–102.

Holland, B. (2002) Sexual selection fails to promote adaptation to a new environment. *Evolution* **56**, 721–730.

Holt, R. D. (2003) On the evolutionary ecology of species' ranges. *Evolutionary Ecology Research* **5**, 159–178.

Holt, R. D. and Gaines, M. S. (1992) The analysis of adaptation in heterogeneous landscapes: implications for the evolution of fundamental niches. *Evolutionary Ecology* **6**, 433–447.

Houle, D. (1992) Comparing evolvability and variability of quantitative traits. *Genetics* **130**, 195–204.

Johnson, T. and Barton, N. H. (2005) Theoretical models of selection and mutation on quantitative traits. *Philosophical Transactions of the Royal Society of London Series B-Biological Sciences* **360**, 1411–1425.

Jones, A. G., Arnold, S. G. and Bürger, R. (2003) Stability of the G-matrix in a population experiencing pleiotropic mutation, stabilising selection, and genetic drift. *Evolution* **57**, 1747–1760.

Jones, A. G., Arnold, S. G. and Bürger, R. (2004) Evolution and stability of the G-matrix on a landscape with a moving optimum. *Evolution* **58**, 1639–1654.

Jump, A. S. and Penuelas, J. (2005) Running to stand still: adaptation and the response of plants to rapid climate change. *Ecology Letters* **8**, 1010–1020.

Kawecki, T. J. and Holt R. D. (2002) Evolutionary consequences of asymmetrical dispersal rates. *American Naturalist* **160**, 333–347.

Kellerman, V. M., van Heerwarden, B., Hoffmann, A. A. and Sgro, C. M. (2006) Very low additive genetic variance and evolutionary potential in multiple populations of two rainforest *Drosophila* species. *Evolution* **60**, 1104–1108.

Kingsolver, J. G., Hoekstra, H. E., Hoekstra, J. M., *et al.* (2001) The strength of phenotypic selection in natural populations. *American Naturalist* **157**, 245–261.

Kirkpatrick, M. and Barton, N. H. (1997) Evolution of a species' range. *American Naturalist* **150**, 1–23.

Kopp, M. and Hermisson, J. (2007) Adaptation of a quantitative trait to a moving optimum. *Genetics* **176**, 715–719.

Kruuk, L. E. B., Merila, J. and Sheldon, B. C. (2001) Phenotypic selection on a heritable size trait revisited. *American Naturalist* **158**, 557–571.

Kruuk, L. E. B., Slate, J., Pemberton, J. M., *et al.* (2002) Antler size in Red Deer: heritability and selection but no evolution. *Evolution* **56**, 1683–1695.

Lande, R. (1976) Natural selection and random genetic drift in phenotypic evolution. *Evolution* **30**, 314–334.

Lande, R. (1979) Quantitative genetic analysis of multivariate evolution, applied to brain-body size allometry. *Evolution* **33**, 402–416.

Lande, R. and Arnold, S. J. (1983) The measurement of selection on correlated characters. *Evolution* **37**, 1210–1266.

Lande, R. and Shannon, S. (1996) The role of genetic variation in adaptation and population persistence in a changing environment. *Evolution* **50**, 434–437.

Lenormand, T. (2002) Gene flow and the limits to natural selection. *Trends in Ecology & Evolution* **17**, 183–189.

Lorch, P. D., Proulx, S., Rowe, L. and Day, T. (2003) Condition-dependent sexual selection can accelerate adaptation. *Evolutionary Ecology* **5**, 867–881.

Lynch, M. and Lande, R. (1993) Evolution and extinction in response to environmental change. In: *Biotic Interactions and Global Change* (ed. P. Kareiva, J. G. Kingsolver and R. B. Huey). Sinauer Associates, MA, USA.

Lynch, M. and Walsh, B. (1998) *Genetics and Analysis of Quantitative Traits.* Sinuaer Associates, MA, USA.

Martins, G. and Lenormand, T. (2006) A general multivariate extension of Fisher's model and the distribution of mutation fitness effects across species. *Evolution* **60**, 893–907.

McGuigan, K. (2006) Studying phenotypic evolution using multivariate quantitative genetics. *Molecular Ecology* **15**, 883–896.

McGuigan, K. and Blows, M. W. (2007) The phenotypic and genetic covariance structure of drosophilid wings. *Evolution* **61**, 902–911.

McGuigan, K., Chenoweth, S. F. and Blows, M. W. (2005) Phenotypic divergence along lines of least resistance. *American Naturalist* **165**, 32–43.

Merila, J., Sheldon, B. C. and Kruuk, L. E. B. (2001) Explaining stasis: microevolutionary studies in natural populations. *Genetica* **112–113**, 199–222.

Mitchell-Olds, T. (1996) Pleiotropy causes long-term genetic constraints on life-history evolution in *Brassica rapa*. *Evolution* **50**, 1849–1858.

Nagylaki, T. (1975) Conditions for the existence of clines. *Genetics* **80**, 595–615.

Nussey, D. H., Clutton-Brock, T. H., Albon, S. D., Pemberton, J. and Kruuk, L. E. B. (2005) Constraints on plastic responses to climate variation in Red Deer. *Biology Letters* **1**, 457–460.

Otto, S. P. and Lenormand, T. (2002) Resolving the paradox of sex and recombination. *Nature Reviews Genetics* **3**, 252–261.

Otto, S. P. and Yong, P. (2002) The evolution of gene duplicates. *Advances in Genetics* **46**, 451–483.

Orr, H. A. (2005) The genetic theory of adaptation: a brief history. *Nature Reviews Genetics* **6**, 119–127.

Parmesan, C. (2006) Ecological and evolutionary responses to recent climate change. *Annual Review of Ecology, Evolution, and Systematics* **37**, 637–669.

Parmesan, C. and Yohe, G. (2003) A globally coherent fingerprint of climate change impacts across natural systems. *Nature* **421**, 37–42.

Parmesan, C., Gaines, S., Gonzalez, L., *et al.* (2005) Empirical perspectives on species' borders: from traditional biogeography to global change. *Oikos* **108**, 58–75.

Pease, C. M., Lande, R. and Bull, J. J. (1989) A model of population growth, dispersal and evolution in a changing environment. *Ecology* **70**, 1657–1664.

Peterson, A. T., Soberon, J. and Sanchez-Cordero, V. (1999) Conservatism of ecological niches in evolutionary time. *Science* **285**, 1265–1267.

Polechová, J., Barton, N. and Marion, G. (in preparation) Species range, adaptation in space and time.

Poon, A. and Otto, S. P. (2000) Compensating for our load of mutations: freezing the meltdown of small populations. *Evolution* **54**, 1467–1479.

Price, T., Turrelli, M. and Slatkin, M. (1993) Peak shifts produced by correlated responses to selection. *Evolution* **47**, 280–290.

Prince, S. D. and Carter, R. N. (1985) The geographical distribution of Prickly Lettuce (*Lactuata serriola*). 3. Its performance in transplant sites beyond its distributional limit in Britain. *Journal of Ecology* **73**, 49–64.

Qvarnstrom, A. (1999) Genotype by environment interactions in the determination of the size of a secondary sexual character in the Collared Flycatcher (*Ficedula albicollis*). *Evolution* **53**, 1564–1572.

Reusch, T. B. H. and Wood, T. E. (2007) Molecular ecology of global change. *Molecular Ecology* **16**, 3973–3992.

Ricklefs, R. E. and Latham, R. E. (1992) Intercontinental correlation of geographical ranges suggests stasis in ecological traits of relict genera of temperate perennial herbs. *American Naturalist* **139**, 1305–1321.

Robinson, M. R., Pilkington, J. G., Clutton-Brock, T. H., Pemberton, J. M. and Kruuk, L. E. B. (2006) Live fast, die young: trade-offs between fitness components and sexually antagonistic selection on weaponry in Soay Sheep. *Evolution* **60**, 2168–2181.

Roff, D. A. (2007a) A centennial celebration for quantitative genetics. *Evolution* **61**, 1017–1032.

Roff, D. A. (2007b) Contributions of genomics to life-history theory. *Nature Reviews Genetics* **8**, 116-125.

Root, T. L., Price, J. T., Hall, K. R., *et al.* (2003) Fingerprints of global warming on wild animals and plants. *Nature* **421**, 57-60.

Roughgarden, J. (1998) *A Primer of Ecological Theory*. Benjamin Cummings, Prentice-Hall, USA.

Roy, K., Jablonski, D., Valentine, J. W. (1998) Marine latitudinal diversity gradients: tests of causal hypotheses. *Proceedings of the National Academy of Sciences of the United States of America* **95**, 3699-3702.

Rundle, H. D., Chenoweth, S. F. and Blows, M. W. (2006) The roles of natural and sexual selection during adaptation to a novel environment. *Evolution* **60**, 2218-2225.

Schluter, D. (1996) Adaptive radiation along lines of least resistance. *Evolution* **50**, 1766-1774.

Schluter, D. (2000) *The Ecology of Adaptive Radiation*. Oxford University Press, UK.

Schwartz, M. W., Iverson, L. R., Prasad, A. M., Matthews, S. N. and O'Connor, R. J. (2006) Predicting extinctions as a result of climate change. *Ecology* **87**, 1611-1615.

Sheldon, B. C., Kruuk, L. E. B. and Merila, J. (2003) Natural selection and inheritance of breeding time and clutch size in the Collared Flycatcher. *Evolution* **57**, 406-420.

Simmons, A. D. and Thomas, C. D. (2004) Changes in dispersal during species' range expansions. *American Naturalist* **164**, 378-395.

Slatkin, M. (1973) Gene flow and selection in a cline. *Genetics* **75**, 733-756.

Slatkin, M. and Maruyama, T. (1975) Influence of gene flow on genetic distance. *American Naturalist* **109**, 597-601.

Smith, F. A. and Betancourt, J. L. (1998) Response of bushy-tailed woodrats (*Neotoma cinerea*) to late Quaternary climatic change in the Colorado plateau. *Quaternary Research* **50**, 1-11.

Smith, F. A., Browning, H., Shepherd, U. L. (1998) The influence of climate change on the body mass of woodrats *Neotoma* in an arid region of New Mexico, USA. *Ecogeography* **21**, 140.

Steppan, S. J., Phillips, P. C. and Houle, D. (2002) Comparative quantitative genetics: evolution of the G matrix. *Trends in Ecology & Evolution* **17**, 320-327.

Stockwell, C. A., Hendry, A. P. and Kinnison, M. P. (2003) Contemporary evolution meets conservation biology. *Trends in Ecology & Evolution* **18**, 94-101.

Thomas, C. D. and Kunin, W. E. (1999) The spatial structure of populations. *The Journal of Animal Ecology* **68**, 647-657.

Thomas, C. D., Bodsworth, E. J., Wilson, R. J., *et al.* (2001) Ecology and evolutionary processes at expanding range margins. *Nature* **411**, 577-581.

Thomas, C. D., Cameron, A., Green, R. E., *et al.* (2004) Extinction risk from climate change. *Nature* **247**, 145-148.

Van Homrigh, V., Higgie, M., McGuigan, K. and Blows, M. W. (2007) The depletion of genetic variance by sexual selection. *Current Biology* **17**, 528-532.

Wagner, W. L. and Funk, V. A. (1995) *Hawaiian Biogeography: Evolution on a Hot Spot Archipelago*. Smithsonian Institution Press, Washington, DC.

Waxman, D. and Peck, J. R. (1999) Sex and adaptation in a changing environment. *Genetics* **153**, 1041-1053.

Wiens, J. J. and Graham, C. H. (2005) Niche conservatism: integrating evolution, ecology, and conservation biology. *Annual Review in Ecology and Systematics* **36**, 519-539.

Willi, Y., Buskirk, J. V. and Hoffmann, A. A. (2006) Limits to the adaptive potential of small populations. *Annual Review in Ecology and Systematics* **37**: 433-458.

Willi, Y. Van Buskirk, J., Schmid, B. and Fischer, M. (2007) Genetic isolation of fragmented populations is exacerbated by drift and selection. *Journal of Evolutionary Biology* **20**, 534-542.

Wilson, R. J., Gutierrez, D., Gutierrez, J., *et al.* (2005). Changes to the elevational limits and extents of species' ranges associated with climate change. *Ecology Letters* **8**, 1138–1346.

Wilson, A. J., Pemberton, J. M., Pilkington, J. G., *et al.* (2006) Environmental coupling of selection and heritability limits evolution. *PloS Biology* **4**, 1270–1275.

Wilson, A. J., Pemberton, J. M., Pilkington, J. G., *et al.* (2007) Quantitative genetics of growth and cryptic evolution of body size in an island population. *Evolutionary Ecology* **21**, 337–356.

Wilson D. S., and Turrelli, M. (1986) Stable underdominance and the evolutionary invasion of vacant niches. *American Naturalist* **127**, 835–50.

Woodward, F. I. and Kelly, C. K. (2003) Why are species not more widely distributed? Physiological and environmental limits. In: *Macroecology: Concepts and Consequences* (ed. T. Blackburn and K. Gaston). Blackwell Science, Oxford, UK.

CHAPTER SEVEN

# Dynamic patterns of adaptive radiation: evolution of mating preferences

SERGEY GAVRILETS AND AARON VOSE

## Introduction

Adaptive radiation is defined as the evolution of ecological and phenotypic diversity within a rapidly multiplying lineage (Simpson 1953; Schluter 2000). Examples include the diversification of Darwin's finches on the Galápagos islands, *Anolis* lizards on Caribbean islands, Hawaiian silverswords, a mainland radiation of columbines, and cichlids of the East African Great Lakes, among many others (Simpson 1953; Givnish & Sytsma 1997; Losos 1998; Schluter 2000; Gillespie 2004; Salzburger & Meyer 2004; Seehausen 2006). Adaptive radiation typically follows the colonization of a new environment or the establishment of a 'key innovation' (e.g. nectar spurs in columbines, Hodges 1997) which opens new ecological niches and/or new paths for evolution.

Adaptive radiation is both spectacular and a remarkably complex process, which is affected by many different factors (genetical, ecological, developmental, environmental, etc.) interweaving in non-linear ways. Different, sometimes contradictory scenarios explaining adaptive radiation have been advanced (Simpson 1953; Mayr 1963; Schluter 2000). Some authors emphasize random genetic drift in small founder populations (Mayr 1963), while others focus on strong directional selection in small founder populations (Eldredge 2003; Eldredge *et al.* 2005), strong diversifying selection (Schluter 2000), or relaxed selection (Mayr 1963). Identifying the more plausible and general scenarios is a highly controversial endeavour. The large timescale involved and the lack of precise data on its initial and intermediate stages even make identifying general patterns of adaptive radiation very difficult (Simpson 1953; Losos 1998; Schluter 2000; Gillespie 2004; Salzburger & Meyer 2004; Seehausen 2006). Further, it is generally unknown if the patterns identified in specific case studies apply to other systems.

The difficulties in empirical studies of general patterns of adaptive radiation, its timescales, driving forces, and consequences for the formation of biodiversity make theoretical approaches important. Perhaps surprisingly, the phenomenon of adaptive radiation remains largely unexplored from a theoretical modelling perspective. Adaptive radiation can be viewed as an extension of

*Speciation and Patterns of Diversity*, ed. Roger K. Butlin, Jon R. Bridle and Dolph Schluter. Published by Cambridge University Press. © British Ecological Society 2009.

the process of speciation (driven by ecological factors and subject to certain initial conditions) to larger temporal and spatial scales. A recent explosion in empirical speciation work (reviewed by Coyne & Orr 2004; Price 2008) has been accompanied by the emergence of a quantitative theory of speciation (Gavrilets 2004). In contrast, there have been few attempts to build genetically based models of large-scale evolutionary diversification. The few existing examples are applicable mostly to asexual populations or do not explicitly treat spatially heterogeneous selection and ecological competition but, instead, focus on stochastic factors (Hubbell 2001; Chow *et al.* 2004; Gavrilets 2004). However, it is thought that diversifying selection is very important in adaptive radiations, and adaptive diversification has been shown to overcome stochasticity and historical contingencies in similar environments (Losos *et al.* 1998). Therefore, diversifying, ecologically based selection needs to be explicitly incorporated into modelling approaches.

Recently, we (Gavrilets & Vose 2005) presented our initial attempts to build a realistic genetically explicit, individual-based model of adaptive radiation driven by ecological selection in a spatially heterogeneous environment. We were able to model evolutionary dynamics of populations with hundreds of thousands of sexual diploid individuals over a time span of 100 000 generations assuming realistic mutation rates and allowing for genetic variation in a large number of both selected and neutral loci. Our approach was built upon a model of speciation via adaptation to a new ecological niche occurring simultaneously with the evolution of genetically based preferences for the niche (Diehl & Bush 1989; Fry 2003). Although this model was originally proposed as a model of non-allopatric speciation in the apple maggot fly *Rhagoletis pomonella* (Feder 1998), it is applicable to other systems where individuals choose an ecological niche in which they mate and raise offspring. Besides its simplicity and generality, this model provides one of the easiest ways to achieve non-allopatric speciation (Gavrilets 2004). Furthermore, it is the only mathematical model of non-allopatric speciation strongly supported by experimental work (Rice & Salt 1990; Rice & Hostert 1993). We used this model within the framework of parapatric speciation (Coyne & Orr 2004; Endler 1977; Gavrilets 2003, 2004) in which spatial heterogeneity in selection, isolation by distance and migration into new patches all play critical roles. Our goal was to help identify potential general patterns of adaptive radiation and evaluate their characteristic timescales.

We consider a scenario in which a few individuals of a sexual diploid species colonize a new environment (e.g. an island or a lake) in which a number of spatially structured empty ecological niches are available. For example, *Anolis* radiation on Caribbean islands is largely driven by adaptation to six ecological niches associated with different parts of vegetation (Losos 1998). While the founders have low fitness, the abundant resources and lack of competitors allow them to seed a population that is able to survive throughout the environment at low densities. The founders have no particular preference for the ecological niches available in the new environment. However, as selection acts on the new genetic variation supplied by mutation, different lineages can become adapted to and simultaneously develop genetic preferences for different ecological niches. The process of ecological and phenotypic diversification driven by selection for local adaptation was accompanied by the growth in the densities of emerging species.

In this chapter, we start by outlining our model and then discuss its behaviour when mating remains random throughout the simulations (Gavrilets & Vose 2005). However, the evolution of mating preferences is a crucial factor in many speciation models and is strongly associated with reproductive isolation in empirical studies. Therefore, we examine next the evolution of non-random mating and its influence on the dynamics of reproductive isolation and diversification.

## Mathematical model

We use a generalization of the model in Gavrilets and Vose (2005) which in turn generalizes and extends those in Diehl and Bush (1989), Johnson *et al.* (1996) and Fry (2003) (see also Kawecki 1996, 1997). The most important extension of our previous model concerns the incorporation of three additional traits controlling (non-random) mating (Gavrilets *et al.* 2007). We will use this model to study the process of invasion by a species into an environment where new ecological niches are available (Kawata 2002; Gavrilets & Vose 2005, 2007; Gavrilets *et al.* 2007). In the new environment, the invading species can be viewed as a low-fitness generalist. In contrast to most modelling work on evolution in a spatially heterogeneous environment which assumes *soft selection* (Kisdi & Geritz 1999; Spichtig & Kawecki 2004 and references therein, but see DeMeeus *et al.* 1993) and does not consider population densities explicitly, selection for local adaptation in our model is not only density-dependent but also *hard* (sensu Christiansen 1975). That is, the contribution of each niche to offspring depends on the fitness of individuals in the niche (which is more biologically realistic under the scenario we study). The following describes the major components of the model.

### Space and environment

Space is subdivided into a rectangular array of 'patches' with each patch supporting a population of a certain size. For example, one can think of different parts of a lake environment, or different types of vegetation or soil. There are $k$ environmental factors $\theta_i$ ($i = 1, 2, \ldots, k$). Each of these factors can take only two discrete values: 0 and 1, corresponding to two contrasting environmental conditions, such as sandy or rocky lake bottom, high or low light level, basalt or calcarenite soil, etc. Consequently, each patch belongs to one of $2^k$ possible types ('ecological niches'). Initially, ecological niches are assigned to patches randomly with equal probabilities.

## Individuals

Individuals are diploid. Each individual has a number of additive quantitative characters:

- $k$ 'ecological' characters $x_1,\ldots, x_k$,
- $k$ 'habitat preference' characters $y_1,\ldots, y_k$,
- three 'mating compatibility' characters $m$, $f$ and $c$.

Ecological and habitat preference characters are expressed in both sexes. Following previous models (Dieckmann & Doebeli 1999; Bolnick 2004, 2006; Gavrilets *et al.* 2007) we assume that male display trait $m$ is expressed in males only, whereas female mating preference traits $f$ and $c$ are expressed in females only. Trait $f$ describes the value of the male trait that the female prefers most. Trait $c$ characterizes the choosiness of the female. All these traits are scaled to [0,1] and are controlled by different unlinked diallelic loci with equal effects. Mutations occur at equal rates across all loci; the probabilities of forward and backward mutations are equal. In addition, there is a number of unlinked neutral loci with a large number of alleles subject to stepwise mutation (Ohta & Kimura 1973). These loci have higher mutation rates and are used to evaluate the levels of genetic divergence within and between species one would observe if using microsatellite markers.

## Life-cycle

Generations are discrete and non-overlapping. The life cycle consists of the following stages, in order:

- density-dependent viability selection within the patch,
- preferential dispersal of surviving adults among neighbouring patches (including the patch of origin),
- non-random mating among individuals within the patch and offspring production.

Note that our description of the third stage implies that mating pairs are formed in the feeding habitat (e.g. as in many plant-feeding insects).

## Viability selection

The $i$-th ecological character $x_i$ controls the fitness component $w_i$ associated with the $i$-th environmental factor. Specifically,

$$w_i = \exp\left[-\frac{(x_i - \theta_i)^2}{2\sigma_s^2}\right],\tag{7.1}$$

where $\theta_i$ is the optimum phenotype (which is given by the value of the $i$-th environmental factor in the niche). That is, each ecological character is subject to directional selection towards an extreme value (0 or 1). Parameter $\sigma_s$ measures the strength of ecological selection; smaller values of $\sigma_s$ mean stronger selection. The overall fitness $w$ is taken to be the product of the fitness components (i.e. $w = w_1, w_2, \ldots, w_k$). Note that fitness of a 'specialist' (i.e. a genotype perfectly adapted to one niche) in a niche differing by $j$ environmental factors ($j \leq k$) is $w_{j,spec} = \exp[-j/(2\sigma_s^2)]$. Fitness of a 'generalist' (i.e. an individual with $x = 1/2$) is $w_{gen} = \exp[-k/(8\sigma_s^2)]$ in all niches. Note that a hybrid between two individuals adapted to the two alternative states of an environmental factor will have an intermediate phenotype and low fitness in both environments (as implied in the scenario of ecological speciation (Schluter 2000; Forister 2005; Rundle & Nosil 2005)).

Within each patch selection acts on viability and is density dependent. The overall fitness $w$ controls a carrying capacity, $K = K_0 w$, associated with the phenotype, where $K_0$ is the maximum carrying capacity (the same for all niches). The probability that an individual survives to the age of reproduction is given by the Beverton–Holt model (Kot 2001):

$$v = \frac{1}{1 + (b-1)\frac{N}{K}},$$ (7.2a)

for monoecious individuals, and

$$v = \frac{1}{1 + (\frac{b}{2} - 1)\frac{N}{K}},$$ (7.2b)

for dioecious individuals, where $b > 0$ is a parameter (the average number of offspring per female; see below), and $N$ is the number of juveniles in the patch.

### Habitat preference and dispersal

The $i$-th preference character $y_i$ controls preference component $p_i$ for the $i$-th environmental factor. Preference component $p_i$ is given by a linear function of $y_i$:

$$P_i = \frac{1}{2} \pm a_i \left( y_i - \frac{1}{2} \right),$$ (7.3)

where $0 \leq a_i \leq 1$ is a parameter measuring the maximum possible preference and the sign is '+' for $\theta_i = 1$ and '−' for $\theta_i = 0$. Note that if $y_i = 1/2$, the individual has equal preference for both states of the environmental factor (i.e. $p_i = 1/2$). The value of $1 - a_i$ can be interpreted as the probability that an individual with the highest preference for one habitat mistakenly goes to the other habitat. The overall preference for an ecological niche characterized by environmental factors $\theta_1, \theta_2, \ldots, \theta_k$ is taken to be the product of the individual preference components (i.e. $p = p_1, p_2, \ldots, p_k$). The probability that an adult enters a patch to mate and raise offspring is proportional to its preference $p$ for the ecological niche present in the patch. Note that a hybrid between two individuals with strong preferences to the two alternative states of an environmental factor will have low preference for either environment (e.g. as experimentally demonstrated in species of maggot flies *Rhagoletis* where hybrids show reduced response to parental host-fruit odours (Linn *et al.* 2004)).

Each adult migrates to one of the eight neighbouring patches or returns back to its native patch with probabilities proportional to its preferences for the corresponding ecological niches. For patches at a boundary, the number of patches available for emigration is reduced according to the number of neighbouring patches.

## Mating preference

The relative probability of mating between a female with traits $f$ and $c$ and a male with trait $m$ is

$$\psi(m,f,c) = \begin{cases} \exp\left[-(2c-1)^2 \frac{(f-m)^2}{2\sigma_a^2}\right], & \text{if } c>0.5, \\ 1, \text{if } c = 0.5, \\ \exp\left[-(2c-1)^2 \frac{(f-(1-m))^2}{2\sigma_a^2}\right], & \text{if } c>0.5, \end{cases} \tag{7.4}$$

where parameter $\sigma_a$ scales the strength of female mating preferences (Gavrilets *et al.* 2007). Under this parameterization, females with $c=0.5$ mate randomly, females with $c>0.5$ prefer males whose trait $m$ is close to the female's trait $f$ (positive assortative mating), and females with $c<0.5$ prefer males whose trait $m$ is close to $1-f$ (negative assortative mating). Note that the absolute value $|2c-1|$, which we will denote as $C$, characterizes the extent of deviation of the female's mate choice from random: females with $C=0$ mate randomly while those with $C=1$ exhibit the strongest possible (negative or positive) assortative mating. Parameter $\sigma_a$ governs the width of the mating probability distribution; the small numerical value used below (0.1) implies that the mating preference loci have very strong effects on the probability of mating (Gavrilets *et al.* 2007).

## Offspring production

Each mating results in a number of offspring drawn from a Poisson distribution with parameter $b$. We assume that all adult females mate. This assumption implies that any costs of mate choice, which can easily prevent divergence and speciation (Pomiankowski 1987; Bolnick 2004; Gavrilets 2004, 2005; Gourbiere 2004; Kirkpatrick & Nuismer 2004; Waxman & Gavrilets 2005), are absent. This assumption also means that the effective population size is increased relative to the actual number of adults (Gavrilets & Vose 2005).

## Local extinction

At the start of every generation, each patch goes extinct with a small probability $\varepsilon$. When this happens, all individuals present there die and the 'niche' assigned to this patch is chosen anew randomly. The later assumption is a simple way to have some turnover of ecological niches. For example, if a patch represents a tree of a certain host species, then when the tree dies its space can be occupied by a tree of a different host species.

## Reproductive isolation and species

We assume that genetic incompatibilities (Coyne & Orr 2004; Gavrilets 2004) are absent. Reproductive isolation can evolve in the model via differentiation in habitat preferences, selection against immigrants and hybridization, and sexual selection. In our model, for each individual, there is a niche, say niche $J$, where it is most fit, and there is a niche, say niche $I$, for which it has the strongest preference. We interpret each individual for whom $J=I$ as a member of 'ecological' species $I$. Each ecological species can be comprised by a number of 'sexual morphs' or 'sexual species', i.e. groups of individuals reproductively isolated by differences in mating preferences. Because fitness in a niche, preference for a niche and mating preferences are controlled genetically, our 'species' also represent distinguishable genetic clusters which are reproductively isolated to a certain degree (Mallet 1995; Pigliucci 2003). The degree of reproductive isolation gets progressively amplified as a by-product of ecological adaptation and strengthening of habitat and mating preferences.

## Initial conditions

$K_{init}$ adults populate a single patch in the upper left corner of the system. All individuals are identical homozygotes with all traits exactly at 1/2. Each microsatellite locus is heterozygous with two intermediate alleles out of $2^8$ possible alleles.

## Parameter values

We varied system size ($8 \times 8$, $16 \times 16$, $32 \times 32$), number of loci per each trait ($L=4$, 8, 16) and the local extinction rate ($\varepsilon=0$, 0.0025, 0.01, 0.04). The following parameters did not change: number of traits $k=3$, $\sigma_s=0.356$ (which corresponds to the fitness of the generalists being 0.05 of the maximum possible value), average number of offspring $b=4$, maximum strength of preference $a_i=0.99$, number of 'microsatellite' loci $M=8$ and mutation probability $\mu=10^{-5}$ for the loci controlling ecological and preference traits and $\mu_n=10^{-3}$ for the 'microsatellite' loci. 30–50 runs were done for each parameter combination. Simulations ran for up to 100 000–150 000 generations or until global extinction.

## Population genetic structure at neutral loci

To estimate the levels of spatial structuring in neutral loci we used the AMOVA framework (Excoffier et al. 1992; Excoffier 2001).

## Diversification under random mating

Here, we briefly summarize our previous results corresponding to the case when mating was forced to remain random throughout the simulations (Gavrilets & Vose 2005). In these simulations, individuals were monoecious, the number of founders was $K_{init} = 10$ individuals, and maximum carrying capacity was $K_0 = 500$ individuals. Under these conditions and for parameter combinations used in our numerical simulations, adaptive radiation into a number of ecological niches often follows the colonization of a new environment. Generally, in the course of the simulations, ecological traits evolve faster, approach their optimum trait values closer, and maintain less genetic variation at (stochastic) equilibrium than the habitat preference traits.

### Area effect

In *Anolis* lizards, empty ecological niches get filled only on islands of sufficiently large area (Losos 1998). In our simulations, larger areas allow for more intensive diversification (see Fig. 2 in Gavrilets & Vose 2005). For parameter values used, eight ecological niches are always available. However, in systems of smaller size (e.g. $8 \times 8$ or $16 \times 16$) not all niches are filled. The area effect has the following explanation. First, larger areas can support larger population sizes which in turn results in more advantageous mutants on which diversifying selection can act. Second and more importantly, in larger areas new locally advantageous genes may become better protected by distance from the diluting effect of locally deleterious genes, which otherwise can easily prevent adaptation to a new niche (Riechert 1993). Isolation by distance allows new advantageous combinations of genes to accumulate in large numbers, promoting further adaptations to new ecological niches.

### Effect of the number of loci

In our simulations, increasing the number of loci underlying the traits decreases diversification (see Fig. 2 in Gavrilets & Vose 2005). This happens because a larger number of loci implies weaker selection per each individual locus and a stronger overall effect of recombination in destroying coadapted gene complexes. Similar effects have been observed in a number of related models (Gavrilets 2004; Gavrilets & Vose 2007; Gavrilets *et al.* 2007).

### Timing of speciation

Typically if more than one species emerges, there is a burst of speciation soon after colonization rather than a more or less continuous process of speciation (see Fig. 3 in Gavrilets & Vose 2005). The explanation can be given in terms of ecological opportunity and genetic constraints (Erwin *et al.* 1987). Initially, the former is much larger (because there are more empty niches, local densities are low and competition is weaker), whereas the latter is much smaller (because the founders are not specialized) than later in the radiation. As a consequence, in our simulations more than 98% of speciation events occurred within the first 10 000 generations.

### Overshooting effect

In some adaptive radiations, the diversity (i.e. the number of species) peaks early in the radiation. For example, the number of species of spiny-legged *Tetragnatha* spiders on younger Hawaii islands is larger than on the older ones (Gillespie 2004). Our simulations provide some support for the generality of this 'overshooting effect'. An explanation of the overshooting effect can be given in terms of the differences between the rates of species extinction and origination. Whereas the former is more or less constant in time (excluding the first few thousand generations), the latter decreases in time because of the effects of ecological opportunity and genetic constraints as discussed above. Note that the overshooting effect was system-size specific; it was not observed in $32 \times 32$ systems. The most likely reason is that larger systems require a longer period of time for a decline in diversity to become apparent.

### Hybridization and neutral gene flow

In our simulations, species can stably maintain their divergence in a large number of selected loci for very long periods of time in spite of substantial hybridization and gene flow that removes differentiation in neutral markers. Similar observations are often made in natural populations. For example, blue butterfly species *Lycaeides idas* and *Lycaeides melissa* utilize different hosts and have diverged significantly in morphology, yet show no differentiation in neutral markers (Nice *et al.* 2002).

### 'Least action effect'

In our simulations, speciation occurring after the initial burst usually involves a change in a single pair of characters. This observation provides theoretical support for a prediction that shifts to radically different hosts will be much less common than shifts to similar hosts (Bush 1969). That is, if a host shift happens, it proceeds, metaphorically speaking, in the direction of least action. A related observation is that when some niches are not filled, the existing species differ in the minimum number of characteristics (1 or 2). These effects are explained by the fact that the deleterious effects of immigration of locally disadvantageous genes on the possibility of accumulation of locally advantageous genes become stronger with genetic difference between immigrating and resident genotypes.

## Evolution of non-random mating

Next we consider individuals with discrete sexes and allow non-random mating to evolve. We have performed additional simulations varying the number $L$ of

(a)    (b)

(c)    (d)

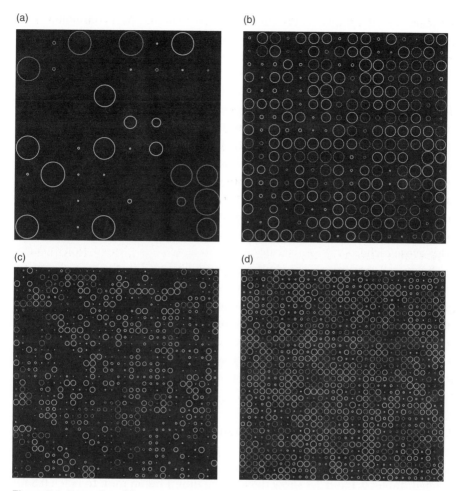

**Figure 7.1** Examples of the system state with three environmental factors ($k = 3$, so that there are eight possible niches). Each square represents a patch. The colour of the square represents the ecological niche assigned to the patch. Each local population is represented by a circle. The radius of the circle is proportional to the population size (the maximum radius corresponds to a population at carrying capacity). The colour of the circle defines the niche preferred by the majority of individuals. Matching of the color of the corresponding square and circle (observed in most cases) means that the majority of individuals in the patch have preference for the ecological conditions they experience. (a) $8 \times 8$ systems with two ecological niches filled, (b) $16 \times 16$ system with eight ecological niches filled, (c) $32 \times 32$ system with four ecological niches filled, and (d) $32 \times 32$ system with eight ecological niches filled. (see colour plate)

loci per trait, the system size and the local extinction rate $\varepsilon$ in the same way as in the previous section. Throughout the simulations we assumed that the carrying capacity was $K = 600$ individuals per deme; set parameter $\sigma_a$ of the preference function (3) at 0.1; and started with 20 founders (10 males and 10 females). The founders were homozygous with ecological, habitat preference and mating preference traits all set at 0.5 for each individual. Thus, all founders were randomly mating generalists. For most parameter combinations, we accumulated 30 runs with the system surviving for about 100 000–150 000 generations. However, in systems of the smallest size (8 × 8) and with non-zero extinction rate, long-term survival was not observed (Gavrilets & Vose 2005). Figure 7.1 illustrates some states reached by the system with $k = 3$ (and, thus, eight possible ecological niches). In the four cases shown, 2, 8, 4 and 8 ecological niches were filled by two to three sexual species each. In the cases shown, a large number of local populations are close to the carrying capacity. However, in some patches populations are very small. These are 'sink populations' (Holt 1997) that cannot adapt to the conditions they experience because of the deleterious effect of migration of locally maladapted genotypes. A gallery of graphical results illustrating the final state of the system can be viewed at http://neko.bio.utk.edu/~avose/niche.

The overall dynamics were similar to those observed in the case of monoecious populations described above. However, because of an increased demographic stochasticity of dioecious populations, successful invasion was much less common. (That was actually the reason why we increased both the number of founders and the population carrying capacity relative to those used in the previous section.) The number of ecological niches filled during the diversification stage was reduced for the same reason.

**Strength of non-random mating**

The strength of non-random mating in the system can be characterized by the average value $\bar{C}$ of trait $C$ $(= |2c - 1|$; see above) in the population. In our simulations, strong non-random mating does evolve often on the timescale studied. Tables 7.1 and 7.2 show the average time $T$ taken for $\bar{C}$ to reach 0.5 for the first time. Also given there are the number of runs in which $C$ did reach 0.5. Notice that time $T$ is typically one order of magnitude larger than the characteristic time for the evolution of local adaptation and habitat preference (see above and Gavrilets & Vose 2005). Simulations also show $C$ does not necessarily evolve to the maximum possible value but can be stably maintained at some intermediate values (cf. Matessi et al. 2001) or can drastically fluctuate in time (cf. Hayashi et al. 2007). Overall, the model predicts that partial sexual isolation (e.g. as between Littorina ecotypes (Rolán-Alvarez et al. 1997) or between Timema walking-stick ecotypes (Nosil et al. 2002) should be common.

**Drift versus selection (against hybridization)**

In principle, non-random mating in our model can evolve by drift and/or selection against hybridization. The deleterious effect of hybridization is not present if only one ecological niche is filled. Therefore, the runs in which $C$ evolved to

**Table 7.1** *The average time to the evolution of non-random mating (in thousands of generations) over the runs with a single 'filled' niche*

| | | System size | | |
|---|---|---|---|---|
| $L$ | $\varepsilon$ | $8 \times 8$ | $16 \times 16$ | $32 \times 32$ |
| 4 | 0 | 61.6(4) | – | – |
| | 0.0025 | – | 50.4(17) | – |
| | 0.01 | – | 33.9(25) | – |
| | 0.04 | – | 70.5(2) | – |
| 8 | 0 | 84.8(5) | – | – |
| | 0.0025 | – | 70.2(20) | – |
| | 0.01 | – | 44.8(22) | – |
| | 0.04 | – | 66.3(7) | – |
| 16 | 0 | 113.1(4) | – | – |
| | 0.0025 | – | 49.5(2) | – |
| | 0.01 | – | 67.7(12) | – |
| | 0.04 | – | 87.8(8) | – |

*Note*: The number of runs is shown in the parentheses.

**Table 7.2** *The average time to the evolution of non-random mating (in thousands of generations) over the runs with more than one 'filled' niche*

| | | System size | | |
|---|---|---|---|---|
| $L$ | $\varepsilon$ | $8 \times 8$ | $16 \times 16$ | $32 \times 32$ |
| 4 | 0 | 52.6(26) | 39.0(30) | 25.0(30) |
| | 0.0025 | – | 25.5(13) | 20.0(30) |
| | 0.01 | – | 13.4(5) | 20.8(30) |
| | 0.04 | – | 32.2(28) | 22.7(30) |
| 8 | 0 | 92.4(13) | 77.6(17) | 83.8(15) |
| | 0.0025 | – | 24.5(1) | 65.9(26) |
| | 0.01 | – | 33.5(1) | 71.3(26) |
| | 0.04 | – | 83.5(16) | 69.4(28) |
| 16 | 0 | 65.0(1) | 69.0(1) | 142.0(1) |
| | 0.0025 | – | 121.1(1) | – |
| | 0.01 | – | 73.5(1) | 123.0(1) |
| | 0.04 | – | – | 114.0(5) |

*Note*: The number of runs is shown in the parentheses.

high values when only one ecological niche was filled (shown in Table 7.1) illustrate the evolution of non-random mating by drift alone. The runs in which $C$ evolved to high values when more than one ecological niche was filled (shown in Table 7.2) can be interpreted as illustrating the evolution of non-random mating by selection against hybridization. Comparing Tables 7.1 and 7.2, one can see that non-random mating often evolved by drift in $16 \times 16$ systems, whereas this never happened in $32 \times 32$ systems. Table 7.2 shows that if selection against hybridization is present, non-random mating evolves faster in larger systems with fewer loci. Non-zero local extinction promotes the evolution of non-random mating (because local extinction causes turnover of populations, more contact between different ecological species, and more hybridization). However, provided local extinction rate $\varepsilon$ is non-zero, the effect of its exact value seems to be small.

## Divergence in mating characters

Evolution of non-random mating does not mean, by itself, divergence in mating characters between or within ecological species as mating preferences can be shared across different 'ecological species'. However, in the simulation we do observe both between- and within-(ecological) species divergence in mating preferences. Figure 7.2 shows examples of the distributions of mating characters observed in simulations with different number of ecological niches filled. For example, in case (a), two ecological niches are filled with two sexual species, each characterized by moderately strong negative assortative mating ($c < 0.5$). In case (d), all eight ecological niches are filled by three sexual species, each characterized by strong positive assortative mating ($c > 0.5$). In case (b), all eight ecological niches are filled by two sexual species; in each pair, one species exhibit strong positive assortative mating ($c > 0.5$) while the other species strong negative assortative mating ($c < 0.5$).

The distribution of mating characters often has a discrete nature. To characterize this quantitatively, we defined the number of (sexual) morphs as the number of local maxima in the distribution of the male mating trait $m$ observed after removing 'odd' trait values. For example, with $L = 4$ the possible trait values are 0, 1/8, 2/8, 3/8, 4/8, 5/8, 6/8, 7/8 and 8/8. After removal of 'odd' values, we are left with five 'even' values 0, 2/8, 4/8, 6/8 and 8/8. Simulations strongly suggest that 'even' values capture the internal cluster structure of the population much better than the 'odd' ones (because peaks in the distribution of the male trait $m$ at 'odd' numbers often represent hybrids). Table 7.3 shows the number of sexual morphs (or 'sexual species') at the end of the run per filled ecological niche. In simulations, we have observed up to 3–4 distinct sexual morphs per ecological species and up to 24 discrete reproductively isolated species. However, in some cases, we observed an almost uniform distribution of the male trait $m$ rather than discrete clusters or 'peaks' (see Fig. 7.2).

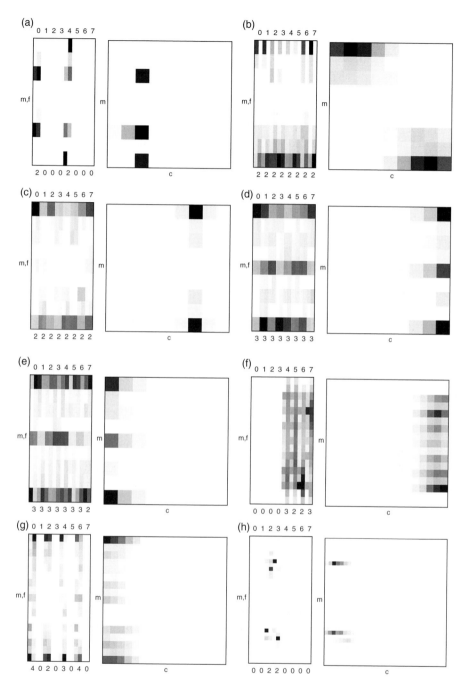

**Figure 7.2** Examples of the distributions of mating characters across different ecological niches and overall in the system. The first graph in each pair shows the distributions of male display trait (m; left bar for each niche) and female preference trait (f; right bar for each niche) within each of the eight niches (0, 1, ... , 7). The traits change between 0 (upper boundary of the graph) and 1 (lower boundary of the graph). The intensity of the black

**Table 7.3** *Average number of mating morphs per 'filled' ecological niche at the end of simulations*

| L | $\varepsilon$ | System size | | |
| --- | --- | --- | --- | --- |
| | | $8 \times 8$ | $16 \times 16$ | $32 \times 32$ |
| 4 | 0 | 1.31 | 1.89 | 1.97 |
| | 0.0025 | – | 1.42 | 2.62 |
| | 0.01 | – | 1.00 | 2.48 |
| | 0.04 | – | 1.10 | 2.26 |
| 8 | 0 | 1.26 | 1.63 | 2.05 |
| | 0.0025 | – | 1.16 | 2.35 |
| | 0.01 | – | 1.10 | 1.58 |
| | 0.04 | – | 1.00 | 1.24 |
| 16 | 0 | 1.31 | 1.63 | 1.78 |
| | 0.0025 | – | 1.33 | 1.36 |
| | 0.01 | – | 1.16 | 1.15 |
| | 0.04 | – | 1.00 | 1.13 |

Table 7.4 shows an alternative diversity index $\gamma$ for the male traits which we defined as twice the standard deviation of trait $m$ in the population. This index can take values between zero and one with the latter being observed when the distribution is symmetric bimodal with peaks at 0 and 1. In both cases, we chose to work with male character $m$ rather than female characters $f$ and $c$ because males are under (sexual) selection whereas female traits can evolve more or less randomly/neutrally (at least as long as selection against hybridization is not too strong).

←————————————————————————————————

**Figure 7.2 (cont)** color is proportional to the frequency of individuals with the corresponding trait value. The numbers at the bottom give the number of 'sexual morphs' per each ecological niche. The second graph in each pair shows the overall distribution of male display trait (m) versus the overall distribution of female mating tolerance trait c. The traits change between 0 (upper left corner of the graph) and 1 (lower right corner of the graph). The intensity of the black colour is proportional to the frequency of individuals with the corresponding trait value. Shown are: (a) run 11 in set 0 ($8 \times 8$, $L=4$, $\varepsilon=0$, 2 filled niches), (b) run 6 in set 12 ($16 \times 16$, $L=4$, $\varepsilon=0$, 8 filled niches), (c) run 23 in set 12 ($16 \times 16$, $L=4$, $\varepsilon=0$, 8 filled niches), (d) run 12 in set 25 ($32 \times 32$, $L=4$, $\varepsilon=0.0025$, 8 filled niches), (e) run 13 in set 25 ($32 \times 32$, $L=4$, $\varepsilon=0.0025$, 8 filled niches), (f) run 0 in set 29 ($32 \times 32$, $L=8$, $\varepsilon=0.0025$, 4 filled niches), (g) run 9 in set 29 ($32 \times 32$, $L=8$, $\varepsilon=0.0025$, 4 filled niches) and (h) run 6 in set 35 ($32 \times 32$, $L=16$, $\varepsilon=0.04$, 2 filled niches). Note that when mating is positive assortative (so that $c > 0.5$; cases c, d, f and, partially, b) the distributions of m and f are fairly similar; if mating is negative assortative (so that $c < 0.5$; the rest of the cases), the distribution of m is a mirror image of the distribution of f around 1/2. See http://neko.bio.utk.edu/~avose/niche for a compete gallery of graphical results.

Table 7.4 *Average diversity* $\gamma$ *in the male mating character* m *at the end of simulations* $(0 \leq \gamma \leq 1)$

| | | System size | | |
|---|---|---|---|---|
| $L$ | $\varepsilon$ | $8 \times 8$ | $16 \times 16$ | $32 \times 32$ |
| 4 | 0 | 0.47 | 0.82 | 0.79 |
| | 0.0025 | – | 0.47 | 0.84 |
| | 0.01 | – | 0.07 | 0.83 |
| | 0.04 | – | 0.13 | 0.82 |
| 8 | 0 | 0.17 | 0.33 | 0.53 |
| | 0.0025 | – | 0.12 | 0.77 |
| | 0.01 | – | 0.06 | 0.59 |
| | 0.04 | – | 0.05 | 0.65 |
| 16 | 0 | 0.08 | 0.14 | 0.16 |
| | 0.0025 | – | 0.08 | 0.11 |
| | 0.01 | – | 0.06 | 0.08 |
| | 0.04 | – | 0.03 | 0.08 |

Importantly, throughout the simulations there is much more variation in female mating preference traits than in the male traits (because there is no cost of choosiness for females, but males are often subject to strong sexual selection). Both Tables 7.3 and 7.4 show that diversification in male mating character is more extensive if the number of loci and the local extinction rate are small but the system size is large. That is, the qualitative effects of these parameters on the diversification in mating characters are the same as on the diversification in ecological characters.

### Does more ecological species mean more sexual species?

Whether species diversity promotes speciation is a controversial topic (Cadena et al. 2005; Emerson & Kolm 2005). Does ecological diversification per se promote the evolution of non-random mating in our simulations? To answer this question, we calculated the correlation between the number of ecological niches filled at the moment when $C$ reached 1/2 for the first time (which can be viewed as a measure of ecological diversity at speciation) and the following four variables: (1) $T$ (i.e. the time to $C=0.5$), (2) the number of sexual morphs per filled niche, (3) diversity index $\gamma$, and (4) the value of $C$. The last three variables were evaluated at the end of simulations. To save space, rather than giving four different tables with correlations and marking statistically significant correlations with, say, stars we decided to collect all statistically significant correlations in a single table (Table 7.5). This table suggests that more filled ecological

**Table 7.5** *Significant correlations (at P = 0.05) of measures of sexual diversification with the number of filled niches at C = 0.5*

| | | System size | | |
|---|---|---|---|---|
| $L$ | $\varepsilon$ | 8 × 8 | 16 × 16 | 32 × 32 |
| 4 | 0 | +1 | | |
| | 0.0025 | | +2,3,4 | −1 |
| | 0.01 | | −1 | −1 |
| | 0.04 | | −1 | −1 + 3 |
| 8 | 0 | | | +3 |
| | 0.0025 | | | +2 |
| | 0.01 | | | −1 + 2 |
| | 0.04 | | | |
| 16 | 0 | | | |
| | 0.0025 | | +1 | |
| | 0.01 | | | |
| | 0.04 | | | |

*Note*: Plus signifies positive correlation and minus negative. The numbers correspond to the correlations with (1) $T$ (i.e. the time to $C = 0.5$), (2) the number of sexual morphs per filled niche, (3) diversity index $\gamma$, and (4) the value of $C$. Empty spots mean statistically insignificant correlations.

niches sometimes result in shorter times to the evolution of non-random mating, more sexual morphs, larger diversity in male characters and larger strength of non-random mating. However, these effects are not necessarily strong.

### The role of reinforcement

A standard argument in evolutionary biology is that a contact between populations adapted to different ecological niches will result in their divergence in mating characters so that the effects of hybridization and resulting deleterious gene flow between them are reduced.

This is a classical reinforcement scenario supported by data and a number of mathematical models (Butlin 1987, 1995; Howard 1993; Servedio & Noor 2003; Coyne & Orr 2004; Gavrilets 2004). How common is between-niche diverge in mating characters without accompanying within-niche diversification in our simulations? The surprising answer is that 'classical' reinforcement is not common at all. Overall, with two ecological niches filled, strong reproductive isolation (with $C > 0.5$ and $\gamma > 0.5$) has evolved in 40 runs under eight different sets of parameters. Within-niche diversification in mating characters was absent in only four of these cases (under two different sets of parameters). With four ecological niches filled, strong reproductive isolation (with $C > 0.5$ and $\gamma > 0.5$) has evolved in

more than 130 runs under 13 different sets of parameters. Within-niche diversification was absent in only 13 of these cases (under two different sets of parameters).
These results can be understood in the following terms. The populations occupying
and adapted to different ecological niches do start diverging in mating characters
upon a contact (as postulated in the reinforcement scenario). However, reproductive
isolation between them is never strong initially. Therefore, whatever new mating
traits emerge in one ecological niche, they readily spread across the (ecological)
species boundary and become represented in other ecological niches. Such 'parallel'
diversification in mating characters does result in a reduction of gene flow between
different ecological niches. Simultaneously, 'parallel' diversification creates within-
niche divergence in mating characters and reduction in within-niche gene flow. At
least in the model studied here, the classical reinforcement appears to be a short-
term process.

Previous modelling work on reinforcement has never been performed at the
level of complexity and biological realism used here. This probably explains
why the within-species diversification in mating characters accompanying reinforcement has not been identified earlier.

## The role of hybridization

Recently, there has been a lot of interest in hybridization in natural populations
(Rieseberg 1995, 1999; Arnold 1997; Dowling and Secor 1997; Barton 2001) with
some researchers forcefully arguing for a very important role of hybridization in
biological diversification, speciation and adaptive radiation of not only plants
but also animals (Seehausen 2004; Arnold & Meyer 2006; Mavarez et al. 2006;
Mallet 2007). Is there a role for hybridization in this model? In the previous
subsection, we already discussed the phenomenon of parallel diversification in
mating characters which occurs as a result of hybridization following the
emergence of new mating traits. A similar behaviour occasionally occurs with
regard to ecological traits. In most modelled cases, all species observed at the
end of a simulation run originated more or less simultaneously (see also
Gavrilets and Vose (2005)). However, occasionally new ecological species origi-
nate significantly after the initial diversification. In such a case, the new ecological trait spreads across the species boundary, which results in essentially
doubling the number of species in the system. We note that these dynamics (e.g.
rapid emergence of multiple species after an extended period of stable species
diversity) resemble those emphasized by punctuated equilibrium theory
(Eldredge 1971; Eldredge & Gould 1972; Gould 2002).

Hybridization may be even more important under the scenario of a secondary
contact. For example, if one starts, say, with two species diverged in all there
ecological and all three habitat preference characters, hybridization between
them can rapidly result in six additional ecological species. The importance and
plausibility of this process will depend on the strength of selection for local

adaptation and the strength of non-random mating present prior to the hybrid-ization event. We have not yet explored these factors systematically.

### Local adaptation and speciation

In our model, more (ecological) species typically means less local adaptation for each individual species. This happens because more species implies more oppor-tunities for hybridization, the consequence of which is locally deleterious gene flow. The correlations between fitness and the number of ecological niches filled and between fitness and the overall number of species present are con-sistently negative and in many cases are statistically significant.

### Stages of diversification

Simulations show that diversification in the model usually occurs in a particular order. First, populations diverge in ecological traits; second, in habitat prefer-ences; and, third, in mating characters. Unless there is a global extinction, the first two steps are observed always and typically occur within several thousand generations after the founders enter the new environment. The first two steps are usually separated by several hundred generations but, rarely, occur one or two thousand generations apart. The third step is not guaranteed. If it does occur, it usually happens tens of thousands of generations after the first two steps. Diversification in mating characters usually occurs in parallel in different ecological niches. Only rarely do new ecological niches become colonized sig-nificantly after the initial bout of diversification. In this case, most mating characters quickly become represented in the new niche. Overall, diversifica-tion occurs in pulses.

## Discussion

Here, we have described an explicit genetic model of adaptive radiation driven by selection for adaptation to discrete multivariate ecological niches. We have shown that strong ecologically based spatially heterogeneous selection coupled with limited migration, genetically based habitat choice and genetically based mate choice can indeed result in rapid phenotypic and ecological diversification and the emergence of multiple species reproductively isolated by a variety of mechanisms (local adaptation, habitat and mating preferences).

In our model, ecological traits evolve faster, approach their optimum trait values closer, and maintain less genetic variation at (stochastic) equilibrium than the habitat preference traits. Mating preference traits evolve slower than the other two sets of traits, maintain more genetic variation, and can fluctuate dramatically in time. Mating preferences can diverge both between and within species utilizing different ecological niches.

Our earlier work (Gavrilets & Vose 2005) has already provided strong theoret-ical support, clarification and explanation for a number of patterns of adaptive radiation including the area effect, the overshooting effect and the least action effect. It also suggested that diversification is strongly promoted if genetics underlying the traits involved are simple and most speciation events occur soon after colonization of a new environment while the rest occur mostly in pulses. We also showed that species can stably maintain their divergence in a large number of selected loci for very long, effectively infinite, periods of time in spite of substantial hybridization and gene flow that removes differentiation in neutral markers. The latter observation strongly supports the idea that genomes can be rather porous (Feder 1998; Wu 2001). The new results presented here support the generality of all these earlier conclusions.

Our new model shows that non-random mating can often evolve. With a non-negligible probability, it can happen by mutation and random drift alone (Gavrilets 2004). As expected, selection against hybridization dramatically increases the likelihood of evolution of strong non-random mating. However, there are a couple of underappreciated patterns. First, reproductive isolation is almost never absolute, and some hybridization and gene flow are always happening. As a result, neutral alleles can pass between different species. Second, this gene flow also results in sharing mating characters (especially during the initial stages when non-random mating is not strong yet) and, less commonly, ecological characters across species boundaries. These processes produce parallel speciation when new mating characters get shared across different ecological niches and/or new ecological characters get shared across different 'sexual morphs'. Third, these processes result in within-niche differentiation in mating preference and in emergence of sister species. Fourth, the model demonstrates that while ecological characters and habitat preference are rather stable in time, traits involved in mating are very dynamic and often vary dramatically both in time and in space. A related observation is that differentiation in mating characters is often of more a continuous nature with clearly defined discrete morphs not necessarily present.

Both the existing empirical data and theory show that genetic, ecological and environmental details have profound effects on the dynamics of speciation and diversification and that no universal rules of speciation exist. At the same time, one should expect that some relatively common trends or tendencies of evolutionary diversification in related groups of organisms can be identified. For example, in birds of the New Guinea mountains the diversification with respect to habitat elevation preceded diversification with respect to other characteristics (Diamond 1986; Schluter 2000; but see Price 2008 for an alternative interpretation). In some fishes, birds, and lizards, a common sequence of events in evolutionary diversification is divergence, first, in habitat, then in food type, and, finally, in mating signals (Streelman & Danley 2003). Earlier, analysis of various speciation models (Gavrilets 2004) has suggested that the following sequence of diversification events should be typically observed: (1) divergence with respect to macrohabitat, (2) evolution of microhabitat choice and divergence with respect to microhabitat, (3) divergence with respect to 'magic traits' (i.e. traits that control simultaneously the degree of local adaptation and non-random mating), and (4) divergence with respect to other traits controlling survival and reproduction. The model studied here does not differentiate between microhabitat and macrohabitat and does not include 'magic traits'. However, it does show that divergence in mating characters occurs much later than divergence in traits that control adaptation to and preference for habitat and, thus, provides partial support for the arguments above.

We note that some of the patterns discussed here have been identified previously in specific biological systems. However, the generality of these patterns cannot be assured on the basis of single cases that are known currently.

The fact that we were able to reproduce these macroevolutionary patterns starting with microevolutionary processes of mutation, random drift, migration, recombination and selection strongly support the generality of these patterns. We expect that these patterns will be observed in most adaptive radiations. Besides being able to capture the essence of adaptive radiation qualitatively, the model studied here allows one to make some quantitative conclusions on the important timescales of diversification, the number of species likely to emerge, and the effects of various parameters.

Due to computing power limitations, we varied only a limited number of parameters. Here we briefly discuss the expected effects of changing other parameters and assumptions (see also Gavrilets 2004; Gavrilets & Vose 2005). For the parameter values used here, no generalists ever survived competition against specialists. Weaker selection for local adaptation may, however, result in generalists being the most dominant type (see Gavrilets & Vose 2007; Gavrilets et al. 2007). Reduced probability of leaving the patch will make divergence slower (because selection against immigrants from diverged habitats will be less effective). Assuming non-equal allelic effects will constrain divergence in mating characters and result in fewer species (because the loci with smaller effects will be less responsive to selection and, thus, will diverge to a smaller degree). The loci with larger effects are expected to diverge first. Introducing costs of being choosy will significantly reduce or completely prevent divergence in mating characters. Linkage will result in differential genetic divergence across parts of genome with genes closely linked to selected loci being more differentiated than loosely linked genes. As a result, diverged genes will tend to form clusters (Wu 2001; Emelianov et al. 2004; Gavrilets 2004). Allowing for other forms of reproductive isolation (e.g. genetic incompatibilities) to evolve is not expected to change dramatically the patterns discussed above because they will evolve at much slower rates than ecological traits, habitat preferences and mating characters (Gavrilets 2004). We also note that assuming environmental factors are distributed along gradients rather than randomly does not seem to reduce neutral gene flow between emerging species (Gavrilets & Vose unpublished).

The current explosive growth in empirical knowledge on different aspects of speciation and adaptive radiation should soon make it possible to evaluate the generality and relevance of the patterns identified here and test the predictions made.

## Acknowledgements

We thank Suzanne Sadedine, Patrik Nosil, Doug Schemske and Roger Butlin for very helpful comments and suggestions and M. D. Vose for help with numerical implementations. Supported by National Institutes of Health grant GM56693.

## References

Arnold, M. L. (1997) *Natural Hybridization and Evolution*. Oxford University Press, Oxford.

Arnold, M. and Meyer, A. (2006) Natural hybridization in primates: one evolutionary mechanism. *Zoology* **109**, 261–276.

Barton, N. H. (2001) The role of hybridization in evolution. *Molecular Ecology* **10**, 551–568.

Bolnick, D. (2004) Waiting for sympatric speciation. *Evolution* **58**, 895–899.

Bolnick, D. (2006) Multi-species outcomes in a common model of sympatric speciation. *Journal of Theoretical Biology* **241**, 734–744.

Bush, G. (1969) Sympatric host race formation and speciation in frugivorous flies of the genus *Rhagoletis* (Diptera, Tephritidae). *Evolution* **23**, 237–251.

Butlin, R. (1987) Speciation by reinforcement. *Trends in Ecology and Evolution* **2**, 8–13.

Butlin, R. K. (1995) Reinforcement: an idea evolving. *Trends in Ecology and Evolution* **10**, 432–434.

Cadena, C. D., Ricklefs, R. E., Jimenez, I. and Bermingham, E. (2005) Is speciation driven by species diversity? *Nature* **438**, E1–E2.

Chow, S. S., Wilke, C. O., Ofria, C., Lenski, R. E. and Adami, C. (2004) Adaptive radiation from resource competition in digital organisms. *Science* **305**, 84–86.

Christiansen, F. (1975) Hard and soft selection in a subdivided population. *American Naturalist* **109**, 11–16.

Coyne, J. and Orr, H. A. (2004) *Speciation*. Sinauer Associates, Inc., Sunderland, MA.

DeMeeus, T., Michalakis, Y., Renaud, F. and Olivieri, I. (1993) Polymorphism in heterogeneous environments, evolution of habitat selection and sympatric speciation – soft and hard selection models. *Evolutionary Ecology* **7**, 175–198.

Diamond, J. (1986) Evolution of ecological segregation in the New Guinea mountane avifauna. In: *Community Ecology* (ed. J. Diamond and T. J. Case), pp. 98–125. Harper and Row, Publishers, New York.

Dieckmann, U. and Doebeli, M. (1999) On the origin of species by sympatric speciation. *Nature* **400**, 354–357.

Diehl, S. R. and Bush, G. L. (1989) The role of habitat preference in adaptation and speciation. In: *Speciation and Its Consequences* (ed. D. Otte and J. A. Endler), pp. 345–365. Sinauer, Sunderland, MA.

Dowling, T. and Secor, C. (1997) The role of hybridization and introgression in the diversification of animals. *Annual Review of Ecology and Systematics* **28**, 593–619.

Eldredge, N. (1971) The allopatric model and phylogeny in Paleozoic invertebrates. *Evolution* **25**, 156–167.

Eldredge, N. (2003) The sloshing bucket: how the physical realm controls evolution. In: *Towards a Comprehensive Dynamics of Evolution:*

*Exploring the Interplay of Selection, Neutrality, Accident, and Function* (ed. J. Crutchfield and P. Schuster), pp. 3–32. Oxford University Press, New York.

Eldredge, N. and Gould, S. J. (1972) Punctuated equilibria: an alternative to phyletic gradualism. In: *Models in Paleobiology* (ed. T. J. Schopf), pp. 82–115. Freeman, Cooper, San Francisco.

Eldredge, N., Thompson, J. N., Brakefield, P. M., *et al.* (2005) Dynamics of evolutionary stasis. *Paleobiology* **31**, 133–145.

Emelianov, I., Marec, F. and Mallet, J. (2004) Genomic evidence for divergence with gene flow in host races of the larch budmoth. *Proceedings of the Royal Society London B* **271**, 97–105.

Emerson, B. and Kolm, N. (2005) Species diversity can drive speciation. *Nature* **434**, 1015–1017.

Endler, J. A. (1977) *Geographic Variation, Speciation and Clines.* Princeton University Press, Princeton, NJ.

Erwin, D. H., Valentine, J. W. and Sepkoski, J. J. (1987) A comparative study of diversification events: the early Paleozoic versus the Mesozoic. *Evolution* **41**, 1177–1186.

Excoffier, L. (2001) Analysis of population subdivision. In: *Handbook of Statistical Genetics* (ed. D. J. Balding, M. Bishop and C. Cannings), pp. 271–307. John Wiley & Sons, Ltd, chichester, UK.

Excoffier, L., Smouse, P. and Quattro, J. (1992) Analysis of molecular variance inferred from metric distances among DNA haplotypes: application to human mitochondrial DNA restriction data. *Genetics* **131**, 479–491.

Feder, J. L. (1998) The apple maggot fly, *Rhagoletis pomonella*: flies in the face of conventional wisdom about speciation? In: *Endless Forms: Species and Speciation* (ed. D. J. Howard and S. H. Berlocher), pp. 130–144. Oxford University Press, New York.

Forister, M. L. (2005) Independent inheritance of preference and performance in hybrids between host races of Mitoura butterflies. *Evolution* **59**, 1149–1155.

Fry, J. D. (2003) Multilocus models of sympatric speciation: Bush versus Rice versus Felsenstein. *Evolution* **57**, 1735–1746.

Gavrilets, S. (2003) Models of speciation: what have we learned in 40 years? *Evolution* **57**. 2197–2215.

Gavrilets, S. (2004) *Fitness Landscapes and the Origin of Species.* Princeton University Press, Princeton, NJ.

Gavrilets, S. (2005) 'Adaptive speciation': it is not that simple. *Evolution* **59**, 696–699.

Gavrilets, S. and Vose, A. (2005) Dynamic patterns of adaptive radiation. *Proceedings of the National Academy of Sciences of the United States of America* **102**, 18040–18045.

Gavrilets, S. and Vose, A. (2007) Case studies and mathematical models of ecological speciation. 2. Palms on an oceanic island. *Molecular Ecology* **16**, 2910–2921.

Gavrilets, S., Vose, A., Barluenga, M., Salzburger, W. and Meyer, A. (2007) Case studies and mathematical models of ecological speciation. 1. Cichlids in a crater lake. *Molecular Ecology* **16**, 2893–2909.

Gillespie, R. (2004) Community assembly through adaptive radiation. *Science* **303**, 356–359.

Givnish, T. and Sytsma, K. (1997) *Molecular Evolution and Adaptive Radiation.* Cambridge University Press, Cambridge.

Gould, S. J. (2002) *The Structure of Evolutionary Theory.* Harvard University Press, Boston.

Gourbiere, S. (2004) How do natural and sexual selection contribute to sympatric speciation? *Journal of Evolutionary Biology* **17**, 1297–1309.

Hayashi, T. I., Vose, M. D. and Gavrilets, S. (2007) Genetic differentiation by sexual conflict. *Evolution* **61**, 516–529.

Hodges, S. (1997) A rapid adaptive radiation via a key innovation in aquilegia. In: Molecular evolution and adaptive radiations. In: *Molecular Evolution and Adaptive Radiations* (ed. T. Givinish and K. Sytsma), pp. 391–405. Cambridge University Press, Cambridge.

Holt, R. D. (1997) On the evolutionary stability of sink populations. *Evolutionary Ecology* **11**, 723–731.

Howard, D. J. (1993) Reinforcement: origin, dynamics, and fate of an evolutionary hypothesis. In: *Hybrid Zones and the Evolutionary Process* (ed. R. G. Harrison), pp. 46–69. Oxford University Press, New York.

Hubbell, S. P. (2001) *The Unified Neutral Theory of Biodiversity and Biogeography*. Princeton University Press, Princeton, NJ.

Johnson, P. A., Hoppensteadt, F. C., Smith, J. J. and Bush, G. L. (1996) Conditions for sympatric speciation: a diploid model incorporating habitat fidelity and non-habitat assortative mating. *Evolutionary Ecology* **10**, 187–205.

Kawata, M. (2002) Invasion of empty niches and subsequent sympatric speciation. *Proceedings of the Royal Society London B* **269**, 55–63.

Kawecki, T. J. (1996) Sympatric speciation driven by beneficial mutations. *Proceedings of the Royal Society London B* **263**, 1515–1520.

Kawecki, T. J. (1997) Sympatric speciation via habitat specialization driven by deleterious mutations. *Evolution* **51**, 1751–1763.

Kirkpatrick, M. and Nuismer, S. L. (2004) Sexual selection can constrain sympatric speciation. *Proceedings of the Royal Society London B* **271**, 687–693.

Kisdi, E. and Geritz, S. A. H. (1999) Adaptive dynamics in allele space: evolution of genetic polymorphism by small mutations in heterogeneous environment. *Evolution* **53**, 993–1008.

Kot, M. (2001) *Elements of Mathematical Ecology*. Cambridge University Press, Cambridge.

Linn, C. E., Dambroski, H. R., Feder, J. L., *et al.* (2004) Postzygotic isolating factor in sympatric speciation in *Rhagoletis* flies:

reduced response of hybrids to parental host-fruit odor. *Proceedings of the National Academy of Sciences of the United States of America* **101**, 17753–17758.

Losos, J. B. (1998) Ecological and evolutionary determinants of the species–area relationship in Carribean anoline lizards. In: *Evolution on Islands* (ed. P. Grant), pp. 210–224. Oxford University Press, Oxford.

Losos, J. B., Jackman, T. R., Larson, A., de Queiroz, K. and Rodriguez-Schettino, L. (1998) Contingency and determinism in replicated adaptive radiations of island lizards. *Science* **279**, 2115–2118.

Mallet, J. (1995) A species definition for the modern synthesis. *Trends in Ecology and Evolution* **10**, 294–299.

Mallet, J. (2007) Hybrid speciation. *Nature* **446**, 279–283.

Matessi, C., Gimelfarb, A. and Gavrilets, S. (2001) Long term buildup of reproductive isolation promoted by disruptive selection: how far does it go? *Selection* **2**, 41–64.

Mavarez, J., Salazar, C., Bermingham, E., *et al.* (2006). Speciation by hybridization in Heliconius butterflies. *Nature* **441**, 868–871.

Mayr, E. (1963) *Animal Species and Evolution*. Belknap Press, Cambridge, MA.

Nice, C. C., Fordyce, J. A., Shapiro, A. M. and French-Constant, R. (2002) Lack of evidence for reproductive isolation among ecologically specialized lycaenid butteflies. *Ecological Entomology* **27**, 702–712.

Nosil, P., Crespi, B. J. and Sandoval, C. P. (2002) Host-plant adaptation drives the parallel evolution of reproductive isolation. *Nature* **417**, 440–443.

Ohta, T. and Kimura, M. (1973) A model of mutation appropriate to estimate the number of electrophoretically detectable alleles in a finite population. *Genetical Research* **22**, 201–204.

Pigliucci, M. (2003) Species as family resemblance concepts: the (dis-)solution of the species problem? *Bioessays* **25**, 596–602.

Pomiankowski, A. (1987) The costs of choice in sexual selection. *Journal of Theoretical Biology* **128**, 195–218.

Price, T. (2008) *Speciation in Birds*. Roberts and Company, Greenwood Village, CO.

Rice, W. R. and Hostert, E. E. (1993) Laboratory experiments on speciation: what have we learned in 40 years? *Evolution* **47**, 1637–1653.

Rice, W. R. and Salt, G. (1990) The evolution of reproductive isolation as a correlated character under sympatric conditions: experimental evidence. *Evolution* **44**, 1140–1152.

Riechert, S. E. (1993) Investigation of potential gene flow limitation of behavioral adaptation in an aridland spider. *Behavioral Ecology and Sociobiology* **32**, 355–363.

Rieseberg, L. H. (1995) The role of hybridization in evolution: old wine in new skins. *American Journal of Botany* **82**, 944–953.

Rieseberg, L. H. (1999) Hybrid origin of plant species. *Annual Review of Ecology and Systematics* **28**, 359–389.

Rolán-Alvarez, E., Johannesson, K. and Erlandson, J. (1997) The maintenance of a cline in the marine snail *Littorina Saxatilis*: the role of home site advantage and hybrid fitness. *Evolution* **51**, 1838–1847.

Rundle, H. D. and Nosil, P. (2005) Ecological speciation. *Ecology Letters* **8**, 336–352.

Salzburger, W. and Meyer, A. (2004) The species flocks of East African cichlid fishes: recent advances in molecular phylogenetics and population genetics. *Naturwissenschaften* **91**, 277–290.

Schluter, D. (2000) *The Ecology of Adaptive Radiation*. Oxford University Press, Oxford.

Seehausen, O. (2004) Hybridization and adaptive radiation. *Trends in Ecology and Evolution* **19**, 198–207.

Seehausen, O. (2006) African cichlid fish: a model system in adaptive radiation research. *Proceedings of the Royal Society London B* **273**, 1987–1998.

Servedio, M. R. and Noor, M. A. F. (2003) The role of reinforcement in speciation: theory and data. *Annual Review of Ecology and Systematics* **34**, 339–364.

Simpson, G. G. (1953) *The Major Features of Evolution*. Columbia University Press, New York.

Spichtig, M. and Kawecki, T. (2004) The maintenance (or not) of polygenic variation by soft selection in heterogeneous environment. *American Naturalist* **164**, 70–84.

Streelman, J. T. and Danley, P. D. (2003) The stages of vertebrate evolutionary radiation. *Trends in Ecology and Evolution* **18**, 126–131.

Waxman, D. and Gavrilets, S. (2005) Issues of terminology, gradient dynamics, and the ease of sympatric speciation in Adaptive Dynamics. *Journal of Evolutionary Biology* **18**, 1214–1219.

Wu, C.-I. (2001) The genic view of the process of speciation. *Journal of Evolutionary Biology* **14**, 851–865.

# CHAPTER EIGHT

# Niche dimensionality and ecological speciation

## PATRIK NOSIL AND LUKE HARMON

The ecological niche plays a central role in the process of 'ecological speciation', in which divergent selection between niches drives the evolution of reproductive isolation (Muller 1942; Mayr 1947, 1963; Schluter & Nagel 1995; Funk 1998; Schluter 2000). Ecological by-product speciation occurs because ecological traits that have diverged between populations via divergent selection, or traits that are genetically correlated with such traits, incidentally affect reproductive isolation. This process can occur under any geographic arrangement of populations (e.g. allopatry, parapatry or sympatry). A central prediction of ecological speciation is that ecologically divergent pairs of populations will exhibit greater levels of reproductive isolation than ecologically similar pairs of populations of similar age. Another prediction is that traits under divergent selection, or those genetically correlated with them, should often incidentally affect reproductive isolation (e.g. mate preference, hybrid fitness). In recent years, these predictions have been supported in a range of taxa (see Feder *et al.* 1994; Funk 1998; Via 1999; Rundle *et al.* 2000; Jiggins *et al.* 2001; Funk *et al.* 2002, 2006; Bradshaw & Schemske 2003; Rundle & Nosil 2005; and Funk, this volume, for review), and processes such as resource competition and predation are now known to be involved (Mallet & Barton 1989; Schluter 1994; Rundle *et al.* 2003; Vamosi 2005; Nosil & Crespi 2006a).

However, ecological divergence often results in patterns inconsistent with the completion of speciation, such as imperfect reproductive isolation and weak genotypic clustering (Drès & Mallet 2002; Coyne & Orr 2004; Nosil *et al.* 2005; Rundle & Nosil 2005; Nosil 2007). Moreover, the collapse of distinct species pairs formed by selection has been documented (Hubbs 1955; Arnold 1997; Seehausen *et al.* 1997; Taylor *et al.* 2006; Richmond & Jockusch 2007). What factors determine the likelihood that divergent selection completes speciation, and that distinct species are maintained? In the broadest sense, divergence during speciation is often continuous in nature (even if the end point of this continuous process is the development of a discontinuity). For example, the magnitude of reproductive isolation can vary quantitatively (Coyne & Orr 2004), as can the degree of genotypic clustering (Jiggins & Mallet 2001), and the extent

*Speciation and Patterns of Diversity*, ed. Roger K. Butlin, Jon R. Bridle and Dolph Schluter. Published by Cambridge University Press. © British Ecological Society 2009.

of monophyly in gene genealogies (Dopman *et al.* 2005). These different degrees of divergence can be thought of as arbitrary 'stages' of the continuous process of speciation (Drès & Mallet 2002; de Queiroz 2005; Mallet *et al.* 2007; Hey, this volume), with greater divergence equating to greater progress towards the completion of speciation. What factors predict this degree of progress?

Three relatively well-considered factors for variability in progress towards speciation are genetic architecture, time since divergence, and levels of gene flow. For example, speciation is promoted by pleiotropic effects on reproductive isolation of genes under selection (Rice & Salt 1990; Rice & Hostert 1993; Boughman 2002; Kirkpatrick & Ravigné 2002; Bradshaw & Schemske 2003), physical linkage of genes under selection and those conferring reproductive isolation (Hawthorne & Via 2001; Noor *et al.* 2001; Rieseberg 2001), one-allele assortative mating mechanisms (Felsenstein 1981; Ortíz-Barrientos & Noor 2005; Rundle & Nosil 2005 for review), increased time since divergence (Coyne & Orr 2004) and geographic barriers to gene flow (Gavrilets 2004; Bridle and Vines, this volume).

An alternative, and less-considered, explanation for the degree of progress towards ecological speciation concerns the number of niche dimensions in which pairs of taxa differ. One aspect of the niche, that dealing with the environmental requirements of species (Chase & Leibold 2003), can be related to fitness landscapes, with species' niches corresponding to peaks on that landscape (Simpson 1944; Schluter 2000). This landscape can vary in any number of ecological dimensions, with divergence in a greater number of dimensions potentially promoting speciation. This 'niche dimensionality hypothesis' is the focus of this chapter, and the alternative causal interpretation, that speciation facilitates increased dimensionality of niche divergence, is addressed in a latter section of the chapter. We also note that geographic, genetic, time-based and ecological hypotheses for the degree of progress towards speciation are not mutually exclusive.

## Dimensionality in niche divergence and speciation

the stronger [a population's] need for local adaptation ... the greater the probability of changes in the components of isolating mechanisms.                    Mayr (1963)

The magnitude, targets and causes of divergent selection are affected by ecological conditions (Endler 1977, 1986; Reimchen 1979; Schluter 2000; Kingsolver *et al.* 2001; Grant & Grant 2002). Thus, in addition to genetic scenarios, there are also specific ecological scenarios that might facilitate speciation. As exemplified by Mayr's quote, it is generally believed that as populations become more locally adapted, the probability of speciation increases. However, what is meant by 'stronger' local adaptation?

One answer is adaptation to a great number of ecological axes (i.e. dimensions). The nature of the ecological niche has received continued study over the last century (Grinnell 1917; Elton 1927; Hutchinson 1957, 1959; Pianka 1978; Schoener 1989; Wiens & Graham 2005). A major aspect of the ecological niche is its dimensionality (Hutchinson 1957; Harmon *et al.* 2005). According to this concept, an organism's niche is defined by numerous biotic and abiotic variables, each of which can be considered an axis in multidimensional space. The dimensionality of the ecological niche can be assessed using measurements of ecological variables (Maguire 1967; Green 1971) or phenotypic variables that interact with the environment to determine fitness (Vandermeer 1972; Harmon *et al.* 2005). For example, in a given species, selection might act on morphology ('dimension 1' – e.g. seed size), physiology ('dimension 2' – e.g. optimal temperature), or both.

Divergent selection between niches is a central component of ecological speciation, but little is known about whether niche dimensionality affects speciation. A distinction between considering niche dimensionality for a species versus for the process of speciation is that *differences in the niche between pairs (or among sets) of taxa* must be examined for the latter (rather than niche properties of a single taxon). Here, the focus is on dimensionality of niche *divergence*, rather than niche dimensionality. This point is critical. For example, sets of populations that each individually exhibit highly dimensional niches would not exhibit highly dimensional niche divergence if their niches were similar to one another in all axes. In essence, as niches differ in more dimensions, they become more 'discrete', thereby promoting phenotypic divergence. The number of niche dimensions that pairs of taxa differ in can be measured and related to speciation using the following three steps (illustrated in Fig. 8.1).

First, axes of niche differentiation must be measured for all taxa. This could be achieved using measurements of differentiation in ecological variables (Maguire 1967; Green 1971), divergence in phenotypic traits that determine fitness (Vandermeer 1972; Harmon *et al.* 2005), or differences in direct measurements of natural selection (Lande & Arnold 1983; Schluter 2000; Kingsolver *et al.* 2001). The focus here is on ecological traits that mediate interactions between organisms and aspects of their biotic or abiotic environment. Figure 8.1 depicts how pairs of taxa can potentially be subject to divergent selection on two different abiotic ecological axes, and can differ in mean phenotype in these two axes.

Second, the dimensionality of niche divergence is quantified. Exactly what is considered a 'dimension' or an 'axis' can vary, but a critical point is that different dimensions should be independent. Ideally, ecologically relevant and statistically independent axes can be identified (Green 1971; Harmon *et al.* 2005), but genetic and functional independence could also be considered. For example, sets of functionally related traits might be pooled to make more general dimensions (e.g. all morphological traits related to foraging used to construct a 'foraging morphology' dimension). When it comes to quantifying the dimensionality

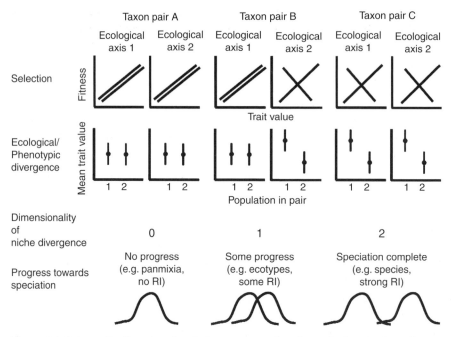

**Figure 8.1** Schematic diagram of variation among pairs of taxa in the number of dimensions (i.e. ecological axes) their niches vary in, and the potential effects on speciation. There are two ecological axes (e.g. measured by the phenotypic surrogates morphology and physiology). Taxon pair A does not differ along either niche axis, and shows no progress towards speciation. Taxon pair B differs along a single niche axis, and exhibits partial progress towards speciation. Taxon pair C differs along both niche axes, and exhibits the greatest progress towards speciation.

of niche divergence, if niche dimensions do not appear highly correlated then one approach is to simply count up the number of niche dimensions that pairs of taxa differ in. Figure 8.1 depicts how pairs of taxa can differ in zero, one or two dimensions. More quantitative descriptions of the degree to which taxa differ along different axes could also be constructed. For example, one could calculate an among-taxa variance–covariance matrix of various niche axes. There are then various statistical techniques that could be used to measure the effective dimensionality of this matrix (Schluter 2000; Herrera *et al.* 2002; Blows & Hoffman 2005; McGuigan *et al.* 2005; Mezey & Houle 2005; Chenoweth & Blows 2006; Hine & Blows 2006; Van Homrigh *et al.* 2007). In general, multivariate statistics can be used to generate independent niche dimensions (as in the example with *Timema* walking-stick insects below).

In considering statistical independence of 'dimensions', it is important to realize that independence of different environmental variables which potentially cause niche divergence may or may not come with independence of the axes of corresponding trait divergence (see Seehausen, this volume). Take the example of fish in a lake: water depth and distance from shore are statistically non-independent, yet they may select divergently on genetically independent traits (e.g. visual system and predator avoidance behaviour, respectively). Conversely, water depth and water turbidity are statistically independent. Yet, they may reinforce each other in selecting divergently on the same trait (e.g. colour vision). Ultimately, it is probably the number of genetically independent axes of trait divergence that determines the progress to ecological speciation

Third, dimensionality of niche divergence is related to measures of progress towards speciation. Examples of such measures are experimental estimates of reproductive isolation, the degree of gene flow or genotypic clustering inferred from molecular markers and traditional taxonomic status. Higher levels of reproductive isolation, lower gene flow, increased genotypic clustering and higher taxonomic status (e.g. species versus ecotype) all indicate greater progress towards speciation (Mallet 1995; Schluter 2000; Jiggins & Mallet 2001; Coyne & Orr 2004). These three steps are best implemented using divergent conspecific populations, or recently formed species, to avoid conflating differences evolving after speciation with those involved in speciation. In a formal analysis, time since the initiation of divergent selection can be taken as a covariate (e.g. Funk et al. 2006).

## Increased total selection strength

We emphasize that a general factor of interest here is the total strength of divergent selection that a population pair is exposed to, which could increase via selection on a greater number of ecological dimensions or simply via stronger selection on a single dimension. We argue that the former mechanism, divergent selection on more dimensions, may be important for generating increased total strength of selection in natural populations. For example, the degree of divergence along any single dimension is limited by the distance between adaptive peaks in that single dimension (Schluter 2000), and may be further limited by lack of suitable genetic variation in that particular dimension (Futuyma et al. 1995) and by physical or functional constraints (Arnold 1992). Notably, if the total strength of selection goes up with greater dimensionality of niche divergence, the per locus selection coefficient does not necessarily decrease as more loci become involved in divergence. Finally, divergence involving more ecological niche dimensions likely also increases the number of genes involved. Because of this, if speciation requires epistasis (Dobzhansky–Muller incompatibilities, Bateson 1909; Dobzhansky 1936, 1937; Muller 1940, 1942; Orr 1995) or divergence in just a few key genes that affect reproductive isolation, then divergence in many dimensions (genes) may be more effective at promoting speciation than greater divergence in just one or a few dimensions (genes).

Substantial phenotypic divergence and speciation may commonly require multidimensional niche divergence.

## Empirical evidence that dimensionality in niche divergence affects speciation

### Laboratory experimental evolution studies

Laboratory experiments collectively indicate that multifarious ... divergent selection can readily lead to complete reproductive isolation, but that single-factor ... divergent selection will typically lead to only incomplete reproductive isolation.

Rice and Hostert (1993, p. 1647)

Laboratory selection experiments provide ideal opportunities for testing how dimensionality in niche divergence affects speciation. Replicate lines could be selected divergently on zero, one, two or more traits (i.e. dimensions), controlling for the total strength of selection. At the end of the experiment, one can test whether lines selected divergently for more traits exhibit greater reproductive isolation (Fig. 8.2). Rice and Hostert (1993) reviewed 38 selection experiments aimed at producing speciation in the lab, and concluded that selection on more traits promotes speciation.

To assess the status of this conclusion fourteen years later, using a new and relevant classification scheme, a literature review was conducted using 59 studies recovered from Rice and Hostert (1993) and a Web of Science search on the terms 'selection and experiment and speciation'. The studies were subdivided into whether they imposed selection directly on premating isolation or on a non-isolation trait (e.g. sternopleural bristle number) (direct and indirect selection, respectively), and according to whether selection was imposed with or without gene flow (following Kirkpatrick & Ravigné 2002). Table 8.1 provides details concerning this scheme.

The most obvious point emerging from the survey is that experiments allowing strong comparisons are lacking. No single study has selected on one versus multiple traits. Only five studies applied selection on more than one trait, but there was some suggestion that these five were more successful at evolving reproductive isolation than those selecting on a single trait. For experiments involving direct selection, all three experiments that selected on more than one trait resulted in the evolution of near complete premating isolation. In contrast, only one of 13 experiments that imposed direct selection on only a single trait resulted in near complete premating isolation. However, comparison within gene-flow categories is not possible, because all studies selecting directly on multiple traits also involved gene flow. For indirect selection, one of two experiments selecting on multiple traits evolved consistent, but partial, premating isolation. In contrast, only 11 of 42 experiments selecting on a single trait resulted in consistent, but partial,

(a) Phylogenetic test

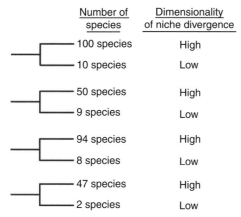

(b) Population genetic test

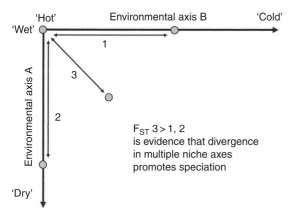

(c) Experimental evolution test

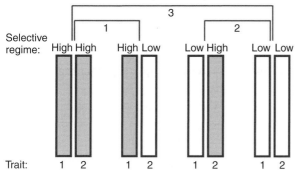

**Figure 8.2** Tests for whether the number of niche dimensions that pairs of taxa differ in affects progress towards speciation. See text for details. (a) Phylogenetic test. (b) Population genetic test. (c) Experimental evolution test.

**Table 8.1** *Summary of 59 laboratory speciation experiments in relation to the type of selection imposed, and number of traits subject to selection.*

|  | No. of studies | Result | No. of traits | References |
|---|---|---|---|---|
| (1) No Gene Flow Direct Selection | 7 | Mixed RI | 1 | 1–7 |
|  | 5 | Partial RI | 1 | 8–12 |
|  | 1 | Strong RI | 1 | 13 |
| (2) No Gene Flow Indirect Selection | 7 | No RI | 1 | 14–20 |
|  | 2 | Mixed RI | 1 | 17, 21 |
|  | 8 | Partial RI | 1 | 22–28 |
|  | 1 | No RI | 4 | 29 |
|  | 1 | Partial RI | 2 | 30 |
| (3) Gene Flow Direct Selection | 3 | Strong RI | 4 | 31–33 |
| (4) Gene Flow Indirect Selection | 21 | No RI | 1 | 10, 24, 34–38 |
|  | 1 | Mixed RI | 1 | 39 |
|  | 3 | Partial RI | 1 | 25, 27, 40 |

*Note:* No RI refers to a lack of significant RI in all cases. Mixed RI refers to cases where partial RI was detected in some cases, but RI varied among years, replicates and strains such that RI was sometimes lacking. Partial RI refers to consistent patterns of incomplete RI. Strong RI refers to over 90 percent assortative mating. (1) No Gene Flow Direct Selection. These experiments destroy-all-hybrids and impose selection directly on reproductive isolation during the course of the experiment. (2) No Gene Flow Indirect Selection. These experiments destroy-all-hybrids and impose selection on a non-isolation trait(s) during the experiment, and then testing for premating isolation at the end of the experiment. (3) Gene Flow Direct Selection Experiments. These impose selection on reproductive isolation during the course of the experiment, but not all hybrids are destroyed. (4) Gene Flow Indirect Selection Experiments. These impose selection on a non-isolation trait(s) during the course of the experiment, not all hybrids are destroyed, and RI is measured at the end of the experiment.

1 = Barker & Karlsson 1974; 2 = Ehrman 1971; 3 = Ehrman 1973; 4 = Ehrman 1979; 5 = Kessler 1966; 6 = Koepfer 1987; 7 = Wallace 1953; 8 = Crossley 1974; 9 = Dobzhansky *et al.* 1976; 10 = Hostert 1997; 11 = Knight *et al.* 1956; 12 = Koopman 1950; 13 = Paterniani 1969; 14 = Barker & Cummins 1969; 15 = Ehrman 1964; 16 = Ehrman 1969; 17 = Markow 1981; 18 = Mooers *et al.* 1999; 19 = Santibanez & Waddington 1958; 20 = van Dijken & Scharloo 1979; 21 = Halliburton & Gall 1981; 22 = de Oliveira & Cordeiro 1980; 23 = del Solar 1966; 24 = Dodd 1989; 25 = Grant & Mettler 1969; 26 = Hurd & Eisenburg 1975; 27 = Rundle *et al.* 2005; 28 = Soans *et al.* 1974; 29 = Rundle 2003; 30 = Killias *et al.* 1980; 31 = Rice 1985; 32 = Rice & Salt 1988; 33 = Rice & Salt 1990; 34 = Chabora 1968; 35 = Scharloo 1971; 36 = Scharloo *et al.* 1967; 37 = Spiess & Wilke 1984; 38 = Thoday & Gibson 1970; 39 = Coyne & Grant 1972; 40 = Thoday & Gibson 1962.

premating isolation. Factors not considered here, such as variation in effective population size (Ödeen & Florin 2000; Florin & Ödeen 2002) or a 'file-drawer effect' from unpublished results, could affect the patterns reported here. Thus, although trends are consistent with the idea that selection on more traits promotes speciation, experimental evolution studies explicitly designed to test this hypothesis would be useful.

## Data from natural populations of *Timema* walking-stick insects

*Dimensionality in divergence of cryptic colour patterns*

*Timema* are wingless insects inhabiting the chaparral of southwestern North America (Vickery 1993; Crespi & Sandoval 2000). Individuals feed and mate exclusively on the host plants upon which they live, and many species exhibit host-plant ecotypes (defined by the host they are found upon). A major axis of host-related differentiation in these insects is cryptic colouration (Sandoval 1994a,b; Crespi & Sandoval 2000; Nosil & Crespi 2006a). The phenotypic dimensionality of divergence in cryptic colouration between populations of *Timema* that had not completely speciated (conspecific host ecotypes using different plants) was compared to that between species ($n = 14$ ecotypes from eight species, where species are designated based upon taxonomic classification). The traits examined are the six colour measurements reported in Nosil and Crespi (2006a; hue, saturation and brightness of the exterior and central parts of the dorsal surface, which are all measured on similar scales, $n = 1032$ individuals). Each of these traits was comprised of three measurements for the exterior body parts (one measurement on each of head, abdomen and thorax) and two measurements for the central body parts (one measurement on each of abdomen and thorax). Using these 15 raw measurements, all pairwise differences in colour measurements between different species and between distinct ecotypes within species were calculated (Kirkpatrick & Lofsvold 1992). These differences were then averaged to yield the six traits examined in Nosil and Crespi (2006a; e.g. differences between the hue measurements from the head, abdomen and thorax were average to calculate 'mean difference in body hue'). The dimensionality of each set of comparisons was then calculated based on the variance–covariance matrix of differences among the colour axes. Since the direction of subtraction for each pair was arbitrary, all variances were calculated with a mean of zero and covariances were forced through the origin.

The concept is visualized in Fig. 8.3, which shows the amount of the variance explained by each of $m$ independent direction, for ecotype versus species pairs. Each direction is a composite trait, made up from a linear combination of the original traits. The dimensionality ('evenness') of each matrix was calculated from its eigenvalues using Levene's index (see Schluter 2000, pp. 220–221, for details)

$$L = \frac{1}{\sum p_i^2} \tag{8.1}$$

where $p_i$ is the proportion of total variance accounted for by the direction $i$. $L = 1$ if all variance is in the first direction, and $L = m$, the number of traits, if variance is equitably distributed. $L$ is therefore a useful measure of 'phenotypic

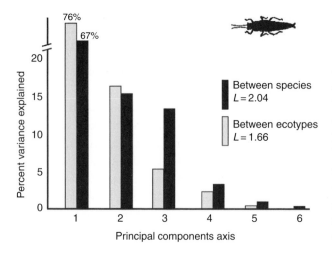

**Figure 8.3** Phenotypic dimensionality of colour-pattern divergence is greater between *Timema* species versus between ecotypes within species. Dimensionality was calculated based on the variance–covariance matrix of differences among the colour axes, using the Levene's index (*L*) (larger values indicate more 'evenness', and thus greater dimensionality).

dimensionality'. Dimensionality was lower for comparisons between ecotypes (*L* = 1.66) versus between distinct species (*L* = 2.04), with marked differences in the distribution of eigenvalues (Fig. 8.3). It is hard to infer causality from these data, which also suffer from lack of replication, but certainly they suggest that distinct species exhibit greater dimensionality in cryptic colouration than do ecotypes that did not fully speciate.

*Dimensionality in divergence in morphology and physiology*
Another avenue is to examine divergence in two different axes of host-related differentiation: cryptic morphology and physiological adaptation to different hosts. Consider two species, *Timema podura* and *Timema cristinae*, who are not sister-species and whose geographic ranges do not overlap (Law & Crespi 2002; Sandoval & Nosil 2005). Both species are composed of two host-plant ecotypes (the '*Ceanothus* ecotype' and the '*Adenostoma* ecotype'), and divergent host-adaptation has driven the evolution of various forms of reproductive isolation between ecotypes in both species (Nosil *et al.* 2002, 2003, 2005, 2006a,b; Nosil 2004; Sandoval & Nosil 2005; Nosil & Crespi 2006b). However, speciation (e.g. reproductive isolation) between ecotypes is incomplete in both species (see Nosil 2007 for details). The main argument for incomplete speciation focuses on the observation that divergence tends to be weaker in parapatry than in allopatry, a classic signature of gene flow, and thus incomplete reproductive isolation, between parapatric populations (Coyne & Orr 2004). In *T. cristinae*, this pattern of weaker divergence in parapatry is observed for experimental estimates of reproductive isolation (Nosil 2007), for mitochondrial and nuclear genetic differentiation (Nosil *et al.* 2003, unpublished), and for morphological traits relevant to host-plant adaptation (Sandoval 1994a; Nosil & Crespi 2004). Moreover, bimodality in

morphological traits is weak and often collapses in parapatry, indicative of incomplete reproductive isolation (Mallet 1995; Jiggins & Mallet 2001). Finally, there is no evidence for F1 hybrid egg inviability in 'hybrids' between the ecotypes (Nosil *et al.* 2007). Similar trends are observed in *T. podura*, although data are scarcer (Vickery 1993; Law & Crespi 2002; Sandoval & Nosil 2005; Nosil unpublished).

In ecotypes of both species, there is evidence for strong divergent selection, but only along the single axis of cryptic morphology. For example, controlled predation trials with jays reveal strong fitness trade-offs between hosts for ecotypes of both species (Sandoval 1994b; Sandoval & Nosil 2005). In *T. cristinae*, a manipulative field experiment revealed strong, host-specific divergent selection on morphology in the presence, but not in the absence, of visual predation (Nosil & Crespi 2006a). Thus, there is no evidence for divergent selection acting on morphology independent of crypsis (e.g. biomechanical/manoeuvreing reasons). When physiology is considered, again divergent selection is lacking. In field reciprocal-transplant experiments conducted in the absence of predation, both ecotypes of both species show higher fecundity when reared on *Ceanothus* (i.e. no physiological trade-offs in host plant use) (Sandoval & Nosil 2005). Perhaps the unidimensionality of divergent selection explains why only partial progress towards speciation has occurred.

Indeed, recent data from a distinct species pair, *T. podura* and *T. chumash*, suggest this is the case (Nosil & Sandoval 2008 for review). Like the ecotype pairs, one species (*T. chumash*) is found on *Ceanothus*, whereas the other on *Adenostoma*. However, the species pair is more genetically, morphologically, and behaviorally distinct than either of the two ecotype pairs, thereby representing a later stage of evolutionary divergence. For example, divergent host-plant preference (a form of reproductive isolation) is more than twice as divergent between the species pair relative to that observed between ecotypes. As for the ecotype pairs, selection on the species pair was estimated for both crypsis and physiology. For crypsis, results similar to those from ecotype pairs were observed; strong divergent selection between hosts was detected. However, the results for physiology differed markedly from those for ecotypes pairs. Specifically, the species pair was subject to strong divergent selection on physiology such that each insect species exhibited higher fitness (i.e. survival and fecundity) on its native host plant (i.e. physiological trade-offs in host-plant use). In summary, the species pair represents a later stage of the speciation process than the ecotype pairs, and was also subject to divergent selection on a greater number of niche dimensions (i.e. crypsis and physiology, rather than crysis alone). The cumulative results suggest that divergent selection on the single niche dimension of cryptic colouration can result in ecotype formation and intermediate levels of reproductive isolation, but that the completion of speciation may involve divergent selection on additional niche dimensions

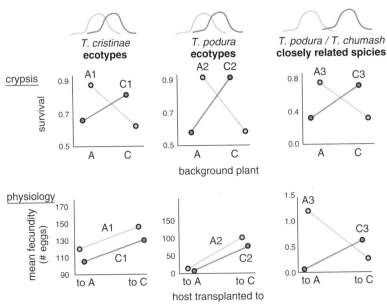

**Figure 8.4** A model of speciation in *Timema* stick-insects, where speciation is promoted by divergent selection on a greater number of niche dimensions. Three taxon pairs are represented, two ecotype pairs that exhibit partial progress towards speciation, and a species pair that represents a later stage of the speciation process. A1 and C1 refer to ecotypes of T. cristinae (A = Adenostoma and C = Ceanothus hereafter). A2 and C2 refer to ecotypes of T. podura. A3 and C3 refer to the species pair T. podura and T. chumash, respectively. The graphs depict fitness (*y*-axis) against trait value (*x*-axis) for two niche dimensions: cryptic morphology and physiology. The ecotype pairs are subject to divergent selection only on crypsis, whereas the species pair is subject to divergent selection on both niche dimensions. Modified from Nosil & Sandoval (2008) and reprinted with permission of the Public Library of Science.

(Fig. 8.4). However, because only a few taxon pairs were examined, further replication is required before robust conclusions can be made.

## Other empirical tests

A number of other tests for an association between dimensionality in niche divergence and speciation are possible. A phylogenetic test would use methods for identifying variability in net diversification (speciation minus extinction) rates (Mitter *et al.* 1988; Slowinski & Guyer 1989; Hey 1992; Nee *et al.* 1992; Harvey *et al.* 1994; Chan & Moore 2002; Ree 2005; Vamosi & Vamosi 2005). The general idea would be to quantify the dimensionality of differences between taxa within clades (Arnqvist 1998; Warheit *et al.* 1999; Cornwell *et al.* 2006), and then test if this is related to variation in diversification rate among clades. For example, a consistent, positive association across sister-clades between the dimensionality of differences between taxa and species richness could mean that dimensionality in niche divergence is associated with net diversification rate (Fig. 8.2a).

A population genetic test would examine the relationship between ecological divergence and neutral genetic differentiation. Specifically, when ecological divergence causes the evolution of reproductive barriers, it can result in a general barrier to even neutral gene flow (Barton & Bengtsson 1986; Charlesworth *et al.* 1997; Pialek & Barton 1997; Gavrilets & Cruzan 1998; see Gavrilets 2004, pp. 147–148, for summary). General barriers reduce the homogenizing effects of gene flow across the genome, thereby allowing these neutral regions to diverge via genetic drift. If neutral gene flow decreases as adaptive divergence increases, the resulting increase in drift may result in a positive correlation between the degree of adaptive phenotypic divergence and differentiation at neutral loci (independent from the geographic distance between populations). Several studies have reported this pattern (MacCallum *et al.* 1998; Lu & Bernatchez 1999; Cooper 2000; Ogden & Thorpe 2002; Vines *et al.* 2003; Rocha *et al.* 2005; Grahame *et al.* 2006; Parchman *et al.* 2006; Pilot *et al.* 2006; Egan *et al.* 2008; Nosil *et al.* 2008), and their basic design could be modified to examine the role of dimensionality in niche divergence (Fig. 8.2b). The modification requires gradients in two or more ecological variables. Population pairs could be sampled along a single (e.g. dry-hot) or multiple gradients (e.g. dry-hot and wet-cold). The prediction is that genetic differentiation would be greater, and thus gene flow lower, in the latter case.

## Niche dimensionality and speciation: alternative mechanisms generating the association

To the extent that increased dimensionality in niche divergence results in greater genetic divergence, it must often promote speciation. This assumption of increased genetic divergence seems reasonable given that increased dimensionality in niche divergence might (1) subject a greater number of traits/genes to divergent selection, thereby increasing adaptive genetic divergence between populations, and (2) result in general barriers to gene flow to facilitate even neutral genetic divergence via genetic drift. However, a number of different mechanisms might generate association between dimensionality in niche divergence, genetic divergence and progress towards speciation. These mechanisms can be viewed as alternative hypotheses concerning why niche dimensionality might affect speciation.

### Increased genetic dimensions and ecologically dependent hybrid fitness

The genotype of a species is an integrated system adapted to the ecological niche in which the species lives. Gene recombination in ... hybrids may lead to ... discordant gene patterns.

Dobzhansky (1951)

Ecological speciation often involves ecologically based reductions in hybrid fitness (Schluter 2000; Rundle & Whitlock 2001; Rundle & Nosil 2005). Hybrids exhibit low fitness because their intermediate phenotypes place them in valleys of low fitness in the fitness landscape. When adaptation to an additional ecological dimension involves divergence in an additional set of loci, hybrid fitness is expected to decrease with the number of dimensions. This can be conceptualized by considering recombinant hybrid genotypes: these genotypes can be 'mismatched' to the environment in one, two, three, etc. dimensions, depending on how many sets of loci are each adapted to different ecological dimensions. Each set of independent loci can be thought of as a 'genetic dimension', with more dimensions resulting in lower fitness (Fig. 8.5a). This mechanism requires genetic dimensions that contribute independently to fitness; the extent to which this occurs is an open empirical question.

### Changing fitness landscapes

A mechanism by which the dimensionality of fitness landscapes reduces hybrid fitness without increased genetic dimensionality involves changes in the shape of the landscape (i.e. the number of loci affecting fitness is held constant as the landscape changes). Consider the circumstance in which each new axis has two distinct optimal phenotypic values. As the number of dimensions that peaks in the landscape differ in rises (Fig. 8.5b versus Fig. 8.5c), the overall distance between peaks increases, as does the depth of the fitness valley that hybrid phenotypes tend to fall within (Fig. 8.5d). Thus using simulations it can be shown that hybrid fitness, relative to parents, decreases as landscape dimensionality increases, even if the number of loci affecting fitness is help constant (Fig. 8.5e; see figure legend 8.5 for details). Notably, the same pattern would occur if peaks were simply moved further apart along any single dimension. Thus this mechanism is most likely to be important when the distance between peaks in any single dimension is limited (i.e. divergent selection along a single axis is weak), or when divergence between distant peaks is limited by lack of suitable genetic variation.

### Pleiotropy

In the scenario directly above, the dimensionality of the fitness landscape itself changes. However, reductions in hybrid fitness can also occur without such changes due to the effects of pleiotropy on genetic variance (Turelli 1985; Wagner 1989). Consider a fitness landscape which has two separate optima that always differ in only a single dimension. Two distinct populations, each adapted to one of these two optima, were created using simulations (Guillaume & Rougemont 2006; see legend of Fig. 8.5 for details). In these simulations, traits were purely additive, and mutations had pleiotropic effects on all traits. Hybrids were formed by randomly selecting one parent from each population, and the average fitness of both hybrids and parents was calculated. The relative fitness distribution of the hybrids has a lower mean and variance when mutation affects two traits compared to when mutations affect a single trait (Fig. 8.5f).

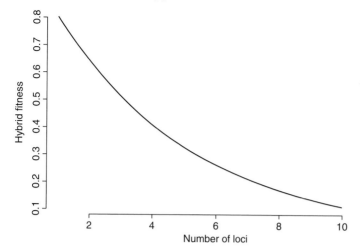

(a) Hybrid fitness with increasing genetic dimensions

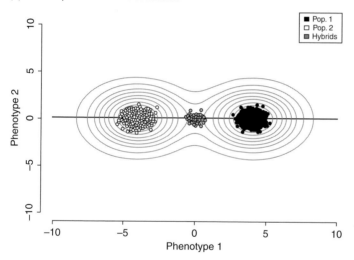

(b) Landscape differs in one dimension

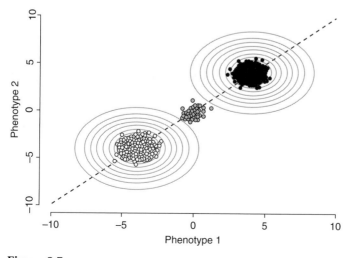

(c) Landscape differs in two dimensions

**Figure 8.5**

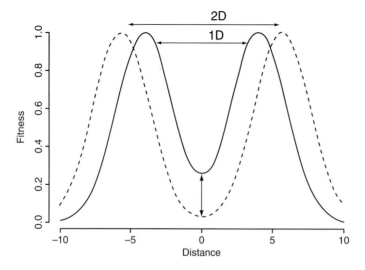

(d) Distance between peaks, 1D versus 2D

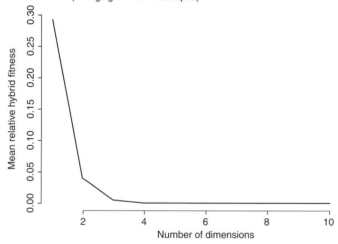

(e) Hybrid fitness as a function of dimensions
(changing fitness landscapes)

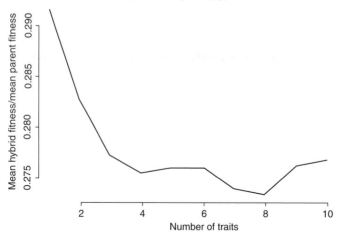

(f) Mean hybrid fitness in relation to number of traits
affected by mutation (pleiotropy)

This occurs despite the fact that the distance between the peaks and their relative width is identical. Extending this result to more traits shows that mean hybrid fitness is negatively related to the number of traits pleiotropically affected by mutation (Fig. 8.5f). This effect can be attributed to a reduction in genetic variation along each axis in the parent population due to the pleiotropic effects of mutations (Turelli 1985; Wagner 1989). As the number of traits increases, hybrids become more concentrated in the 'saddle' of the selective valley. The realism of this scenario might be questioned (as for models invoking pleiotropy in general, see below). However, less extreme scenarios should produce the same effect (i.e. mutations need not affect all traits), and it illustrates how hybrid fitness can be reduced without changing the fitness landscape itself.

## Dobzhansky–Muller incompatibility models

A general class of speciation models involves the evolution of genetic incompatibilities in hybrids due to negative epistasis between alleles at different loci ('Dobzhansky–Muller incompatibilities'; Bateson 1909; Dobzhansky 1936, 1937; Muller 1940, 1942; Orr 1995; Orr & Turelli 2001; Coyne & Orr 2004; Gavrilets 2004; Dettman *et al.* 2007). The underlying logic is that genes within

---

**Figure 8.5** The effects of dimensionality of fitness landscapes on the fitness of hybrids relative to parental taxa. (a), total fitness is determined by the independent effects of an increasing number of loci which are each under selection; at each locus, heterozygote fitness is equal to 80% of the parental homozygotes. In (b) and (c), populations evolving on a fitness landscape are considered; in (b), peaks differ in their position along only one phenotypic axis ($x$; $-4$ and 4; $y = 0$); in (c), peaks differ in their position along both the $x$ and $y$ axes. (d) compares the shape of the fitness surface along a straight line connecting both peaks in (b) and (c). To investigate hybrid fitness, simulations were conducted using the software Nemo (Guillaume & Rougemont 2006). For each simulation, two populations were created. Each population was made up of 10 000 individuals whose traits were determined by 50 unlinked loci. Mutations occurred at a low rate (0.0002) and had pleiotropic effects on all traits; mutational effects were drawn from a multivariate normal distribution with mutational variance $\sigma^2_m = 0.05$ for each trait and zero expected genetic covariance among traits. Thus, each of the 50 loci contributed to all of the phenotypic traits. Populations were evolved for 10 000 generations, at which point mean fitness and genetic variances had stabilized due to mutation-selection-drift equilibrium. The fitness surface was described by two peaks, each determined by a multivariate normal distribution with variance of four and no selective covariance among traits. Hybrids were formed by drawing one parent from each population. For one set of simulations (e), the populations evolved on multidimensional fitness landscapes similar to (c), where peaks differ at their position along all axes. In another set of simulations, (f), peaks on the fitness landscape with peaks differing only along one axis ('Phenotype 1'; b). These simulations show the effects of the number of traits pleiotropically affected by mutations on the fitness of hybrids relative to parental taxa.

a population are tested to work well together, whereas genes from different populations have not been tested together. Thus, alleles that are advantageous (or neutral) within populations might be incompatible when combined in between-population hybrids. Under this class of models, increased genetic divergence translates to increased probability of genetic incompatibilities, and higher likelihood of speciation (see Orr 1995 for details). For example, a first substitution at locus A cannot cause a between-locus incompatibility, a second substitution at locus B might cause one incompatibility (with locus A), a third substitution at locus C might cause two incompatibilities (with A and B), and in general the Kth substitution can be incompatible with K-1 loci. Thus for incompatibilities between pairs of the loci, the number of incompatibilities increases with the square of the number of substitutions (Orr 1995). For complex incompatibilities involving three or more loci, the rise of the number of incompatibilities with the number of substitutions is even faster. The Dobzhansky–Muller model generates the prediction that increased dimensionality in niche divergence promotes speciation if the following assumption is made: as dimensionality in niche divergence increases, genetic divergence between populations (i.e. number of substitutions separating them) increases.

### Models invoking magic traits or linkage disequilibrium

In another class of models, genes under divergent selection have a pleiotropic effect on forms of reproductive isolation such as mate or habitat preference ('single-variation' models of Rice & Hostert 1993, 'direct selection' models of Kirkpatrick & Ravigné 2002, 'no-gene' models of Coyne & Orr 2004, 'magic-trait' models of Gavrilets 2004). Speciation via this mechanism might become more likely as the dimensionality of divergence increases, because as niches become divergent in more dimensions, more traits differ between taxa, and genetic divergence in the key genes (traits) that incidentally affect reproductive isolation becomes more likely. A similar argument could apply to models using the build up of linkage disequilibrium between selected and reproductive isolation loci (Felsenstein 1981; Kirkpatrick & Ravigné 2002; see Gavrilets 2004 for review); when dimensionality increases, the probability that a gene involved in adaptation will be in linkage disequilibrium (e.g. physically linked) with a gene involved in reproductive isolation might increase. In general though, there is not much explicit theoretical work relating niche dimensionality to speciation. Thus further theoretical work, particularly work focusing on analytical theory and macro-evolutionary models (see Gavrilets & Vose 2005 and this volume), could be fruitful.

### Alternatives summarized: many genes or the right genes?

The alternative mechanisms noted above can be divided into two main classes: those that invoke divergence in a greater number of genes that each more or less equally affect reproductive isolation (e.g. genetic dimensions and ecological selection against hybrids) and those that invoke a greater sampling of the genome such that divergence is more likely in the few key genes affecting reproductive isolation (e.g. Dobzhansky–Muller and Magic Trait models). These can be thought of as 'accumulative' and 'sampling' models, respectively. An analogy can be drawn with the association between ecosystem function (e.g. productivity, nutrient cycling) and species diversity: does this association arise because a greater number of more or less equally important species better fills ecosystem requirements, or because greater species number increases the probability that a few key (e.g. keystone) species are represented within the ecosystem (Hooper & Vitousek 1997; Tilman *et al.* 1997)?

## Niche dimensionality and peak shifts during speciation

Throughout this article we have implicitly assumed that populations will reach different peaks in the adaptive landscape, and thus that the major problem of speciation is the build up of reproductive isolation, particularly in the face of homogenizing gene flow (see Bridle & Vines 2006, this volume; see Gavrilets & Vose, this volume, for consideration of the limits to adaptation and the costs of gene flow when divergence is multidimensional). In contrast, it has been argued that peak shifts themselves are a central problem for speciation, because large peak shifts are difficult, but small ones result in only weak reproductive isolation (Wright 1932; Barton & Rouhani 1987; Fear & Price 1998; Gavrilets 2004; but see Whitlock 1997). It could be that large valleys between peaks in a single dimension do not generate speciation because they cannot be crossed. In contrast, with many dimensions each separated by a smaller valley, perhaps each dimension can be crossed, gradually building up reproductive isolation as more dimensions are adapted to.

Alternatively, by increasing separation between peaks, increasing niche dimensionality might decrease the probability of peak shifts (but increase the probability of speciation should those valleys be crossed). Additionally, peak shifts might be further affected by frequency-dependent selection. For example, niche dimensionality might start out low, with a narrow valley that can be crossed in a single dimension. As adaptation proceeds, species interact in new ways, thus gradually building up niche dimensionality (and the potential for reproductive isolation) among different sets of adaptive characters (Streelman & Danley 2003). Thus, if peak shifts are a central issue in speciation, a correlation between dimensionality of niche divergence and speciation might be expected, but only under certain scenarios. The net outcome of these counteracting forces is unclear, and deserves further study.

## Causality and the role of time since divergence

A final issue to consider is causality. Not only can gene flow decrease (i.e. reproductive isolation increase) with increasing dimensionality of niche divergence, but, conversely, the dimensionality of niche divergence can grow with decreasing levels of potentially homogenizing gene flow. Thus when it comes to an association between dimensionality of niche divergence and levels of gene flow, it may often be difficult to determine the direction of causality. Moreover, a feedback loop can exist between the two processes, such that niche divergence reduces gene flow, which then allows further niche divergence, which then further reduces gene flow, and so on (Hendry & Taylor 2004; Nosil & Crespi 2004). A minimum dimensionality of niche divergence could be required for divergent selection to be sufficiently strong to initiate speciation, but once gene flow becomes sufficiently reduced, divergence in other traits that are under weaker selection becomes possible (Streelman & Danley 2003). Such a scenario might be occurring in the African cichlids studied by Seehausen and colleagues (Seehausen, this volume).

The issue of causality further raises the question of the role of time since divergence. It is accepted that time since divergence is positively related to the degree of progress towards speciation (reviewed in Coyne & Orr 2004). It could be that adaptation to a greater number of niche dimensions requires more time than adaptation to one or a few dimensions. In that case, a critical role for time in speciation could arise through the association between time and dimensionality of niche divergence. Alternatively, the dimensionality of niche divergence could be positively related to the degree of progress towards speciation independent of time. In this case, population pairs of equal age that differ in the dimensionality of their niche divergence would differ in their degree of progress towards speciation. In both scenarios, dimensionality of niche divergence could play a role in speciation, but with time sometimes playing an important role and sometimes not. The types of reproductive barriers involved might determine whether time plays a role. For barriers that evolve quickly and early in the speciation process (e.g. premating isolation), the dimensionality of niche divergence may play a role independent from time, whereas for more slowly evolving barriers (e.g. intrinsic incompatibilities) a role for time may be more likely. Empirical data are required to determine the incidence of these scenarios.

## Conclusions

Divergent selection between niches can promote speciation. However, progress towards ecological speciation is often incomplete, or species pairs collapse. A number of ecological, genetic and geographic factors might explain the degree of progress towards complete speciation, and the maintenance of species. The focus here was on whether dimensionality in niche divergence affects speciation. Although existing information is suggestive of a potential role for such divergence, direct data are almost completely lacking. Once more definitive evidence for a positive association between dimensionality in niche divergence and speciation is established, it will be of interest to determine causality, and to distinguish among the alternative mechanisms generating the association.

## Acknowledgements

We thank D. Schluter, R. Butlin and J. Bridle for organizing the Symposium, and inviting us to participate. The manuscript benefited from discussions about speciation with A. Hendry, J. Galindo, S. Egan, J. Lee-Yaw, D. Funk, S. Rogers, C. P. Sandoval, O. Seehausen and B. J. Crespi. We especially thank T. Reimchen for suggesting the link between niche dimensionality and speciation, and D. Schluter for forcing us to be more precise in our thinking on the topic. J. Losos, D. Schluter, T. Reimchen, T. Price, J. Bridle, J. Mallet, O. Seehausen and two anonymous reviewers provided constructive criticism on previous versions. The analogy with ecosystem function was raised by D. Schluter, the role of peak shifts by T. Price, and the issue of causality by O. Seehausen. F. Guilliame modified the NEMO software package for use in these analyses. P. Nosil was funded by a post-doctoral fellowship from the Natural Sciences and Engineering Research Council of Canada (NSERC). L. Harmon was funded by the Biodiversity Centre at the University of British Columbia.

## References

Arnold, S. J. (1992) Constraints on phenotypic evolution. *American Naturalist* **140**, S85–S107.

Arnold, M. L. (1997) *Natural Hybridization and Evolution*. Oxford University Press, Oxford.

Arnqvist, G. (1998) Comparative evidence for the evolution of genitalia by sexual selection. *Nature* **393**, 784–786.

Barker, J. S. F. and Cummins, L. J. (1969) The effect of selection for sternopleural bristle number on mating behavior in *Drosophila melanogaster*. *Genetics* **61**, 713–719.

Barker, J. S. F. and Karlsson, L. J. E. (1974) Effects of population size and selection intensity on responses to disruptive selection in *Drosophila melanogaster*. *Genetics* **78**, 715–735.

Barton, N. and Bengtsson, B. O. (1986) The barrier to genetic exchange between hybridizing populations. *Heredity* **57**, 357–376.

Barton, N. H. and Rouhani, S. (1987) The frequency of shifts between alternative equilibria. *Journal of Theoretical Biology* **125**, 397–418.

Bateson, W. (1909) Heredity and variation in modern lights. In: *Darwin and Modern Science* (ed. A. C. Seward), pp. 85–101. Cambridge University Press, Cambridge.

Blows, M. W. and Hoffman, A. A. (2005) A reassessment of genetic limits to evolutionary change. *Ecology* **86**, 1371–1384.

Boughman, J. W. (2002) How sensory drive can promote speciation. *Trends in Ecology and Evolution* **10**, 1–7.

Bradshaw, H. D. and Schemske, D. W. (2003) Allele substitution at a flower colour locus produces a pollinator shift in monkeyflowers *Nature* **426**, 176–178.

Bridle, J. R. and Vines, T. H. (2006) Limits to evolution at range margins: when and why does adaptation fail? *Trends in Ecology and Evolution* **22**, 140–147.

Chabora, A. J. (1968) Disruptive selection for sternopleural chaeta number in various strains of *Drosophila melanogaster*. *American Naturalist* **102**, 525–532.

Chan, K. M. A. and Moore, B. R. (2002) Whole-tree methods for detecting differential diversification rates. *Systematic Biology* **51**, 855–865.

Charlesworth, B., Nordborg, M. and Charlesworth, D. (1997) The effects of local selection, balanced polymorphism and background selection on equilibrium patterns of genetic diversity in subdivided populations. *Genetical Research* **70**, 155–174.

Chase, J. M. and Leibold, M. A. (2003) *Ecological Niches: Linking Classical and Contemporary Approaches.* University of Chicago Press, Chicago, IL.

Chenoweth, S. F. and Blows, M. (2006) Dissecting the complex genetic basis of mate choice. *Nature Review Genetics* **7**, 681–692.

Cooper, M. L. (2000) Random amplified polymorphic DNA analysis of southern brown bandicoot (*Isoodon obesulus*) populations in Western Australia reveals genetic differentiation related to environmental variables. *Molecular Ecology* **9**, 469–479.

Cornwell, W. K., Schwilk, D. W. and Ackerly, D. D. (2006) A trait-based test for habitat filtering: convex hull volume. *Ecology* **87**, 1465–1471.

Coyne, J. A. and Grant, B. (1972) Disruptive selection on I-maze activity in *Drosophila melanogaster. Genetics* **11**, 185–188.

Coyne, J. A. and Orr, H. A. (2004) *Speciation.* Sinauer Associates, Inc. Sunderland, MA.

Crespi, B. J. and Sandoval, C. P. (2000) Phylogenetic evidence for the evolution of ecological specialization in *Timema* walking-sticks. *Journal of Evolutionary Biology* **13**, 249–262.

Crossley, S. A. (1974) Changes in mating behavior produced by selection for ethological isolation between ebony and vestigial mutants of *Drosophila melanogaster. Evolution* **28**, 631–647.

de Oliveira, A. K. and Cordeiro, A. R. (1980) Adaptation of *Drosophila willistoni* experimental populations to extreme pH medium. *Heredity* **44**, 123–130.

de Queiroz, K. (2005) Ernst Mayr and the modern concept of species. *Proceedings of National Academy of Sciences of the United States of America* **102**, 6600–6607.

del Solar, E. (1966) Sexual isolation caused by selection for positive and negative phototaxis and geotaxis in *Drosophila pseudoobscura. Proceedings of National Academy of Sciences of the United States of America* **56**, 484–487.

Dettman, J. R., Sirjusingh, C., Kohn, L. M. and Anderson, J. B. (2007) Incipient speciation by divergent adaptation and antagonistic epistasis in yeast. *Nature* **447**, 585–588.

Dobzhansky, T. (1936) Studies on hybrid sterility. II. Localization of sterility factors in *Drosophila pseudoobscura* hybrids. *Genetics* **121**, 113–125.

Dobzhansky, T. (1937) *Genetics and the Origin of Species.* Columbia University Press, New York.

Dobzhansky, T. (1951) *Genetics and the Origin of Species.* 3rd edn. Columbia University Press, New York.

Dobzhansky, T., Pavlovsky, O. and Powell, J. R. (1976) Partially successful attempt to enhance reproductive isolation between semispecies of *Drosophila paulistorum. Evolution* **30**, 201–212.

Dodd, D. M. B. (1989) Reproductive isolation as a consequence of adaptive divergence in *Drosophila pseudoobscura. Evolution* **43**, 1308–1311.

Dopman, E. B., Perez, L., Bogdanowicz, S. M. and Harrison, R. (2005) Consequences of reproductive barriers for genealogical discordance in the European corn borer. *Proceedings of National Academy of Sciences of the United States of America* **102**, 14706–14711.

Drès, M. and Mallet, J. (2002) Host races in plant-feeding insects and their importance in sympatric speciation. *Philosophical Transactions of the Royal Society of London B* **357**, 471–492.

Egan, S. P., Nosil, P. and Funk, D. J. (2008) Selection and genomic differentiation during ecological speciation: isolating the contributions of host-association via a comparative genome scan of Neochlamisus bebbianae leaf beetles. *Evolution* **62**, 1162–1181.

Ehrman, L. (1964) Genetic divergence in M. Vetukhiv's experimental populations of *Drosophila pseudoobscura*. *Genetical Research* **5**, 150–157.

Ehrman, L. (1969) Genetic divergence in M. Vetukhiv's experimental populations of *Drosophila pseudoobscura*. 5. A further study of the rudiments of sexual isolation. *American Midland Naturalist* **82**, 272–276.

Ehrman, L. (1971) Natural selection for the origin of reproductive isolation. *American Naturalist* **105**, 479–483.

Ehrman, L. (1973) More on natural selection for the origin of reproductive isolation. *American Naturalist* **107**, 318–319.

Ehrman, L. (1979) Still more on natural selection for the origin of reproductive isolation. *American Naturalist* **113**, 148–150.

Elton, C. S. (1927) *Animal Ecology*. Sidwich & Jackson, London.

Endler, J. A. (1977) *Geographic Variation, Speciation, and Clines*. Princeton University Press, Princeton, NJ.

Endler, J. A. (1986) *Natural Selection in the Wild*. Princeton University Press, NJ.

Fear, K. K. and Price, T. (1998) The adaptive surface in ecology. *Oikos* **82**, 440–448.

Feder, J. L., Opp, S. B., Wlazlo, B., *et al.* (1994) Host fidelity is an effective premating barrier between sympatric races of the apple maggot fly. *Proceedings of the National Academy of Sciences of the United States of America* **91**, 7990–7994.

Felsenstein, J. (1981) Skepticism towards Santa Rosalia, or why are there so few kinds of animals? *Evolution* **35**, 124–38.

Florin, A.-B. and Ödeen, A. (2002) Laboratory experiments are not conducive for allopatric speciation. *Journal of Evolutionary Biology* **15**, 10–19.

Funk, D. J. (1998) Isolating a role for natural selection in speciation: host adaptation and sexual isolation in *Neochlamisus bebbianae* leaf beetles. *Evolution* **52**, 1744–1759.

Funk, D. J., Filchak, K. E. and Feder, J. L. (2002) Herbivorous insects: model systems for the comparative study of speciation ecology. *Genetica* **116**, 251–267.

Funk, D. J., Nosil, P. and Etges, W. (2006) Ecological divergence exhibits consistently positive associations with reproductive isolation across disparate taxa. *Proceedings of the National Academy of Sciences of the United States of America* **103**, 3209–3213.

Futuyma, D. J., Keese, M C. and Funk, D. J. (1995) Genetic constraints on macroevolution: the evolution of host affiliation in the leaf beetle genus *Ophraella*. *Evolution* **49**, 797–809.

Gavrilets, S. (2004) *Fitness Landscapes and the Origin of Species*. Princeton University Press, Princeton, NJ.

Gavrilets, S. and Cruzan, M. B. (1998) Neutral gene flow across single locus clines. *Evolution* **52**, 1277–1284.

Gavrilets, S. and Vose, A. (2005) Dynamic patterns of adaptive radiation. *Proceedings of the National Academy of Sciences of the United States of America* **102**, 18040–18045.

Grahame, J. W., Wilding, C. S. and Butlin, R. K. (2006) Adaptation to a steep environmental gradient and an associated barrier to gene exchange in *Littorina saxatilis*. *Evolution* **60**, 268–278.

Grant, P. R. and Grant, B. R. (2002) Unpredictable evolution in a 30-year study of Darwin's finches. *Science* **296**, 707–711.

Grant, B. and Mettler, L. E. (1969) Disruptive and stabilizing selection on the 'escape' behavior of *Drosophila melanogaster*. *Genetics* **62**, 625–637.

Green, R. H. (1971) A multivariate statistical approach to the Hutchinsonian niche: bivalve mollusks of central Canada. *Ecology* **52**, 543–556.

Grinnell, J. (1917) The niche relationships of the California thrasher. *Auk* **34**, 427–433.

Guillaume, G. and Rougemont, J. (2006) Nemo: an evolutionary and population genetics programming framework. *Bioinformatics Applications Note* **22**, 2556–2557.

Halliburton, R. and Gall, G. A. E. (1981) Disruptive selection and assortative mating in *Tribolium castaneum*. *Evolution* **35**, 829–843.

Harmon, L. J., Kolbe, J. J., Cheverud, J. M. and Losos, J. B. (2005) Convergence and the multidimensional niche. *Evolution* **59**, 409–421.

Harvery, P. H., May, R. M. and Nee, S. (1994) Phylogenies without fossils. *Evolution* **48**, 523–529.

Hawthorne, D. J. and Via, S. (2001) Genetic linkage of ecological specialization and reproductive isolation in pea aphids. *Nature* **412**, 904–907.

Hendry, A. P. and Taylor, E. B. (2004) How much of the variation in adaptive divergence can be explained by gene flow? An evaluation using lake-stream stickleback pairs. *Evolution* **58**, 2319–2331.

Herrera, C. M., Cerdaâ, X., Garciaâ, M. B., *et al.* (2002) Floral integration, phenotypic covariance structure and pollinator variation in bumblebee-pollinated *Helleborus foetidus*. *Journal of Evolutionary Biology* **15**, 108–121.

Hey, J. (1992) Using phylogenetic trees to study speciation and extinction. *Evolution* **46**, 627–640.

Hine, E. and Blows, M. (2006) The effective dimensionality of the genetic variance-covariance matrix. *Genetics* **173**, 1135–1144.

Hooper, D. U. and Vitousek, P. M. (1997) The effects of plant composition and diversity on ecosystem processes. *Science* **277**, 1302–1305.

Hostert, E. E. (1997) Reinforcement: a new perspective on an old controversy. *Evolution* **51**, 697–702.

Hubbs, C. L. (1955) Hybridization between fish species in nature. *Systematic Zoology* **4**, 1–20.

Hurd, L. E. and Eisenburg, R. M. (1975) Divergent selection for geotactic response and evolution of reproductive isolation in sympatric and allopatric populations of house flies. *American Naturalist* **109**, 353–358.

Hutchinson, G. E. (1957) Concluding remarks. *Cold Spring Harbor Symposia on Quantitative Biology* **22**, 415–427.

Hutchinson, G. E. (1959) Homage to Santa Rosalia or why are there so many kinds of animals? *American Naturalist* **93**, 145–159.

Jiggins, C. D. and Mallet, J. (2001) Bimodal hybrid zones and speciation. *Trends in Ecology and Evolution* **15**, 250–255.

Jiggins, C. D., Naisbit, R. E., Coe, R. L. and Mallet, J. M. (2001) Reproductive isolation caused by colour pattern mimicry. *Nature* **411**, 302–305.

Kessler, S. (1966) Selection for and against ethological isolation between *Drosophila pseudoobscura* and *Drosophila persimilis*. *Evolution* **20**, 634–645.

Killias, G., Alahiotis, S. N. and Pelecanos, M. (1980) A multifactorial genetic investigation of speciation theory using *Drosophila melanogaster*. *Evolution* **34**, 730–737.

Kingsolver, J. G., Hoekstra, H. E., Hoekstra, J. M., *et al.* (2001) The strength of phenotypic selection in natural populations. *American Naturalist* **157**, 245–261.

Kirkpatrick, M. and Lofsvold, D. (1992) Measuring selection and constraint in the evolution of growth. *Evolution* **46**, 954–971.

Kirkpatrick, M. P. and Ravigné, V. (2002) Speciation by natural and sexual selection: models and experiments. *American Naturalist* **159**, S22–S35.

Knight, G. R., Robertson, A. and Waddington, C. H. (1956) Selection for sexual isolation within a species. *Evolution* **10**, 14–22.

Koepfer, H. R. (1987) Selection for sexual isolation between geographic forms of *Drosophila mojavensis*. I. Interactions between the selected forms. *Evolution* **41**, 37–48.

Koopman, K. F. (1950) Natural selection for reproductive isolation between *Drosophila pseudoobscura* and *Drosophila persimilis*. *Evolution* **4**, 135–148.

Lande, R. and Arnold, S. J. (1983) The measurement of selection on correlated characters. *Evolution* **37**, 1210–1226.

Lu, G. Q. and Bernatchez, L. (1999) Correlated trophic specialization and genetic divergence in sympatric lake whitefish ecotypes (*Coregonus clupeaformis*): support for the ecological speciation hypothesis. *Evolution* **53**, 1491–1505.

Law, J. H. and Crespi, B. J. (2002). The evolution of geographic parthenogenesis in *Timema* walking-sticks. *Molecular Ecology* **11**, 1471–1489.

MacCallum, C. J., Nürnberger, B., Barton, N. H. and Szymura, J. M. (1998) Habitat preference in a *Bombina* hybrid zone in Croatia. *Evolution* **52**, 227–239.

Maguire, B., Jr. (1967) A partial analysis of the niche. *American Naturalist* **101**, 515–523.

Mallet, J. (1995) A species definition for the modern synthesis. *Trends in Ecology and Evolution* **10**, 294–299.

Mallet, J. and Barton, N. H. (1989) Strong natural selection in a warning-color hybrid zone. *Evolution* **43**, 421–431.

Mallet, J., Beltran, M., Neukirchen, W. and Linares, M. (2007) Natural hybridization in heliconiine butterflies: the species boundary as a continuum. *BioMedCentral Evolutionary Biology* **7**, 28.

Markow, T. A. (1981) Mating preference are not predictive of the direction of evolution in experimental populations of *Drosophila*. *Science* **213**, 1405–1407.

Mayr, E. (1947) Ecological factors in speciation. *Evolution* **1**, 263–288.

Mayr, E. (1963) *Animal Species and Evolution*. Harvard University Press, Cambridge.

McGuigan, K., Chenoweth, S. F. and Blows, M. W. (2005) Phenotypic divergence along lines of genetic variance. *American Naturalist* **165**, 32–43.

Mezey, J. G. and Houle, D. (2005) The dimensionality of genetic variation for wing shape in *Drosophila melanogaster*. *Evolution* **59**, 1027–1038.

Mitter, C., Farrell, B. and Wiegmann, B. (1988) The phylogenetic study of adaptive zones: has phytophagy promoted insect diversification? *American Naturalist* **132**, 107–128.

Mooers, A., Rundle, H. D. and Whitlock, M. C. (1999) The effects of selection and bottlenecks on male mating success in peripheral isolation. *American Naturalist* **153**, 437–444.

Muller, H. J. (1940) Bearings of the *Drosophila* work on systematics. In: *The New Systematics* (ed. J. S. Huxley), pp. 185–268. Clarendon Press, Oxford.

Muller, H. J. (1942) Isolating mechanisms, evolution and temperature. *Biological Symposia* **6**, 71–125.

Nee, S., Mooers, A. O. and Harvey, P. H. (1992) Tempo and mode of evolution revealed from molecular phylogenies. *Proceedings of the National Academy of Sciences of the United States of America* **89**, 8322–8326.

Noor, M. A. F., Grams, K. K., Bertucci, L. A. and Reiland, J. (2001) Chromosomal inversions and the reproductive isolation of species. *Proceedings of the National Academy of Sciences of the United States of America* **98**, 12084–12088.

Nosil, P. (2004) Reproductive isolation caused by visual predation on migrants between divergent environments. *Proceedings of the Royal Society of London B* **271**, 1521–1528.

Nosil, P. (2007) Divergent host-plant adaptation and reproductive isolation between ecotypes of *Timema cristinae* walking-sticks. *American Naturalist* **169**, 151–162.

Nosil, P. and Crespi, B. J. (2004) Does gene flow constrain trait divergence or vice-versa? A test using ecomorphology and sexual isolation in *Timema cristinae* walking-sticks. *Evolution* **58**, 101–112.

Nosil, P. and Crespi, B. J. (2006a) Experimental evidence that predation promotes divergence during adaptive radiation. *Proceedings of the National Academy of Sciences of the United States of America* **103**, 9090–9095.

Nosil, P. and Crespi, B. J. (2006b) Ecological divergence promotes the evolution of cryptic reproductive isolation. *Proceedings of the Royal Society of London B* **273**, 991–997.

Nosil, P., Crespi, B. J., Gries, R. and Gries, G. (2007) Natural selection and divergence in mate preference during speciation. *Genetica* **129**, 309–327.

Nosil, P., Crespi, B. J. and Sandoval, C. P. (2002) Host-plant adaptation drives the parallel evolution of reproductive isolation. *Nature* **417**, 440–443.

Nosil, P., Crespi, B. J. and Sandoval, C. P. (2003) Reproductive isolation driven by the combined effects of ecological adaptation and reinforcement. *Proceedings of the Royal Society of London B* **270**, 1911–1918.

Nosil, P., Crespi, B. J. and Sandoval, C. P. (2006a) The evolution of host preferences in allopatric versus parapatric populations of *Timema cristinae*. *Journal of Evolutionary Biology* **19**, 929–942.

Nosil, P., Crespi, B. J., Sandoval, C. P. and Kirkpatrick, M. (2006b) Migration and the genetic covariance between habitat preference and performance. *American Naturalist* **167**, E66–E78.

Nosil, P., Egan, S. P. and Funk, D. J. (2008) Heterogeneous genomic differentiation between walking-stick ecotypes: 'isolation by adaptation' and multiple roles for divergent selection. *Evolution* **62**, 316–336.

Nosil, P. and Sandoval, C. P. (2008) Ecological niche dimensionality and the evolutionary diversification of stick insects. *Public Library of Science ONE* **3**, e1907.

Nosil, P., Vines, T. H. and Funk, D. J. (2005) Perspective: reproductive isolation caused by natural selection against immigrants from divergent habitats. *Evolution* **59**, 705–719.

Ödeen, A. and Florin, A.-B. (2000) Effective population size may limit the power of laboratory experiments to demonstrate sympatric and parapatric speciation. *Proceedings of the Royal Society of London B* **267**, 601–606.

Ogden, R. and Thorpe, R. S. (2002) Molecular evidence for ecological speciation in tropical habitats. *Proceedings of the National Academy of Sciences of the United States of America* **99**, 13612–13615.

Orr, H. A. (1995) The population genetics of speciation: the evolution of hybrid incompatibilities. *Genetics* **139**, 1805–1813.

Orr, H. A. and Turelli, M. (2001) The evolution of postzygotic isolation: accumulating Dobzhansky-Muller incompatibilities. *Evolution* **55**, 1085–1094.

Ortíz-Barrientos, D. and Noor, M. A. F. (2005) Evidence for a one-allele assortative mating locus. *Science* **310**, 1467–1467.

Parchman, T. L., Benkman, C. W. and Britch, S. C. (2006) Patterns of genetic variation in the adaptive radiation of New World crossbills (Aves: Loxia). *Molecular Ecology* **15**, 1873–1887.

Paterniani, E. (1969) Selection for reproductive isolation between two populations of Maize, *Zea mays*, L. *Evolution* **23**, 534–547.

Pialek, J. and Barton, N. H. (1997) The spread of an advantageous allele across a barrier: The effects of random drift and selection against heterozygotes. *Genetics* **145**, 493–504.

Pianka, E. (1978) *Evolutionary Ecology*. 3rd ed. Harper & Row, New York.

Pilot, M., Jedrzejewski, W., Branicki, W., *et al.* (2006) Ecological factors influence population genetic structure of European grey wolves. *Molecular Ecology* **15**, 4533–4553.

Ree, R. H. (2005) Detecting the historical signature of key innovations using stochastic models of character evolution and cladogenesis. *Evolution* **59**, 257–265.

Reimchen, T. E. (1979) Substratum heterogeneity, crypsis, and colour polymorphism in an intertidal snail. *Canadian Journal of Zoology* **57**, 1070–1085.

Rice, W. R. (1985) Disruptive selection on habitat preference and the evolution of reproductive isolation: an exploratory experiment. *Evolution* **39**, 645–656.

Rice, W. R. and Hostert, E. E. (1993) Laboratory experiments in speciation: what have we learned in 40 years? *Evolution* **47**, 1637–1653.

Rice, W. R. and Salt, G. W. (1988) Speciation via disruptive selection on habitat preference: experimental evidence. *American Naturalist* **131**, 911–917.

Rice, W. R. and Salt, G. W. (1990) The evolution of reproductive isolation as a correlated character under sympatric conditions: experimental evidence. *Evolution* **44**, 1140–1152.

Richmond, J. Q. and Jockusch, E. L. (2007) Body size evolution simultaneously creates and collapses species boundaries in a clade of scincid lizards. *Proceedings of the Royal Society of London B* **274**, 1701–1708.

Rieseberg, L. H. (2001) Chromosomal rearrangements and speciation. *Trends in Ecology and Evolution* **16**, 351–358.

Rocha, L. A., Robertson, D. R., Roman, J. and Bowen, B. W. (2005) Ecological speciation in tropical reef fishes. *Proceedings of the Royal Society of London B* **272**, 573–579.

Rundle, H. D. (2003) Divergent environments and population bottlenecks fail to generate premating isolation in *Drosophila pseudoobscura*. *Evolution* **57**, 2557–2565.

Rundle, H., Chenoweth, S. F., Doughty, P. and Blows, M. W. (2005) Divergent selection and the evolution of signal traits and mating preferences. *PLoS Biology* **11**, e368.

Rundle, H. D., Nagel, L., Boughman, J. W. and Schluter, D. (2000) Natural selection and parallel speciation in sympatric sticklebacks. *Science* **287**, 306–308.

Rundle, H. and Nosil, P. (2005) Ecological speciation. *Ecology Letters* **8**, 336–352.

Rundle, H. D., Vamosi, S. M. and Schluter, D. (2003) Experimental test of predation's effect on divergent selection during character displacement in sticklebacks. *Proceedings of the National Academy of Sciences of the United States of America* **100**, 14943–14948.

Rundle, H. D. and Whitlock, M. (2001) A genetic interpretation of ecologically dependent isolation. *Evolution* **55**, 198–201.

Sandoval, C. P. (1994a) The effects of relative geographic scales of gene flow and selection on morph frequencies in the walking stick *Timema cristinae*. *Evolution* **48**, 1866–1879.

Sandoval, C. P. (1994b) Differential visual predation on morphs of *Timema cristinae* (Phasmatodeae: Timemidae) and its consequences for host range. *Biological Journal of the Linnean Society* **52**, 341–356.

Sandoval, C. P. and Nosil, P. (2005) Counteracting selective regimes and host preference evolution in ecotypes of two species of walking-sticks. *Evolution* **59**, 2405–2413.

Santibanez, S. K. and Waddington, C. H. (1958) The origin of sexual isolation between different lines within a species. *Evolution* **12**, 485–493.

Scharloo, W. (1971) Reproductive isolation by disruptive selection: did it occur? *American Naturalist* **105**, 83–86.

Scharloo, W., Hoogmoed, M. S. and Kuile, A. T. (1967) Stabilizing selection on a mutant character in *Drosophila*. I. The phenotypic variance and its components. *Genetics* **56**, 709–726.

Schluter, D. (1994) Experimental evidence that competition promotes divergence in adaptive radiation. *Science* **266**, 798–800.

Schluter, D. (2000) *The Ecology of Adaptive Radiation*. Oxford University Press, Oxford.

Schluter, D. and Nagel, L. M. (1995) Parallel speciation by natural selection. *American Naturalist* **146**, 292–301.

Schoener, T. W. (1989) The ecological niche. In: *Ecological concepts* (ed. J. M. Cherrett), pp. 79–113. Blackwell Scientific, Oxford.

Seehausen, O., van Alphen, J. J. M. and Witte, F. (1997) Cichlid fish diversity threatened by eutrophication that curbs sexual selection. *Science* **277**, 1808–1811.

Simpson, G. G. (1944) *Tempo and Mode in Evolution*. Columbia University Press, New York.

Slowinski, J. B. and Guyer, C. (1989) Testing the stochasticity of patterns of organismal diversity: an improved null model. *American Naturalist* **134**, 907–921.

Soans, B. A., Pimentel, D. and Soans, J. S. (1974) Evolution of reproductive isolation in allopatric and sympatric populations. *American Naturalist* **108**, 117–124.

Spiess, E. B. and Wilke, C. M. (1984) Still another attempt to achieve assortative mating by disruptive selection in *Drosophila*. *Evolution* **38**, 505–515.

Streelman, J. T. and Danley, P. D. (2003) The stages of vertebrate adaptive radiation. *Trends in Ecology and Evolution* **18**, 126–131.

Taylor, E. B., Boughman, J. W., Groenenboom, M., *et al.* (2006) Speciation in reverse: morphological and genetic evidence of the collapse of a three-spined stickleback (*Gasterosteus aculeatus*) species pair. *Molecular Ecology* **15**, 343–355.

Thoday, J. M. and Gibson, J. B. (1962) Isolation by disruptive selection. *Nature* **193**: 1164–1166.

Thoday, J. M. and Gibson, J. B. (1970) The probability of isolation by disruptive selection. *American Naturalist* **104**, 219–230.

Tilman, D., Knops, J., Wedlin, D., *et al.* (1997) The influence of functional diversity and composition on ecosystem processes. *Science* **277**, 1300–1302.

Turelli, M. (1985) Effects of pleiotropy on predictions concerning mutation-selection balance for polygenic traits. *Genetics* **111**: 165–195.

Vamosi, S. M. (2005) On the role of enemies in divergence and diversification of prey: a review and synthesis. *Canadian Journal of Zoology* **83**, 894–910.

Vamosi, S. M. and Vamosi, J. C. (2005) Endless tests: guidelines for analyzing non-nested sister-group comparisons. *Evolutionary Ecology Research* **7**, 567–579.

Vandermeer, J. H. (1972) Niche theory. *Annual Review of Ecology and Systematics* **3**, 107–132.

van Dijken, F. R. and Scharloo, W. (1979) Divergent selection on locomotor activity in *Drosophila melanogaster*. II. Tests for reproductive isolation between selected lines. *Behavioral Genetics* **9**, 555–561.

Van Homrigh, A., Higgie, M., McGuigan, K. and Blows, M. W. (2007) The depletion of genetic variance by sexual selection. *Current Biology* **17**, 528–532.

Via, S. (1999) Reproductive isolation between sympatric races of pea aphids. I. Gene flow restriction and habitat choice. *Evolution* **53**, 1446–1457.

Vickery, V. R. (1993) Revision of *Timema* Scudder (Phasmatoptera: Timematodea) including three new species. *Canadian Entomologist* **125**, 657–692.

Vines, T. H., Köhler, S. C., Thiel, M., *et al.* (2003) The maintenance of reproductive isolation in a mosaic hybrid zone between the fire-bellied toads *Bombina bombina* and *B. variegata*. *Evolution* **57**, 1876–1888.

Wagner, G. P. (1989) Multivariate mutation-selection balance with constrained pleiotropic effects. *Genetics* **122**: 223–234.

Wallace, B. (1953) Genetic divergence of isolated populations of *Drosophila melanogaster*. *Proceedings of the Ninth International Congress of Genetics* **9**, 761–764.

Warheit, K. I., Forman, J. D., Losos, J. B. and Miles, D. B. (1999) Morphological diversification and adaptive radiation: a comparison of two diverse lizard clades. *Evolution* **53**, 1226–1234.

Whitlock, M. (1997) Founder effects and peak shifts without genetic drift: adaptive peak shifts occur easily when environments fluctuate slightly. *Evolution* **51**, 1044–1048.

Wiens, J. J. and Graham, C. H. (2005) Niche conservatism: integrating evolution, ecology, and conservation biology. *Annual Review of Ecology Evolution and Systematics* **36**, 519–539.

Wright, S. (1932) The roles of mutation, inbreeding, crossbreeding, and selection in evolution. *Proceedings of the Sixth International Congress of Genetics* **1**, 356–366.

# CHAPTER NINE

# Progressive levels of trait divergence along a 'speciation transect' in the Lake Victoria cichlid fish *Pundamilia*

OLE SEEHAUSEN

## Introduction and outline

Identifying mechanisms of speciation has proven one of the most challenging problems in evolutionary biology, perhaps mainly for two reasons, speciation is not readily accessible to experimental approaches, and rarely to time series analyses. Any one case of speciation can usually be investigated only at a single stage of completion. Cases of parallel ecological speciation driven repeatedly within the same taxon by divergent selection along replicate environmental gradients, have therefore received considerable attention (Schluter & Nagel 1995; Rundle *et al.* 2000). Several such systems have become major model systems in evolutionary ecology research, including sticklebacks in postglacial lakes (Rundle *et al.* 2000), *Heliconius* butterflies (Mallet *et al.* 1998), leaf beetles (Funk 1998) and *Timema* walking sticks (Nosil *et al.* 2002). They provide powerful means of identifying causes of divergence and may lend themselves to examining associations between variation in the environments and variation in the progress towards speciation (Nosil & Harmon, this volume). However, variation in the progress towards speciation among disconnected populations undergoing parallel speciation may be due to different contingency as much as different environments (Taylor & McPhail 2000). Ideally, to trace the correlates of the transition from panmixis to incipient speciation, one would want to study variation in the progress towards speciation in exactly the same pair of species, and along a continuous progress series, to minimize the potential confounding effect of variable historical contingency.

The cichlid fish species *Pundamilia pundamilia*, *Pundamilia nyererei*, their hybrids and intermediate colour morphs occur at various stages of differentiation at different islands placed along an environmental gradient of water clarity in Lake Victoria. I will subsequently refer to this as a 'speciation transect'. At the same time, populations occurring at different islands do exchange genes as evident from microsatellite-allele frequencies (Seehausen *et al.* 2008), and also from several observed cases of invasion of, and establishment in, previously unoccupied islands within the 15 years time window that I have collected community

*Speciation and Patterns of Diversity*, ed. Roger K. Butlin, Jon R. Bridle and Dolph Schluter. Published by Cambridge University Press. © British Ecological Society 2009.

data for. Hence, I propose that the *Pundamilia* 'speciation transect' is close to the ideal case where variation in the progress towards speciation can be studied in the same pair of species and along a continuous series. As such, it lends itself to the identification of the sequence of events in population divergence that lead to the emergence of non-random mating and built-up of genetic differentiation between non-allopatric incipient species. In this Chapter, I review our current knowledge on this system, and use it to test predictions of alternative theoretical models of non-allopatric speciation.

I start by describing the phenotypes and their distribution patterns, then describe geographical variation in genetic and phenotypic differentiation, introducing the 'speciation transect', and review knowledge on causes of gene flow restriction. This will lead me to the experimental identification of traits involved in mate choice, their inheritance, the form of selection on male nuptial colouration and the sources of selection on female mating preferences. I will use current knowledge on the sources of selection to test alternative models of speciation. Finally, I will show male–male aggression is based on the same traits that female choice is based on; is a frequency-dependent source of selection; and facilitates the coexistence of incipient species. I conclude with a short summary of the main points and perspectives for future research.

## Distribution and phenotypic characterization
## of *P. pundamilia* and *P. nyererei*

The genus *Pundamilia* is endemic to Lake Victoria (Seehausen *et al.* 1998a), including the upper Victoria Nile (pers. obs.). The two best studied species are *P. pundamilia* and *P. nyererei*. Both are found exclusively over rocky substrates. *P. pundamilia* has been recorded from almost every sampled patch of rocky habitat along the mainland shores of the lake, and also from most offshore islands (Seehausen *et al.* 1998a). The geographical range of *P. nyererei* is nested within that of *P. pundamilia*. *P. nyererei* has only been found in places where *P. pundamilia* is present (Seehausen & van Alphen 1999), and it is absent from large stretches of mainland shore and islands despite apparently suitable habitat (Seehausen 1996; Fig. 9.1).

Both species are of small to medium size, of generalized haplochromine shape, but with somewhat specialized unicuspid and recurved teeth, and they are morphologically highly similar (Seehausen 1996; Magalhaes *et al.*, in press). Females of both species are cryptically yellowish, brownish or greyish depending on the population, with a number of distinct darker vertical bars on the flanks. The vertical bars in the much larger males are black. The flanks of male *P. pundamilia* are blue-grey between the bars, with a bright metallic blue spinous dorsal fin and red soft dorsal, caudal and anal fins. Males of *P. nyererei* instead are bright yellow between the bars on the lower flanks and bright crimson red above the lateral line. Their entire dorsal fin is bright crimson, and the other

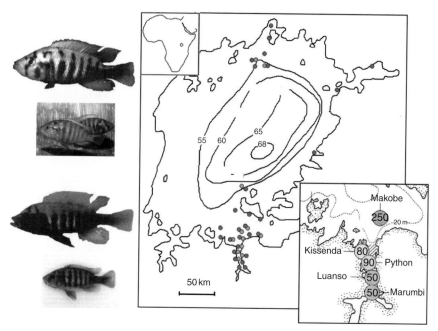

**Figure 9.1** Male and female of *P. pundamilia* (top) and *P. nyererei* (bottom), both from Makobe Island (DS4). Large map: distribution of known records of *P. pundamilia* and *P. nyererei* in Lake Victoria. Small inset map: the 'speciation transect' with location of the five islands: Marumbi (DS1), Luanso (DS1), Python (DS3), Kissenda (DS2) and Makobe (DS4). Water transparencies at the islands (cm Secchi disk readings) are given inside the circles (colours: turbid = brown, clear = blue). Note that Kissenda Island lies inside a lateral embayment, hence the lower transparency than at Python Island despite Kissenda's greater proximity to the clear water areas. (see colour plate)

unpaired fins are pale orange to red (Fig. 9.1). Adult males of both species are two to three times larger and heavier than adult females.

## Geographical variation in the degree of genetic and phenotypic differentiation

I will review in this section population genetic, quantitative genetic and phenotypic evidence which suggest that, in different parts of their shared distribution range, fish of *P. pundamilia-* and *P. nyererei*-like phenotype are either well-differentiated sibling species, weakly differentiated incipient species, or merely extremes along a continuum of phenotypic variation.

Variation in male colour phenotypes can be described by a colour-based hybrid index (Fig. 9.2). The frequency distribution of hybrid index scores changes along a transect through increasingly clear waters, from unimodal in turbid water with predominantly intermediate phenotypes and complete absence of class 4 phenotypes to distinctly bimodal with predominantly

**Figure 9.2** Colour-based hybrid index. Class 0 (*P. pundamilia*-like) = blue-grey lower flanks (lf), blue-grey upper flanks (uf), blue spinous dorsal fin. Class 1 = yellow lower flanks, grey upper flanks, blue spinous dorsal fin. Class 2 = yellow lf, red along upper lateral line, blue spinous dorsal fin. Class 3 = yellow lf, red uf except grey crest, blue dorsal fin. Class 4 (*P. nyererei*-like) = yellow lf, red uf, red df. (a) wild males from Luanso Island. (b) F2 hybrid males generated by crossing F1 hybrids between *P. nyererei* and *P. pundamilia* from Python Islands (1992 lines, DS3). (see colour plate)

phenotype classes 0 and 4 and very few individuals of intermediate colour in clear water (Seehausen 1997; Seehausen *et al.* 1997; Seehausen *et al.* 2008; Fig. 9.3a). Because the transition occurs along a fairly straight line from the highly turbid southern Mwanza Gulf to the relatively clear Speke Gulf, I refer to this transect as 'speciation transect'. Other than productivity and associated water turbidity and ambient light, no other environmental variables are known to correlate with this transition.

To assess gene flow, we genotyped between 20 and 50 individuals each of the red and the blue phenotypes at each of the five islands along 'speciation transect' using 11 microsatellite loci (Seehausen *et al.* 2008). These data confirm the inferences made from phenotypic data. Significant differentiation is observed in the clear water sites and $F_{ST}$ decays with increasing turbidity from 0.03 (Makobe Island) to 0.000 (Luanso and Marumbi Islands). At the high turbidity end of the transect (Luanso and Marumbi Islands), the differentiation was not significantly different from zero, but it was significantly different from zero at the islands Kissenda and Python with intermediate clarity. All 11 loci behaved similar, none showing any differentiation at Marumbi or Luanso, yet 7 of 11 showing significant differentiation at Makobe. Patterns of $F_{ST}$ between island populations of the same species revealed significant isolation by distance in *P. pundamilia* but not in *P. nyererei* where $F_{ST}$ were generally low and some not significant (Seehausen *et al.* 2008).

We generated F2 hybrids between *P. nyererei* and *P. pundamilia* in the laboratory (Haesler & Seehausen 2005) to acquire independent evidence that the variation around an intermediate hybrid index mode in turbid water populations was due to allelic segregation at those loci that are fixed for alleles with opposite sign in

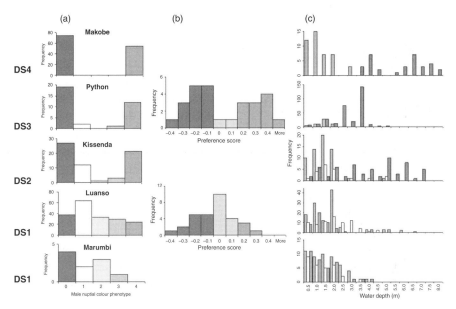

**Figure 9.3(a)** Frequency distributions of hybrid index scores along 'speciation transect'. (b) Frequency distributions for female mating preference phenotypes at two stations along 'speciation transect'. (c) Frequency distributions of males of different colour phenotypes against water depth. Males of *P. pundamilia*-like appearance (blue bars), *P. nyererei*-like appearance (red bars), reddish intermediates (hybrid index classes 2 and 3; orange bars), yellow intermediates (hybrid index class 1; pale yellow bars) DS1 to DS4 = Divergence score 1 to 4. (a) Modified from Seehausen *et al.* (2008). (b) and (c) reproduced with permission from Seehausen *et al.* (2008). (see colour plate)

the clear water areas. We used laboratory lines derived from populations of the two species collected in 1992 at Python Islands in the Mwanza Gulf (Seehausen 1996). By 1992, the frequency of phenotypic intermediates at Python Island was below 1% (Seehausen 1997). Our 1992 laboratory lines bred true. No intermediate male has appeared in the lines of either species and several hundred males have been bred and phenotyped in each line. More than 90% of the F1 hybrid males were intermediate in colour, resembling hybrid index class 3 (Fig. 9.2a), but some families contained a few blue males resembling class 0 or 1. The F2 hybrid males bred from both types of F1 male phenotypes (classes 0 and 3) segregated for colour. The phenotype classes in these F2 hybrid males resembled in great detail those in the wild males from turbid water areas (Fig. 9.2b). The close resemblance of wild phenotypes from turbid waters and laboratory hybrids between the species from clear water makes it likely that the same loci and alleles are responsible for the segregation of colour in both cases.

Based on the combined phenotypic and genetic evidence, I am henceforth assigning a divergence score (DS) to each island: DS1, Marumbi and Luanso; DS2, Kissenda; DS3, Python; and DS4, Makobe (Fig. 9.3).

## Causes of gene flow restriction between incipient species

In the following I will show absence of intrinsic postzygotic isolation, and that divergent spawning site choice may restrict gene flow but strongly only at Makobe (DS4). In contrast, divergent female mating preferences are suitable to restrict gene flow at all stages of differentiation, but apparently only do so where a minimum water transparency is given.

### No intrinsic postzygotic isolation

All Lake Victoria cichlid species that have been tested in the laboratory produce viable and fertile hybrid offspring (Crapon de Caprona & Fritzsch 1984; Seehausen *et al.* 1997). *P. pundamilia* and *P. nyererei* are fully interfertile too, and no intrinsic hybrid disadvantage could be detected in experiments in which a number of fitness-related variables were studied using Python Island fishes (DS3; Van der Sluijs *et al.* 2008b). Gene flow restrictions between the species in nature must hence arise from prezygotic or extrinsic postzygotic mechanisms.

### Spawning habitat choice

Males of *Pundamilia* are highly territorial and spawning takes place within the territories of the males (Seehausen *et al.* 1998b; Maan *et al.* 2004). Both direct mate choice by females and microhabitat selection by territorial males can potentially restrict gene flow between the species. At the clear water Makobe Island (DS4), *P. nyererei* males breed deeper than *P. pundamilia* males and overlap is very limited (Seehausen 1997). However, at Python (DS3) and Kissenda Islands (DS2), where the species are phenotypically strongly, but genetically only weakly (but significantly), differentiated, the microdistributions largely overlap. *P. nyererei* breeds from very shallow to deep, whereas *P. pundamilia* breeds only in the uppermost 3 m (Seehausen 1997; Seehausen *et al.* 2008). Microhabitat selection by males can then only have a minor effect on restricting gene flow. Similarly, at Makobe Island, the spawning site distribution of *P. nyererei* is fully overlapping with that of a third *Pundamilia* species, the blue-black *P.* 'pink anal'. Microhabitat selection by males cannot account for the restriction of gene flow between these phenotypically divergent species either (Seehausen *et al.* 1998b).

### Direct female mate choice

Direct female mating preferences have been measured for females from the very clear island (Makobe Island, DS4), the very turbid one (Luanso Island, DS1) and the one with intermediate turbidity (Python Island, DS3) along 'speciation transect'. Six females tested from the clear Makobe Island had behavioural preferences for conspecific males; four of them spawned, all with conspecific males (Seehausen 1997). A larger number of females were tested from Python Island, and most females actively preferred conspecific males; however, a few had no preference. For Fig. 9.3b I calculated the frequency distribution of female mating preference phenotypes, combining data from several studies and 27 different females (Seehausen & van Alphen 1998; Haesler & Seehausen 2005; Seehausen *et al.* 2008). The distribution is strongly bimodal, with most females having significant preferences either for *P. nyererei* (exclusively *P. nyererei* females) or for *P. pundamilia* males (exclusively *P. pundamilia* females) and very few having no preferences. It is hence likely that divergent female mating preferences restrict gene flow between the incipient species at Python Islands (DS3). However, it also seems likely that divergent selection must be acting on female preferences because the strong bimodality in the trait variance despite the presence of hybrids seems unlikely to be maintained in the absence of divergent selection.

Two studies have investigated female mating preferences in the turbid water population of Luanso Island (DS1), where the male phenotype frequency distribution is unimodal and dominated by males of intermediate colouration (Seehausen *et al.* 2008). The first study was conducted in 1997, using 16 females collected in 1996 (Seehausen 1999). The second study was conducted in 2005–2006 using 30 females collected in 2003 and 2005 (Van der Sluijs *et al.* 2008). Both studies found significant between-female variation in mating preferences. In contrast to the population from Python Island (DS3), the frequency distribution of preference phenotypes was unimodal, with most females lacking preferences, several having a preference for blue males and fewer having a preference for red males. The two studies conducted with a 9-year interval between them and by different researchers yielded highly congruent results. The coincidence of the absence of bimodality in the frequency distributions of both female mating preference and male colouration in the turbid waters is consistent with the hypothesis that divergent female mating preferences maintain non-random mating between the incipient species at the islands with clearer waters.

## The mate choice traits and their inheritance

Two very different manipulation experiments implicate divergent male nuptial colouration as a target of the divergent female mating preferences. Females of the Python Island populations (DS3) that preferred conspecific males under broad spectrum illumination failed to behaviourally discriminate between con- and heterospecific males under narrow spectrum orange light illumination (Seehausen & van Alphen 1998). Recently, Stelkens *et al.* (2008) used a different approach. In their experiment, second-generation hybrid males, segregating for nuptial colour, competed for matings with *P. pundamilia* and *P. nyererei* females from Python Island (DS3) in a partial partition mate choice design (Turner *et al.* 2001). Comparing mating success between the more blueish and the more reddish hybrid males, using molecular paternity tests, Stelkens *et al.* showed that when competing for females of *P. nyererei*, reddish hybrid males had significantly higher mating success than blueish hybrid males, whereas a strong trend approaching significance ($p = 0.07$) in the opposite direction was found when males were competing for *P. pundamilia* females. Overall, hybrid males had significantly higher mating success with females of the species whose male nuptial colouration they resembled than with the females of the other species (Fisher's combined probability test, $p = 0.02$). Importantly, the hybrid males were not produced by backcrossing but by mating F1 hybrids with each other. The result would only be expected if females of both species when choosing mates use male nuptial colouration, but prefer different colours or a trait that tightly cosegregates with colour through linkage or pleiotrophy.

Three investigations have addressed the mechanism of mate preference acquisition in *Pundamilia*. Haesler and Seehausen (2005), using a quantitative genetics approach, were able to demonstrate a heritable basis of the divergence between *P. pundamilia* and *P. nyererei* in their mating preferences; 20 F1 hybrid females expressed inconsistent mate choice decisions and 19 of these showed no significant bias to males of either species. In contrast, significant preferences for either blue or red male phenotypes segregated among 30 F2 hybrid females. Using the Castle–Wright biometric estimator (Castle 1921; Wright 1968; Lande 1981), between 1 and 5 genes were estimated to contribute to the variation in preference. Van der Sluijs *et al.* (2008) studied 21 additional F2 hybrid females and obtained closely corresponding results, again suggesting oligofactorial inheritance of the female mating preference. Both studies used the weakly differentiated populations of Python Islands, which differ mainly in male colouration and female preferences (DS3).

It appears that these inherited preferences can be modulated by imprinting on as yet unknown, but perhaps olfactorial traits. Verzijden and ten Cate (2007), using a cross-fostering design, found a significant effect of species of foster mother on female mating preference. The effect was strong enough to reverse a preference in *P. nyererei* for conspecific males into one for heterospecific males. For this experiment the well-divergent populations of Makobe Island (DS4) were used, which besides differing in male colour and female preference also differ in depth distribution, diet and parasites (see p. 166, 168, Fig. 9.3c).

As discussed above, the differences in male nuptial colouration between *P. pundamilia* and *P. nyererei* are heritable too. Heritability has been shown also for the intraspecific variation in the brightness of the red colouration between populations of *P. nyererei* (Seehausen *et al.* 1997). In a quantitative genetics approach using hybridization between the incipient species from Python Islands (DS3), the frequencies of different phenotype classes in the F2 suggested epistasis between one major gene with two alleles ('blue-grey flank' and 'yellow flank') and at least two quantitative trait loci for the presence and extension of red upwards from the lateral line, which act epistatically on the 'yellow flank' allele Seehausen (unpublished data).

Maan *et al.* (2006c) investigated a similar (blue/yellow) colour polymorphism in a population of the closely related species *Neochromis omnicaeruleus* from Lake Victoria. They also found evidence for heritability, but additionally evidence for ontogenetic switches from yellow to blue in many males, with some males lacking the yellow phase altogether, and a few staying yellow through most, perhaps all of their lives. Similar ontogenetic colour changes have not been observed in *Pundamilia*, but the findings in *Neochromis* raise the possibility that divergence between species or genotypes with yellow-red and blue male colouration may have come about through selection on heterochrony of ontogenetic colour change.

## The form of selection on male nuptial colouration

The experimental and observational data discussed until here suggest that the maintenance of the phenotypic divergence between *P. pundamilia* and *P. nyererei* involves divergent or disruptive selection on male nuptial colouration within island populations. It is possible, and perhaps plausible, that the same applies to the origin of these divergent phenotypes. Geographical distribution patterns of species and of colours in *Pundamilia* suggest both geographical and non-geographical speciation may have occurred. Besides its major peak on close to zero range overlap, the frequency distribution of species' range overlap has a second minor peak on close to 100% overlap whereas intermediate categories of range overlap are less frequent (Seehausen & van Alphen 1999). Moreover, the distribution of male colour patterns is significantly overdispersed, species of similar colour co-occur less often, and species of different colour more often than expected by chance, consistent with the hypothesis that colour pattern diversity is under negative frequency-dependent selection (Seehausen & Schluter 2004).

Negative frequency-dependent selection on male colour could arise from male–male aggression (Seehausen & Schluter 2004). Dijkstra *et al.* (2007a), investigating effects of male colouration on male–male interactions, found that males of clear water populations (DS4) of *P. pundamilia*, *P. nyererei* and *P.* 'pink anal' (another closely related blue species) direct more aggression to conspecific than to heterospecific intruder males, and that these aggression preferences can be eliminated by masking of colour differences under narrow-spectrum light. Dijkstra *et al.* (2005) found that males with red colouration had a significant advantage over blue males in dyadic interactions between *P. nyererei* and *P. pundamilia* males from Python Island (DS3), and this effect too was at least partially eliminated by masking of colour differences under narrow-spectrum light. Finally, Dijkstra *et al.* (in press), studying experimental assemblages of red and blue males from Kissenda island, a population with very weak differentiation at neutral loci and relatively frequent phenotypic hybrids (DS2), found that red males received significantly less aggression in mixed assemblages than in assemblages consisting only of red males.

This is the most direct evidence of negative frequency-dependent seletion on male colouration yet.

Laboratory experiments and field-based quantification of female mating preferences suggest that (1) intraspecific female mate choice within populations of *P. nyererei* exerts positive directional selection on the extension of the red area of males; male mating success in the clear waters of Makobe Island (DS4) correlates significantly with the extension of the red area on the body of males (Maan *et al.* 2004). In the more turbid waters at Kissenda Island (DS2) it correlates with the extension of red and yellow area (Maan 2006). (2) The strength of female mating preferences in *P. nyererei* (Maan 2006) and/or the evolutionary response in the male colouration (Seehausen *et al.* 1997) correlate positively with water clarity. (3) Female mate choice in the turbid water population of Luanso Island (DS1) can exert divergent or disruptive selection on male nuptial colouration; while most females lack any significant preference for red or blue male colouration and mate at random, some females prefer red males and some prefer blue males (van der Sluijs *et al.* 2008). Measuring mating success of F2 hybrid males with F2 hybrid females (both generated from Python populations, DS3) in a laboratory experiment, Stelkens *et al.* (2008) show that such intraspecific variation in female mating preferences can exert disruptive selection, whereby the more blue and the more red males obtain more matings than males of intermediate coloration. Hence, even though the random mating behaviour of F1 hybrid females will tend to cause bimodality in the male trait to decay (Bridle *et al.* 2006), the circumstance that distinct preferences for red and blue males segregate in higher-generation hybrid females will facilitate the build-up and/or maintenance of bimodality and positive associations between preference and trait genes.

### The sources of selection on female mating preferences and testing predictions of speciation models

Complete allopatric speciation without gene flow upon secondary contact can be excluded for *P. pundamilia* and *P. nyererei* as should be apparent from the above. Independent of whether the current levels of gene flow are a primary or a secondary phenomenon (see discussion), the speciation transect allows us to identify the minimum set of traits that have to be under divergent selection to recruit assortative female mating preferences. Hence, I concentrate here on models of speciation with gene flow. Alternative theoretical models of speciation with gene flow often differ primarily in the sources of divergent selection on female mating preferences (Kirkpatrick & Ravigné 2002; but see Gavrilets 2004 for a different classification). I propose that data from the *Pundamilia* 'speciation transect' are suitable to test predictions from four distinct models: classical reinforcement, adaptive speciation, sensory drive speciation and speciation by parasite-mediated divergent sexual selection.

## Classical reinforcement

Classical reinforcement of mating preferences can occur when recombination between divergent genomes produces gene combinations, which incur reduced intrinsic fitness to their bearer relative to the fitness of parental genotypes (Dobzhansky 1937, 1940; Muller 1939; Wilson 1965; Butlin 1987; Coyne & Orr 1997; Kirkpatrick & Ravigné 2002). Two studies have experimentally addressed components of intrinsic hybrid fitness in *Pundamilia*. Seehausen *et al.* (1997) found no evidence for reduced fertility through four generations of hybridization between *P. nyererei* and *Platytaeniodus degeni*. Van der Sluijs *et al.* (2008b) found no evidence for any reduction in survival, growth rate, fecundity or fertility, nor for skewed sex ratios in hybrids between *P. pundamilia* and *P. nyererei*. Given the geological youth of Lake Victoria (reviewed in Stager & Johnson 2008), and the very short speciation intervals required to account for its endemic species diversity (Seehausen 2006), intrinsic hybrid incompatibilities may not be expected. However, sex determining genes of variable strengths have been found segregating in other Lake Victoria cichlid fish, which can cause strongly skewed sex ratios (Seehausen *et al.* 1999), and perhaps the evolution of assortative mating (Lande *et al.* 2001). No such effects are apparent in interspecific hybrids within *Pundamilia*.

Our results do not rule out that Dobzhansky–Muller incompatibilities exist because if reduced fitness affects only some genotypes in F2 and higher hybrid generations, we might have failed to detect them. Molecular marker-aided studies are required to identify Muller–Dobzhansky incompatibilities through comparison of observed with expected frequencies of genotypes in hybrid families. However, if rare, such incompatibility mutations may become purged soon after hybridization started, and before selection would have effectively reinforced mating preferences. Our laboratory experiments do not rule out reinforcement through selection against ecological hybrid intermediacy (see next paragraph). To the extent that the relationship between the degree of habitat overlap and the extent of divergence in mating preferences can be taken as a test, the observed negative relationship (Fig. 9.3) is consistent with gene flow constraining mate preference evolution as opposed to the positive relationship predicted by reinforcement.

## Adaptive speciation

In adaptive speciation driven by negative frequency-dependent selection on resource utilization exerted by competition (Rosenzweig 1978; chapters in Dieckmann *et al.* 2004), variation in mating preferences is recruited by selection to generate assortative mating within resource use types. In species with sexual selection and sexually dimorphic traits, this may happen when variation in a mate choice signal (e.g. nuptial colour) becomes a marker trait for variation in traits related to resource utilization (e.g. the shape of the feeding structures).

Assortative mating between females with a preference and males with the matching trait might then evolve through ecological reinforcement of mating preferences (Kondrashov & Kondrashov 1999; Kirkpatrick 2001). The species *P. pundamilia* and *P. nyererei* typically differ in several ecological traits when they live in clear water, but are ecologically very similar in turbid water. To test one prediction of the adaptive speciation model, Mrosso et al. (manuscript) studied dietary variation using the stable isotope ratios $^{13}C/^{12}C$ and $^{15}N/^{14}N$, phenotypic variation in male colouration, and variation at neutral genetic loci using 11 microsatellites. The ratio $^{13}C/^{12}C$ is a measure of the extent to which an individual fish fed from the benthic versus the limnetic food chain; the ratio $^{15}N/^{14}N$ is a measure of the trophic level that an individual fish fed at. This was done in a DS1, a DS3, and in a DS4 *Pundamilia* community along the 'speciation transect'. Adaptive speciation makes the prediction that dietary differentiation coincides with, or precedes, phenotypic and neutral genetic differentiation.

In turbid waters, where variation in male colour phenotypes was unimodal (DS1), all individuals fell into a single tight cluster in the two-dimensional isotope space, as well as in multilocus genotype space. At Python Island with intermediate water clarity (DS3), there was considerably more individual variation in both isotope ratios, yet the ranges of variation of reds and blues on both axes completely overlapped with no significant differentiation between them. Despite the lack of dietary differentiation, red and blue males had significantly different microsatellite allele frequencies, and intermediate colour phenotypes were rare. In clear water (DS4), red and blue males formed non-overlapping clusters along the carbon isotope axis, highly significantly different allele frequencies at microsatellites, and there were no intermediate colour phenotypes. In line with the differences in microdistribution (see above and below), blue males had less and red males more strongly negative $^{13}C/^{12}C$ ratios, indicating that blues fed more on the benthic food chain and reds more on the limnetic food chain. Their trophic levels ($^{15}N/^{14}N$) were not different. These data suggest assortative mating is sufficiently strong to permit species differentiation at neutral loci even in the absence of significant dietary differences. It seems hence unlikely that species differences in female mating preferences for male nuptial colour variants evolved or are maintained because colouration was a marker trait of feeding specialization at the incipient stage of speciation.

**Sensory drive speciation**
In sensory drive speciation, assortative mating emerges as a by-product of adaptive divergence in a sensory system that is used in mate choice (Endler 1992; Boughman 2001, 2002). We measured the microdistribution (water depth and distance off shore) of males with different colours at several islands along

'speciation transect'. Differentiation in microdistribution between red and blue coincided closely with divergence in colouration and in allele frequencies at neutral loci. At islands where phenotypic variation in male colour was unimodal and genetic differentiation absent (DS1), more reddish and more bluish males had identical microdistributions. At the islands with intermediate water clarity (DS2, 3) – where variation in male colour was bimodal and neutral genetic differentiation weak but significant – red males extended into much greater depth than blue males. Finally, at the clear water island (DS4), where allele frequencies at microsatellites and colour were well differentiated, red males were restricted to deep water and blue males to shallow water with little overlap (Fig. 9.3c; Seehausen *et al.* 2008, see also Seehausen 1997).

These data are consistent with the sensory drive model of speciation. The ambient light spectrum in Lake Victoria shifts towards red with increasing water depth (Seehausen *et al.* 1997). At sites where the *Pundamilia* species are significantly differentiated (DS2–4), the red males hence tend to be found in relatively more red-shifted light environments. Sensory drive makes the prediction that the species differ in their visual properties such that *P. nyererei* sees red light better than *P. pundamilia* and vice versa for blue light. Until very recently two studies had addressed this question. Carleton *et al.* (2005) determined the wavelength of maximum absorbance ($\lambda_{max}$) of cones and rods by microspectrophotometry in seven populations of four closely related species of *Pundamilia*, varying in visual environment and male nuptial colour. They also sequenced the six by then known opsin genes of the same individuals (subsequently, a seventh opsin gene was discovered; Spady *et al.* 2003). Microspectrophotometry determined that the $\lambda_{max}$ of the rod pigment and of two of the three expressed cone pigments were similar in all five species and all populations. However, the long wavelength sensitive (LWS) cone pigment varied among species, with 3–4 nm shifts in $\lambda_{max}$. The shift coincided with two amino acid substitutions in the part of the LWS opsin protein that is directed into its retinal binding pocket and predicted to affect absorbance properties. The subtle shifts in $\lambda_{max}$ coincided with large shifts in male body colour, with longer LWS pigments being confined to individuals of red species and shorter ones largely confined to individuals of blue species. Furthermore, there were indications that *P. nyererei* may have a higher proportion of red/red versus red/ green double cones, which might bias its absorbance further towards red.

Maan *et al.* (2006) used the optomotor response paradigm to compare the sensitivity of *P. pundamilia* and *P. nyererei* (Makobe populations, DS4) to narrow spectrum light of blue and red colour. They found that *P. pundamilia* had a significantly lower detection threshold for blue light but a significantly higher detection threshold for red light than *P. nyererei*. Hence, as predicted by the sensory drive hypothesis, the species that has red nuptial colouration was more sensitive to red light, whereas the species with blue nuptial colouration was more sensitive to blue light. A first comprehensive test of the sensory drive hypothesis has just been published. Comparing the divergence in alleles at opsin genes and nuptial colouration is currently at five sites along the 'speciation transect', Seehausen *et al.* (2008) found evidence consistent with speciation through sensory drive without geographical isolation.

## Parasite-mediated divergent sexual selection

Good genes models of sexual selection are often thought of as constraining population divergence and speciation (Kirkpatrick & Nuismer 2004). However, in a heterogeneous environment, this need not be true because local adaptation may select for different good genes indicators, and hence different preferences in different parts of the environment (Edelaar *et al.* 2004; Reinhold 2004). The red dorsal body colouration of male *P. nyererei* is carotenoid-based, and correlational evidence indicated that it may be a signal of parasite resistance (Maan *et al.* 2004, 2006). Supportive evidence derives from a laboratory study involving experimental infection of males (Dijkstra *et al.* 2007b). Thus, carotenoids may mediate a trade-off between sexual signalling and immune defense, which would perhaps make male red colouration in *P. nyererei* an honest signal of individual quality.

Maan *et al.* (2008) investigated the relationship between male nuptial colouration and parasite load in *P. pundamilia* from Makobe Island (DS4) and found that the extent of iridescent blue on the dorsal fin is negatively correlated with parasite load too. They further found that parasite infestation rates differed quantitatively between the species, in a way that correlates well with species differences in diet and microhabitat, *P. pundamilia*, the shallow water species with the benthic carbon isotope signature, mostly carries the Nematod *Contracaecum*. *P. nyererei*, the deeper water species with the limnetic carbon signature, predominantly carries parasitic copepods. At Makobe Island, egrets and cormorants aggregate in large numbers on the shoreline, covering the rocks with guano. As a result, the abundance of infectious *Contracaecum* stages on the rocky bottom is likely to be very high in shallow waters and to decrease with depth. In contrast, parasitic copepod loads (Ergasilus *lamellifer* and Lamproglena *monodi*) were higher in *P. nyererei*. Due to its zooplanktivorous, limnetic feeding style at Makobe Island, *P. nyererei* may experience increased exposure to these free-living, pelagic copepods relative to the shallower-dwelling and more benthic-feeding *P. pundamilia*. It is hence conceivable that parasite-mediated divergent sexual selection strengthens reproductive isolation between *P. pundamilia* and *P. nyererei* at Makobe Island. However, whether resistance trade-offs exist remains to be tested. To answer the question whether parasite-mediated sexual selection is divergent in the initial stage of speciation or only latches on later when the species have already diverged more strongly ecologically, parasitation patterns will have to be investigated at multiple sites along the 'speciation transect'.

## Colour-based aggression biases in males help stabilize incipient speciation

Own-type aggression biases of males can importantly contribute to the stabilization of coexistence between ecologically similar species with well-developed male territoriality (Mikami *et al.* 2004; Seehausen & Schluter 2004). Theoretical modelling suggests that if the traits under selection by male–male competition and female mate choice are the same, negative frequency-dependent selection exerted by male–male competition could be very important also for the likelihood of sympatric speciation (Van Doorn *et al.* 2004, cf. *magic trait*, Gavrilets 2004). Dijkstra *et al.* 2006 tested predictions of this idea by investigating male aggression biases in the blue species *P. pundamilia* at different stages of speciation along the 'speciation transect'. In the very turbid water site Luanso (DS1), where *Pundamilia* forms a single fully admixed population with a single blue-shifted mode in the phenotype frequency distribution, all males tended to be more aggressive towards blue males than to red males. This would favour the relatively rare more reddish phenotypes.

At sites with intermediate water clarity (Kissenda and Python Islands), two modes in the male phenotype frequency distribution, but persisting low levels of hybridization (DS2 and 3), both red and blue males biased aggression towards red males. This was not predicted by the hypothesis, which had predicted own-type aggression biases. However, red males are more abundant at these islands (Seehausen 1997). Moreover, the disadvantage that red males would experience due to aggression bias against them was offset by a significant behavioural dominance advantage of red over blue males in dyadic contests (Dijkstra *et al.* 2005). It is possible that the joined action of these forces does indeed stabilize coexistence at this intermediate stage of species divergence. Consistent with this, Dijkstra *et al.* (in press) found significantly lower levels of aggression in mixed than in entirely red experimental assemblages. They used males from Kissenda Island (DS2).

Finally, in two sites with clear water and well-differentiated species (DS4), males of both species biased their aggression to conspecifics as predicted by the theory. Hence, even though the distribution of patterns in male–male competition and emergence of own-type biases along the 'speciation transect' are complex, divergent nuptial colours consistently have effects on male–male interactions. In turn, the male–male interactions facilitate both the invasion of a new colour and the coexistence of existing colours.

## Discussion and perspectives

In this chapter, I have developed the argument that the *Pundamilia* 'speciation transect' comes fairly close to the wishful case where variation in the progress towards speciation can be studied in the same pair of species and along a continuous series of progress levels. As such, the *Pundamilia* 'speciation transect' is well suited to identify the sequence of events that lead to the emergence of non-random mating and subsequent built-up of genetic differentiation between non-allopatric incipient species. We do not know at present to what extent the lack of differentiation in turbid waters is a primary feature of those populations or is due to recent breakdown of reproductive isolation (Seehausen *et al.* 1997; Taylor *et al.* 2006). It is, however, almost certainly a combination of both. Turbidity has increased dramatically in the past century, and this has certainly led to increased hybridization. On the other hand, historical data (Graham 1929) show that even before the recent anthropogenic eutrophication, Lake Victoria had large inshore areas of turbid waters, including the southern Mwanza Gulf. There is also evidence that the lake underwent several cycles of eutrophication and re-oligotrophication since the beginning of the Holocene (Schmidt 2003). Finally, turbid water conditions were probably also prevalent during the early stage of refilling of Lake Victoria through flooding of a productive grassland (Stager & Johnson 2008), and haplochromine populations with blue/red variation in male nuptial colouration, similar to DS1 on the *Pundamilia* 'speciation transect', are known from several rivers and small turbid lakes in Tanzania. Hence, rather than being deterred by the uncertainty about the extent to which the 'speciation transect' is a primary or a secondary phenomenon, it should be seen primarily as a rare opportunity to investigate the sequence of divergence/convergence in different traits relative to that in mating patterns. It allows us to identify the minimum set of traits that have to be under divergent selection to recruit and/or maintain non-random mating.

A number of conclusions can be drawn from the investigations carried out to date.

1.  All evidence suggests that the southern end of 'speciation transect' is inhabited by a single species of variable colour: colour and preference phenotypes have a unimodal frequency distribution; there is no association between colour and microdistribution or diet and there is no neutral genetic differentiation between the phenotypes. All evidence suggests that the northern end of 'speciation transect' is inhabited by two different species: colour and preference distributions are strongly bimodal with few or no intermediates; colour is strongly associated with microdistribution and diet, such that red and blue males have almost non-overlapping depth ranges and non-overlapping isotope signatures of diet. Finally, they have significantly different allele frequencies at many neutral loci.

2.  The mate choice traits that mediate assortative mating at the differentiated (north) end of the transect (female preferences and male nuptial colours) are already variable at the non-differentiated (south) end of the transect. *Pundamilia* is not unusual in this regard. There is a large number of cichlid populations in Lake Victoria that segregate for the different nuptial colours

and in fact at various levels of water transparency (Seehausen & van Alphen 1999; Maan *et al.* 2006; Terai *et al.* 2006).

3. Populations from two intermediate sites on the transect are intermediate in that they show strongly bimodal frequency distributions of the mate choice traits, male nuptial colour and female preference, and weakly but significantly different allele frequencies between red and blue males at neutral loci. These populations also show an intermediate level of association between colour and microdistribution; reds have a much wider depth range, while blues are restricted to shallow waters. On the other hand, these populations show no evidence for an association between colour and isotope signature of diet.

4. The data are compatible with models of speciation by sensory drive, where divergent light conditions at different water depths and divergent adaptations of the visual system affect mating preferences and the fitness of male nuptial colour morphs. Detailed analyses of the visual system of populations from all stages of differentiation on the transect have very recently confirmed this (Seehausen *et al.* 2008).

5. The same nuptial colour cues that females use for choosing mates are also used by males when defending territories against competitors. Evidence gathered from a number of different experiments are consistent with the hypothesis that colour-mediated competition between males contributes a negative frequency-dependent selection component that may stabilize the coexistence of the incipient species.

6. Parasitological data have yet to be collected from intermediate sites on the transect to ask whether sexual selection for parasite resistance in different parts of the environment may contribute to divergent selection on mating preferences and colours. Imprinting experiments ought to be conducted with fish from intermediately differentiated and non-differentiated populations to better understand the role of imprinting in the early stages of speciation.

7. The accumulation of divergence in a successively larger number of traits along the *Pundamilia* 'speciation transect' can be equated with increasing dimensionality of niche divergence (Nosil & Harmon, this volume). One potentially important observation then is that the dimensionality of niche divergence can grow rapidly with decreasing gene flow or vice versa, gene flow can rapidly decrease with increasing dimensionality of niche divergence. That divergent adaptation and gene flow can be reciprocally constraining has been shown on other incipient species systems too (Hendry *et al.* 2002; Nosil & Crespi 2004). It will be difficult to decide who is driving who. If the analogy of the different stages of differentiation along the 'speciation transect' with successive stages in speciation is valid, the observations in *Pundamilia* may imply that dimensionality of niche divergence and gene flow restriction co-evolve. A minimum dimensionality of niche

divergence is required for divergent selection to be sufficiently strong to initiate speciation (here the occupation of a sufficiently large range of water depth to be exposed to different light environments). However, as soon as gene flow becomes sufficiently reduced, divergence in other traits becomes possible, which may be under less strong divergent selection. In turn again, the increased dimensionality of niche divergence is likely to reduce gene flow further.

Similar gradients of species differentiation that lend themselves to investigations of 'speciation transects' might exist also in other taxa to the extent that non-geographical speciation occurs. The highest likelihood of discovering them is in sister species pairs with fully sympatric or nested geographical distributions that are sufficiently large to generate some population structure and to cover a range of different environments.

## Acknowledgements

I wish to thank the students who worked with me on the *Pundamilia* system: Peter Dijkstra, Andy Gould, Marcel Haesler, Nellie Konijnendijk, Martine Maan, Isabel Magalhaes, John Mrosso, Linda Söderberg, Rike Stelkens and Inke van der Sluijs. Several collaborators and colleagues have shaped and sharpened my views on this system, including Karen Carleton, Ton Groothuis, Masakado Kawata, Nori Okada, Dolph Schluter, Ola Svensson, Carel ten Cate, Yohey Terai, George Turner, Jacques van Alphen and Kyle Young. I owe many thanks to Erwin Schaeffer, Alan Smith, Kees Hofker and Peter Snelderwald for maintaining the fish stocks for the *Pundamilia* speciation research programme through many years with great care, and to Patrik Nosil, Luke Harmon and Sergey Gavrilets for very good comments on the manuscript.

## References

Bridle, J. R., Saldamando, C. I., Koning, W. and Butlin, R. K. (2006) Assortative preferences and discrimination by females against hybrid male song in the grasshoppers Corthippus brunneus *and* C. jacobsi *(Orthoptera, Acrididae)*. *Journal of Evolutionary Biology* **19**, 1248–1256.

Boughman, J. W. (2001) Divergent sexual selection enhances reproductive isolation in sticklebacks. *Nature* **411**, 944–948.

Boughman, J. W. (2002) How sensory drive can promote speciation. *Trends in Ecology & Evolution* **17**, 571–577.

Butlin, R. (1987) Speciation by reinforcement. *Trends in Ecology & Evolution* **2**, 8–13.

Carleton, K. L., Parry, J. W. L., Bowmaker, J. K., Hunt, D. M. and Seehausen O. (2005) Color vision and speciation in Lake Victoria cichlids of the genus Pundamilia. *Molecular Ecology* **14**, 4341–4353.

Castle, W. E. (1921) An improved method of estimating the number of genetic factors concerned in cases of blending inheritance. *Science* **54**, 223.

Coyne, J. A. and Orr, H. A. (1997) 'Patterns of speciation in Drosophila' revisited. *Evolution* **51**, 295–303.

Coyne, J. A. and Orr, H. A. (2004) Speciation. Sinauer associates, Sunderland.

Crapon de Caprona, M. D. and Fritysch, B. (1984) Interspecific fertile hybrids of haplochromine Cichlidae (Teleostei) and their possible importance for specification. *Netherlands Journal of Zoology* **34**, 503–538.

Dieckmann, U., Metz, J. A. J., Doebeli, M. and Tautz, D. (eds) (2004) *Adaptive Speciation.* Cambridge University Press, Cambridge, UK.

Dijkstra, P. D., Seehausen, O. and Groothuis, T. G. G. (2005) Direct male-male competition can facilitate invasion of new colour types in Lake Victoria cichlids. *Behavioural Ecology and Sociobiology* **58**, 136–143.

Dijkstra, P. D., Seehausen, O., Pierotti M. E. R. and Groothuis, T. G. G. (2007a) Male–male competition and speciation, aggression bias towards differently coloured rivals varies between stages of speciation in a Lake Victoria cichlids species complex. *Journal of Evolutionary Biology* **20**, 496–502.

Dijkstra, P. D., Hekman, R., Schulz, R. W. and Groothuis, T. G. G. (2007b) Social stimulation, nuptial colouration, androgens and immunocompetence in a sexual dimorphic cichlid fish. *Behavioral Ecology and Sociobiology* **61**, 599–609.

Dijkstra, P. D., Hemelrijk, C., Seehausen, O. and Groothuis, T. G. G. (in press) Colour polymorphism and intrasexual competition in assemblages of cichlid fish. *Behavioural Ecology.*

Dijkstra, P. D., Seehausen, O., Gricar, B. L. A., Maan, M. E. and Groothuis, T. G. G. (2006) Can male-male competition stabilize speciation? A test in Lake Victoria haplochromine cichlid fish. *Behavioral Ecology and Sociobiology* **59**, 704–713.

Dobzhansky, T. (1937) *Genetics and the Origin of Species.* Columbia University Press, New York.

Dobzhansky, T. (1940) Speciation as a stage in evolutionary divergence. *American Naturalist* **74**, 312–321.

Edelaar, P., van Doorn, G. S. and Weissing, F. J. (2004) Sexual selection on good genes facilitates sympatric speciation. In: *Sexual Selection and Sympatric Speciation*, PhD Thesis of G. S. van Doorn, pp. 205–210, University of Groningen.

Endler, J. A. (1992) Signals, signal conditions, and the direction of evolution. *American Naturalist* **139**, S125–S153.

Funk, D. J. (1998) Isolating a role for natural selection in speciation, Host adaptation and sexual isolation in Neochlamisus bebbianae leaf beetles. *Evolution* **52**, 1744–1759.

Gavrilets, S. (2004) *Fitness Landscapes and the Origin of Species.* Princeton University Press, Princeton, NJ.

Graham M. M. A. (1929) The Victoria Nyanza and its fisheries. A report on the fishing survey of Lake Victoria 1927–28, and appendices. Crown Agents, London.

Haesler, M. and Seehausen, O. (2005) The inheritance of female mating preference in a sympatric sibling species pair of Lake Victoria cichlids, implications for speciation. *Proceedings of the Royal Society of London Series B* **272**, 237–245.

Hendry, A. P., Taylor, E. B. and McPhail, J. D. (2002) Adapative divergence and the balance between selection and gene flow, lake and stream stickleback in the Misty system. *Evolution* **56**, 1199–1216.

Kirkpatrick, M. (2001) Reinforcement during ecological speciation. *Proceedings of the Royal Society of London Series B* **268**, 1259–1263.

Kirkpatrick, M. and Nuismer, S. L. (2004) Sexual selection can constrain sympatric speciation. *Proceedings of the Royal Society of London Series B* **271**, 687–693.

Kirkpatrick, M. and Ravigné, V. (2002). Speciation by natural and sexual selection, models and experiments. *American Naturalist* **159**, S22–S35.

Kondrashov, A. S. and Kondrashov, F. A. (1999) Interactions among quantitative traits in the course of sympatric speciation. *Nature* **400**, 351–354.

Lande, R. (1981) The minimum number of genes contributing to quantitative variation between and within populations. *Genetics* **99**, 541–553.

Lande, R., Seehausen, O. and van Alphen, J. J. M. (2001) Rapid sympatric speciation by sex reversal and sexual selection in cichlid fish. *Genetica* **112–113**, 435–443.

Maan, M. E. (2006) Sexual selection and speciation, mechanisms in Lake Victoria cichlid fish. PhD Thesis, Leiden. ISBN-10 90.9020523.3; ISBN-13 978.90.9030523.6

Maan, M. E., Haesler, M. P., Seehausen, O. and van Alphen, J. J. M. (2006c) Heritability and heterochrony of polychromatism in a Lake Victoria cichlid fish, stepping stones for speciation? *Journal of Experimental Zoology. Molecular and Developmental Evolution* **306**B, 168–176.

Maan, M. E., Hofker, K. D., van Alphen, J. J. M. and Seehausen, O. (2006b) Sensory drive in cichlid speciation. *The American Naturalist* **167**, 947–954.

Maan, M. E., van Rooijen, A. M. C., van Alphen, J. J. M. and Seehausen, O. (2008) Parasite-mediated sexual selection and species divergence in Lake Victoria cichlid fish. *Biological Journal of the Linnean Society* **94**, 53–600.

Maan, M. E., van der Spoel, M., Quesada Jimenez, P., van Alphen, J. J. M. and Seehausen, O. (2006a) Fitness correlates of male coloration in a Lake Victoria cichlid fish. *Behavioural Ecology*, doi:10.1093/beheco/ark020.

Maan, M., Seehausen, O., Söderberg, L., *et al.* (2004) Intraspecific sexual selection on a speciation trait, male coloration, in the Lake Victoria cichlid Pundamilia nyererei. *Proceedings of the Royal Society of London Series B* **271**, 2445–2452.

Magalhaes, I. S., Mwaiko, S., Schneider, M. V. and Seehausen, O. (in press). Divergent selection and phenotypic plasticity during speciation in Lake Victoria cichlid fish. *Journal of Evolutionary Biology*.

Mallet, J., Jiggins, C. D. and McMillan, W. O. (1998) Mimicry and warning colour at the boundary between races and species. In:

*Endless Forms, Species and Speciation* (ed. D. J. Howard and S. H. Berlocher), pp. 390–403. Oxford University Press, New York.

Mikami, O. K., Kohda, M. and Kawata, M. (2004) A new hypothesis for species co-existence, male–male repulsion promotes co-existence of competing species. *Population Ecology.* **46**, 213–217.

Muller, H. J. (1939) Reversibility in evolution considered from the standpoint of genetics. *Biological Reviews of the Cambridge Philosophical Society* **14**, 261–280.

Nosil, P. and Crespi, B. J. (2004) Does gene flow constrain adaptive divergence or vice versa? A test using ecomorphology and sexual isolation in Timema cristinae walking sticks. *Evolution* **58**, 102–112.

Nosil, P. and Harmon, L. (2008) Niche Dimensionality and Ecological Speciation. This volume.

Nosil, P., Crespi B. J. and Sandoval C. P. (2002) Host-plant adaptation drives the parallel evolution of reproductive isolation. *Nature* **417**, 440–443.

Reinhold, K. (2004) Modeling a version of the good-genes hypothesis, female choice of locally adapted males. *Organisms Diversity & Evolution* **4**, 157–163.

Rosenzweig, M. L. (1978) Competitive speciation. *Biological Journal of the Linnean Society* **10**, 275–289.

Rundle, H. D., Nagel, L., Wenrick Boughman, J. and Schluter, D. (2000) Natural Selection and parallel speciation in sympatric sticklebacks. *Science* **287**, 306–308.

Schluter, D. and Nagel, L. (1995) Parallel speciation by natural selection. *American Naturalist* **146**, 292–301.

Schmidt, P. (2003) Deep time landscape histories and the improvement of environmental management in Africa. In: *Conservation, Ecology, and Management of African Freshwaters* (ed. T. L. Crisman, L. J. Chapman, C. A. Chapman and L. S. Kaufman), pp. 20–37. University Press of Florida, Florida.

Seehausen, O. (1996) *Lake Victoria Rock Cichlids, Taxonomy, Ecology, and Distribution*, p. 304. Verduijn Cichlids, Zevenhuizen, Netherlands.

Seehausen, O. (1997) Distribution of and reproductive isolation among color morphs of a rock dwelling Lake Victoria cichlid. *Ecology of Freshwater Fish* **6**, 59–66.

Seehausen, O. (1999) Speciation and species richness in African cichlid fish, the role of sexual selection. University of Leiden, Netherlands. ISBN 90-9012711-9.

Seehausen, O. (2006) African cichlid fish, a model system in adaptive radiation research. *Proceedings of the Royal Society of London Series B* doi:10.1098/rspb.2006.3539.

Seehausen, O. and Schluter, D. (2004) Male–male competition and nuptial-colour displacement as a diversifying force in lake Victoria cichlid fishes. *Proceedings of the Royal Society of London Series B* **271**, 1345–1353.

Seehausen, O. and van Alphen, J. J. M. (1998) The effect of male coloration on female mate choice in closely related Lake Victoria cichlids (Haplochromis nyererei complex). *Behavioral Ecology and Sociobiology* **42**, 1–8.

Seehausen, O. and van Alphen, J. J. M. (1999) Can sympatric speciation by disruptive sexual selection explain rapid evolution of cichlid diversity in Lake Victoria? *Ecology Letters* **2**, 262–271.

Seehausen, O., van Alphen, J. J. M. and Lande, R. (1999) Colour polymorphism and sex ratio distortion in a cichlid fish as an incipient stage in sympatric speciation by sexual selection. *Ecology Letters* **2**, 367–378.

Seehausen, O., van Alphen, J. J. M. and Witte, F. (1997) Cichlid fish diversity threatened by eutrophication that curbs sexual selection. *Science* **277**, 1808–1811.

Seehausen, O., Lippitsch, E., Bouton, N. and Zwennes, H. (1998a) Mbipi, the rock-dwelling cichlids of Lake Victoria, description of three new genera and fifteen new species (Teleostei). *Ichthyological Exploration of Freshwaters* **9**, 129–228.

Seehausen, O., Witte, F., van Alphen, J. J. M. and Bouton, N. (1998b) Direct mate choice maintains diversity among sympatric cichlids in Lake Victoria. *Journal of Fish Biology* **53** (Suppl. A), 37–55.

Seehausen, O., Terai, Y., Magalhaes, I. S., *et al.* (2008) Speciation through sensory drive in cichlid fish. *Nature*, In press.

Spady T. C., Parry, J. W. L. and Robinson, P. R. (2003) Evolution of the cichlid visual palette through ontogenetic subfunctionalization of the opsin gene array. *Molecular Biology and Evolution* **23**, 1538–1547.

Stager, J. C. and Johnson, T. C. (2008) The Late Pleistocene desiccation of Lake Victoria and the origin of its endemic biota. *Hydrobiologia* **596**, 5–16.

Stelkens, R. B., Pierotti, M. E. R., Joyce, D. A., *et al.* (2008) Disruptive sexual selection on male nuptial colouration in an experimental hybrid population of cichlid fish. *Philosophical Transactions of the Royal Society of London Series B*, doi:10.1098/rstb.2008.0049.

Taylor, E. B. and McPhail, J. D. (2000) Historical contingency and ecological determinism interact to prime speciation in sticklebacks, Gasterosteus. *Proceedings of the Royal Society of London Series B* **267**, 2375–2384.

Taylor, E. B., Boughman, J. W., Groenenboom, M., *et al.* (2006) Speciation in reverse, morphological and genetic evidence of the collapse of a three-spined stickleback (Gasterosteus aculeatus) species pair. *Molecular Ecology* **15**, 343–355.

Terai, Y., Seehausen, O., Sasaki, T., *et al.* (2006) Divergent selection on opsins drives incipient speciation in Lake Victoria cichlids. *PloS Biology* **4**, e433–440.

Turner, G. F., Seehausen, O., Knight, M. E., Allender, C. J. and Robinson, R. L. (2001) How many species of cichlid fishes are there in African lakes? *Molecular Ecology* **10**, 793–806.

Van der Sluijs, I. (2008) Divergent mating preferences and nuptial coloration in sibling species of cichlid fish. University of Leiden, Netherlands. ISBN 978-90-90-9023197-6.

Van der Sluijs, I., van Alphen, J. J. M. and Seehausen, O. (2008a) Preference polymorphism for coloration but no speciation in a population of Lake Victoria cichlids. *Behavioral Ecology* **19**, 177–183.

Van der Sluijs, I., Van Dooren, T. J. M., Seehausen, O. and Van Alphen, J. J. M. (2008b). A test of fitness consequences of hybridization in sibling species of Lake Victoria cichlid fish. *Journal of Evolutionary Biology* **21**, 480–491.

Van der Sluijs, I., Van Dooren, T. J. M., Hofker, K. D., *et al.* Female mating preference functions predict sexual selection against hybrids between sibling species of cichlid fish. *Philosophical Transactions of the Royal Society of London Series B*, doi:10.1098/rstb.2008.0045.

van Doorn, G. S., Dieckmann, U. and Weissing, F. J., (2004) Sympatric speciation by sexual selection: a critical reevaluation. *American Naturalist* **163**, 709–725.

Verzijden, M. N. and ten Cate, C. (2007) Early learning influences species assortative mating preferences in Lake Victoria cichlid fish. *Biology Letters*, doi:10.1098/rsbl.2006.0601

Wilson, E. O. (1965) The challenge from related species. In: *The Genetics of Colonizing Species* (ed. H. G. Baker and G. L. Stebbins), 7–24. Academic Press, New York.

Wright, S. (1968) *Evolution and the Genetics of Populations*, vol. 1. University of Chicago Press, London.

# CHAPTER TEN

# Rapid speciation, hybridization and adaptive radiation in the *Heliconius melpomene* group

## JAMES MALLET

In 1998 it seemed clear that a pair of 'sister species' of tropical butterflies, *Heliconius melpomene* and *Heliconius cydno* persisted in sympatry in spite of occasional although regular hybridization. They speciated and today can coexist as a result of ecological divergence. An important mechanism in their speciation was the switch in colour pattern between different Müllerian mimicry rings, together with microhabitat and host-plant shifts, and assortative mating produced as a side effect of the colour pattern differences. An international consortium of *Heliconius* geneticists has recently been investigating members of the *cydno* superspecies, which are in a sense the 'sisters' of one of the original 'sister species', *cydno*. Several of these locally endemic forms are now recognized as separate species in the eastern slopes of the Andes. These forms are probably most closely related to *cydno*, but in several cases bear virtually identical colour patterns to the local race of *melpomene*, very likely resulting from gene transfer from that species; they therefore can and sometimes do join the local mimicry ring with *melpomene* and its more distantly related co-mimic *Heliconius erato*. I detail how recent genetic studies, together with ecological and behavioural observations, suggest that the shared colour patterns are indeed due to hybridization and transfer of mimicry adaptations between *Heliconius* species. These findings may have general applicability: rapidly diversifying lineages of both plants and animals may frequently share and exchange adaptive genetic variation.

## Introduction

Speciation is increasingly recognized to be associated with ecological divergence, and may indeed occur in the presence of gene flow (see other contributions in this volume). But the extent to which species formation is driven by ecology and the possible role of hybridization in adaptive radiation are still debated (Schluter 2000; Coyne & Orr 2004; Mallet 2007). Here I analyse speciation in *Heliconius* butterflies, which has provided a somewhat classical story of ecology-driven speciation. There are some surprising recent discoveries which cast doubt on the generality of some of our earlier ideas on diversification in this group.

*Speciation and Patterns of Diversity*, ed. Roger K. Butlin, Jon R. Bridle and Dolph Schluter. Published by Cambridge University Press. © British Ecological Society 2009.

Ten years ago, we argued on somewhat slender evidence that speciation in *Heliconius* was largely due to ecological divergence, particularly in mimicry as well as in host and microhabitat choice, and that this was coupled with the evolution of assortative mating (Mallet *et al.* 1998b). Since then, evidence for ecological, especially mimicry-mediated, speciation has become firmer, but in the course of this work additional modes of speciation have been revealed. I begin by assessing *Heliconius* speciation work since the 1990s. I next review evidence on hybridization and introgression between species. Finally, I discuss the possibility that adaptive genes, particularly for mimicry, are transferred among members of rapidly radiating lineages.

## Speciation driven by mimicry

Müllerian mimicry among unpalatable *Heliconius* species and between *Heliconius* and another unpalatable butterfly group, the Ithomiinae, has been known since the work of Henry Walter Bates (Bates 1862). Fritz Müller (1879) provided the first theoretical analysis of Müllerian mimicry, but Bates was the first to explain mimicry in Darwinian terms. Bates was also the first to recognize the existence of mimicry now designated 'Müllerian' between pairs of unpalatable 'co-mimic' species, as well as 'Batesian' mimicry between palatable and unpalatable 'model' species. Bates' most convincing evidence of mimicry was provided by parallel geographic variation in colour patterns of different lineages. He argued that colour-pattern mimicry acted as an important means of local protection for rarer unpalatable *Heliconius*, and it is thus clear he had an inkling that frequency-dependent selection was involved. He argued that divergence in mimicry was a likely means whereby natural selection led to speciation (Bates 1862).

For most of the 20th century, evolutionists saw mimicry as an interesting visual demonstration of natural selection, but no longer related it to speciation (Fisher 1930; Futuyma 1998). Natural selection, such as that for mimicry, was seen mainly as a source of 'microevolution' within populations and species, whereas geographic isolation, and incidental hybrid inviability and sterility were more heavily emphasized as important for speciation. Yet by the 1990s, evidence was accumulating for divergent ecological selection as causes of speciation of a variety of organisms (Mallet 2001b).

By 1998, the majority of our work on contacts between the species pair *Heliconius himera* and *H. erato* had already been published. *Heliconius himera* and *H. erato* differ in colour pattern, tolerance of dry versus wet forest and mate choice (McMillan *et al.* 1997; Mallet *et al.* 1998b). Hybridization rates in narrow ~5 km hybrid zones are low (6%) (Mallet *et al.* 1998a). The two species lack larval food-plant and microhabitat differences (Jiggins *et al.* 1997b), and competition thus probably explains the narrowness of the overlap. All hybrids, backcrosses and further crosses, when formed, are viable and fertile (McMillan *et al.* 1997), but for poorly understood ecological reasons, probably in part due to non-mimetic colour pattern combinations, do not go on to establish successful hybrid swarms in the overlap zone. Furthermore, mtDNA and allozyme profiles of the two species remain distinct (Jiggins *et al.* 1997a). The two forms are considered species on the basis of this coexistence of separate genotypic clusters within the zone of overlap, even though ~10% of the population in the centre of the zone are F1 and backcross hybrids. Divergently patterned geographic races of *erato* also form narrow hybrid zones where they meet, but, in these contact zones, hybrids are always abundant, while the pure forms do not differ at mitochondrial or nuclear markers, and so are considered members of the same species. In conclusion, the sister species *himera* and *erato* have diverged in colour pattern and mate choice, but have no genomic incompatibilities (Mallet *et al.* 1998b).

Work on the *H. cydno* and *H. melpomene* 'sister species' pair was embryonic in 1998, but 10 years later has become a major area of investigation. The two species are sympatric throughout Central America and in the Western Andes. They hybridize regularly, albeit at a very low rate of one in a thousand or less (Mallet *et al.* 2007). This hybridization explains shared polymorphisms at (only some) nuclear genes (Bull *et al.* 2006; Kronforst *et al.* 2006b).

Each of these species mimics different, divergent members of an unrelated *Heliconius* lineage: *cydno* is black with white or yellow bands, and normally mimics *Heliconius sapho* or *Heliconius eleuchia*; in West Ecuador and parts of Colombia it can also be polymorphic and mimic both (Kapan 2001). *H. melpomene* instead has forms with monomorphic 'postman' patterns (black with broad red forewing band, sometimes with a yellow hindwing bar) that mimic postman-patterned *erato*. Other races of *melpomene* also have various 'dennis-ray' patterns (black with a forewing yellow band, a basal forewing orange patch known as 'dennis', and hindwing rays) that mimic 'dennis-ray' *erato* races and those of a number of other species in the Amazon basin. In only the Cauca Valley of Colombia, which lacks *H. melpomene* altogether, does *Heliconius cydno weymeri* have a morph 'gustavi' that mimics the local *Heliconius erato chestertonii* (Linares 1997). Probably, it is not a coincidence that *chestertonii* is also the only race of *erato* that lacks red markings: it has a postman-like yellow hindwing bar, but has entirely blueish-black iridescent forewings. However, whether the other *cydno* races are congenitally unable to match red-marked *erato*, or *H. e. chestertonii* has converged to a pattern like that of *cydno* only in this one valley, is not known.

The colour pattern differences between *cydno* and *melpomene* are controlled by many of the same loci involved in geographic mimicry differences within each species (Linares 1996; Naisbit *et al.* 2003; Joron *et al.* 2006; Kronforst *et al.* 2006a). Because *cydno* and *melpomene* are considerably closer genetically to one another than *sapho* or *eleuchia* are to *erato*, *cydno* and *melpomene* diverged more recently, probably as mimics of *erato*, *sapho* and *eleuchia* (the already divergent 'models'), rather than vice versa (Jiggins *et al.* 2001; Flanagan *et al.* 2004). In addition, comparative evidence strongly implicates *melpomene* and *cydno* as mimics of *erato* and allies, rather than vice versa (Eltringham 1917; Mallet 2001a).

Colour pattern divergence between the two species leads to a number of interesting evolutionary consequences: (1) the rare hybrids that are produced are non-mimetic, and are likely to suffer strongly reduced survival due to bird attacks (Mallet *et al.* 1998b; Jiggins *et al.* 2001); (2) mate location is partly visual and has apparently co-evolved with mimicry, leading to strong assortative mating that can be demonstrated with paper models as well as with natural wing patterns (Jiggins *et al.* 2001; Naisbit *et al.* 2001); (3) Panama *melpomene* are sympatric with *cydno* and mate more assortatively than Guianan *melpomene*, which do not overlap with *cydno* (Jiggins *et al.* 2001). This is consistent with 'reinforcement', i.e. adaptive evolution of more discriminating mate choice to avoid production of low-fitness hybrids with another species in sympatry; (4) the two mimetic pairs also differ in microhabitat – co-mimics *cydno* and *sapho* inhabit forest understory, whereas co-mimics *melpomene* and *erato* are normally found in more open areas (Estrada & Jiggins 2002). Nocturnal gregarious roosts are also stratified by mimicry (Mallet & Gilbert 1995). It is probable that the mimicry switch took place as a response to a habitat shift, but the reverse is possible. In any case, the mimicry shift also provides an additional ecological dimension which further enhances assortative mating due to habitat segregation; (5) hybrids between *cydno* and *melpomene* of both sexes are discriminated against in mate choice experiments, a form of disruptive sexual selection against hybrids (Naisbit *et al.* 2001). Thus, mimicry has important direct and indirect effects which result in manifold, pleiotropic effects on reproductive isolation (Mallet *et al.* 1998b; Jiggins *et al.* 2005).

Not all speciation-related differences between *cydno* and *melpomene* can be attributed to mimicry: hybrid females are usually sterile, an example of Haldane's Rule, because in butterflies it is the females that are heterozygous for sex-determining chromosomes (Naisbit *et al.* 2002, 2003). Like other 'Dobzhansky–Muller incompatibilities', Haldane's Rule sterility is thought to be caused by divergence at sex-linked loci with pleiotropic effects on gamete production (Coyne & Orr 2004). Female hybrid sterility is a major form of 'post-mating isolation' between these two species (~70%), but it is only about as strong as mimicry selection against hybrids (~64%), and host-plant and microhabitat differences are additional selective effects that have not been estimated. However, all these are weaker barriers than the 'pre-mating isolation' due to assortative mating (> 99.9%) (Naisbit *et al.* 2002; Mallet 2006). Thus we became convinced of what might be termed the classical story: that ecological divergence, especially mimicry and its co-evolved effects on mate choice and microhabitat use, was the major cause of *Heliconius* speciation.

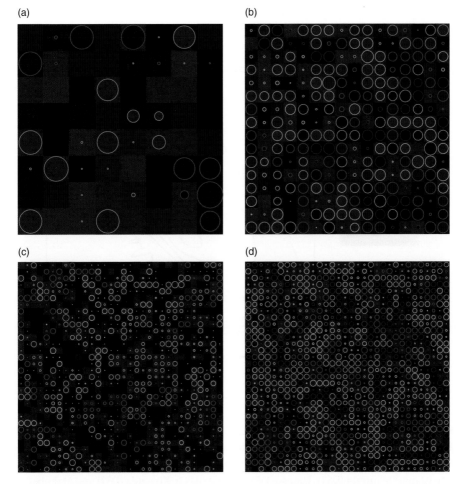

**Figure 7.1** Examples of the system state with three environmental factors ($k = 3$, so that there are eight possible niches). Each square represents a patch. The colour of the square represents the ecological niche assigned to the patch. Each local population is represented by a circle. The radius of the circle is proportional to the population size (the maximum radius corresponds to a population at carrying capacity). The colour of the circle defines the niche preferred by the majority of individuals. Matching of the color of the corresponding square and circle (observed in most cases) means that the majority of individuals in the patch have preference for the ecological conditions they experience. (a) 8 × 8 systems with two ecological niches filled, (b) 16 × 16 system with eight ecological niches filled, (c) 32 × 32 system with four ecological niches filled, and (d) 32 × 32 system with eight ecological niches filled.

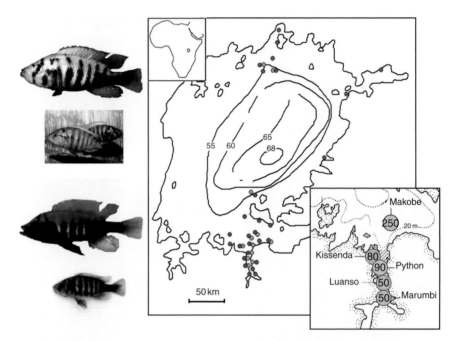

**Figure 9.1** Male and female of *P. pundamilia* (top) and *P. nyererei* (bottom), both from Makobe Island (DS4). Large map: distribution of known records of *P. pundamilia* and *P. nyererei* in Lake Victoria. Small inset map: the 'speciation transect' with location of the five islands: Marumbi (DS1), Luanso (DS1), Python (DS3), Kissenda (DS2) and Makobe (DS4). Water transparencies at the islands (cm Secchi disk readings) are given inside the circles (colours: turbid = brown, clear = blue). Note that Kissenda Island lies inside a lateral embayment, hence the lower transparency than at Python Island despite Kissenda's greater proximity to the clear water areas.

(a)    (b)

4

3

3

1

0

**Figure 9.2** Colour-based hybrid index. Class 0 (*P. pundamilia*-like) = blue-grey lower flanks (lf), blue-grey upper flanks (uf), blue spinous dorsal fin. Class 1 = yellow lower flanks, grey upper flanks, blue spinous dorsal fin. Class 2 = yellow lf, red along upper lateral line, blue spinous dorsal fin. Class 3 = yellow lf, red uf except grey crest, blue dorsal fin. Class 4 (*P. nyererei*-like) = yellow lf, red uf, red df. (a) wild males from Luanso Island. (b) F2 hybrid males generated by crossing F1 hybrids between *P. nyererei* and *P. pundamilia* from Python Islands (1992 lines, DS3).

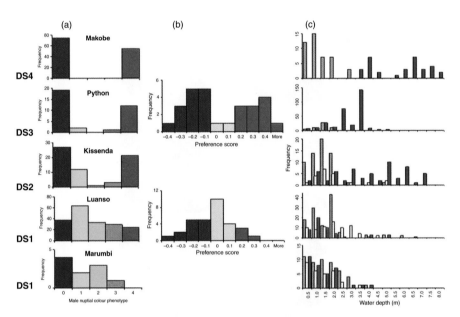

**Figure 9.3(a)** Frequency distributions of hybrid index scores along 'speciation transect'. (b) Frequency distributions for female mating preference phenotypes at two stations along 'speciation transect'. (c) Frequency distributions of males of different colour phenotypes against water depth. Males of *P. pundamilia*-like appearance (blue bars), *P. nyererei*-like appearance (red bars), reddish intermediates (hybrid index classes 2 and 3; orange bars), yellow intermediates (hybrid index class 1; pale yellow bars) DS1 to DS4 = Divergence score 1 to 4. (a) Modified from Seehausen *et al.* (2008). (b) and (c) reproduced with permission from Seehausen *et al.* (2008).

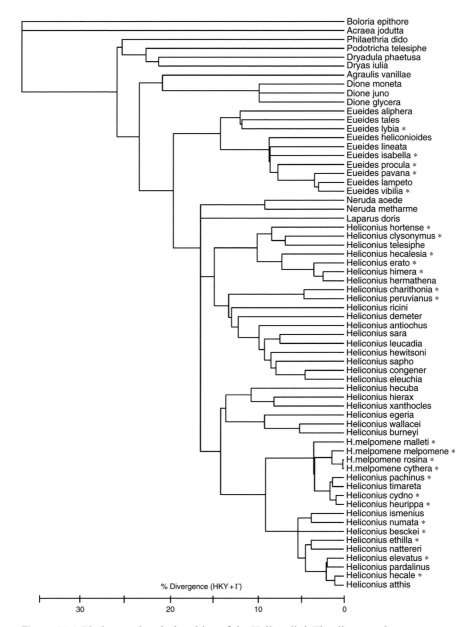

**Figure 10.1** Phylogenetic relationships of the Heliconiini. The diagram shows an mtDNA-based Yang rate-smoothed estimate of phylogenetic relationships among species of the Heliconiini. Species that hybridize in nature (starred) tend to be most closely related, and are in the most recent radiations of *Heliconius* and *Eueides*, rather than from older species in the 'basal genera' (Beltrán *et al.* 2007; after Mallet *et al.* 2007).

## Speciation not driven by mimicry

In *Heliconius*, closely related species often belong to different mimicry rings, and co-mimics in the same ring usually belong to different lineages, as though species in each lineage were adaptively radiating onto different mimicry rings (Turner 1976; Mallet *et al.* 1998b). However, recent evidence suggests that this classical story of mimicry-driven speciation holds for by no means all speciation events among Müllerian mimics. Dasmahapatra *et al.* (in preparation) have recently discovered via DNA sequencing a cryptic species most closely related to *Heliconius demeter*, with which it shares a nearly identical rayed mimicry pattern. Not only are these two scarce sister species difficult to distinguish from one other using morphology, but are also hard to distinguish from the related sympatric rayed *H. erato* throughout the Amazon basin. As another example, in *Ithomia*, from a different subfamily, the Ithomiinae, sister species are frequently Müllerian co-mimics (Jiggins *et al.* 2006), suggesting that speciation in this group can only sometimes be attributed to mimetic divergence, if at all.

Furthermore, although our group were using the term 'sister species' to refer to the *cydno* and *melpomene* species pair (Mallet *et al.* 1998b; Jiggins *et al.* 2001), the actual situation is more complex: one of the 'sister species' itself has sisters. A number of local endemic segregate taxa, such as *Heliconius pachinus*, *Heliconius heurippa*, *Heliconius tristero* and *Heliconius timareta*, thought to be more closely related to *cydno* than to *melpomene*, were interpreted 10 years ago as subspecies or 'semi-species' of *cydno* (Mallet *et al.* 1998b). In all cases, these local endemics overlap with *melpomene*, but occur outside the range of *H. cydno* (*sensu stricto*). DNA sequence data have mostly confirmed closer relationships of these segregate taxa to *cydno* (Brower 1996a,b; Salazar *et al.* 2005; Mavárez *et al.* 2006; Kronforst *et al.* 2006b; Beltrán *et al.* 2007), although the picture is clouded by possible paraphyly of *melpomene* (Fig. 10.1) as well as ongoing allele sharing and autosomal introgression between *cydno* (including its segregates) and *melpomene* (Bull *et al.* 2006; Kronforst *et al.* 2006b).

Whether these disjunct segregate taxa are regarded as full species or geographic races of *cydno* is to some extent a matter of taste. Early systematists, lacking detailed distribution data or large series from zones of sympatry, tended to confuse the red, yellow and black colour patterns of *heurippa*, *timareta* and *tristero* with those found in hybrid zones between mimetic races, and interpreted the segregate taxa as forms of *melpomene* (Eltringham 1917; Emsley 1964). However, it is now becoming usual to treat *heurippa*, *tristero* and *timareta* as separate species, due to their sympatry with *melpomene*; their pronounced morphological divergence from the black and white or yellow mimetic patterns typical of *cydno*; their often prominent red markings; and their assortative mating in tests with both *cydno* and *melpomene* (Brown 1981; Brower 1996a; Mavárez *et al.* 2006; Kronforst *et al.* 2006c). In the most recent checklist, the

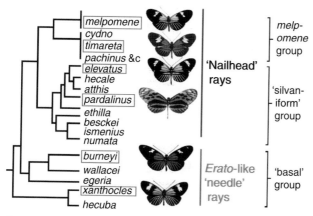

**Figure 10.2** Homoplasy of colour patterns in the *melpomene* group. 'Nailhead' rays, as in mimetic Amazonian races of *H. melpomene*, appear to provide a novel method of producing rays, and are characteristic only of the most recent radiation of the *melpomene/cydno/* silvaniform group of *Heliconius* (upper four 'boxed' species). Plesiomorphic 'needle' rays are the norm in unrelated, rayed co-mimics such as *erato*, *Neruda* and *Laparus*, and even in related *H. burneyi* and *H. xanthocles* (lower two 'boxed' species). The species having 'nailhead rays', are not, however, necessarily each other's closest relatives. Furthermore, red forewing band 'postman' patterns in *besckei*, the *cydno/timareta* group, and *melpomene* also show a patchy distribution across the phylogeny (not shown). Unless the ancestral *melpomene*/silvaniform species was polymorphic for both rays and postman patterns (which is extremely unlikely), either postman or nailhead ray patterns must be homoplasious. A likely source of such homoplasy is hybridization and introgression, as virtually all species in the *melpomene* and silvaniform clades can hybridize and backcross in nature.

entirely yellow and black *pachinus* is treated as a subspecies of *cydno*, while red-marked *heurippa*, *tristero* and *timareta* are regarded as full species (Lamas 2004).

Renewed interest in the *melpomene–cydno* group has been triggered by the suggestion that hybridization is important in the evolution of mimetic and non-mimetic colour patterns (Gilbert 2003; Salazar *et al.* 2005; Mavárez *et al.* 2006), as well as further discoveries of new segregate forms related to *cydno* and *melpomene* in the Eastern Andes.

## Hybridization, mimicry and speciation

My former PhD supervisor, Prof. Lawrence E. Gilbert Jr. proposed that *Heliconius* may gain much of their mimicry adaptability because they have available via hybridization, a 'shared genetic toolbox' allowing acquisition of mimicry colour patterns (Gilbert 2003). Often 1–3 genetic loci ('supergenes') can cause major mimetic switches in colour pattern (Sheppard *et al.* 1985; Joron *et al.* 2006). Adaptive introgression could explain the repeated re-evolution of the same apomorphic mimicry traits, such as 'nailhead rays' or the 'postman' pattern (see Fig. 10.2). This was the first time the proposal for the importance of

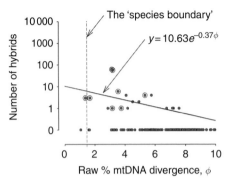

**Figure 10.3** Hybridization between species of *Heliconius*. The graph shows the numbers of wild-caught interspecific hybrids in *Heliconius* and *Eueides* from a survey of world museum collections, plotted against genetic distance (mitochondrial DNA sequence data, based on *CoI* and *CoII*). Each point represents a different pair of species; points ringed with a halo represent pairs of species for which backcrosses, as well as F1 hybrids, are known in nature. The regression line of best fit is also shown. Genetic distances for the same loci tend to be less than ~1.8% within species; this distance therefore marks the approximate 'species boundary'. The numbers of hybrids, and of hybrids that backcross, as well as the fraction of species pairs that hybridize all tend to be higher for species nearest the species boundary (after Mallet *et al.* 2007).

hybridization and introgression had appeared in print, although another Gilbert PhD student, Mauricio Linares, had advocated similar ideas in his PhD thesis (Linares 1989). While I recognized that introgression could explain some *Heliconius* colour pattern evolution, at that time I felt that the idea had limited applicability. In my view, a more pressing problem was the evolution of novel *Heliconius* mimicry patterns, which could not be generated readily using existing adaptations. However, this skepticism about the evolution of true novelty via introgression underestimated the degree to which recombination of existing patterns can generate a much larger array of colour patterns than exists today. For example, many of Gilbert's lab-reared hybrids show how novel patterns can be created via hybridization and recombination in the *melpomene* group to produce convincing mimics of existing unpalatable model species, even some never mimicked by *Heliconius* in nature (Gilbert 2003). Hybridization might be an important source of variation for rapid adaptation to novel mimetic opportunities, even though the components needed to create the new mimicry pattern may be old.

Hybridization between species of *Heliconius* in nature is rare (< 0.1% of specimens), but regular and ongoing, as well as widespread (Figs. 10.1 and 10.3). It results in documented cases of introgression between species such as *melpomene* and *cydno* (Mallet *et al.* 1998b, 2007; Bull *et al.* 2006; Dasmahapatra *et al.* 2007). It might be imagined that only sister species hybridize, but F1 hybrids and in some cases backcrosses are known to occur regularly between non-sister species (e.g. *melpomene* and silvaniforms such as *Heliconius ethilla* and *Heliconius numata*,

Figs. 10.1 and 10.3) (Dasmahapatra *et al.* 2007; Mallet *et al.* 2007). In captivity, the 'silvaniforms' *H. ethilla, hecale, ismenius, atthis* and *numata* can be mated with *cydno, pachinus* and *melpomene* to produce viable male and female offspring. Females are usually sterile (see above), again according to Haldane's Rule, but backcrosses of hybrid males can be used to transmit colour pattern genes between species (Gilbert 2003; Mavárez *et al.* 2006; Mallet *et al.* 2007). Thus hybridization is a likely conduit for flow of mimetic adaptations among species of *Heliconius*, provided that initial disadvantages of hybrid sterility and non-mimetic patterns can be overcome.

### *H. heurippa* and *H. pachinus*

*H. heurippa* is found in the foothills of the Andes near Villavicencio, Colombia, up to about 1800 m, where it overlaps but rarely hybridizes with *Heliconius melpomene melpomene* (Fig. 10.4). *H. heurippa* itself is not obviously mimetic, though it is somewhat reminiscent of the local black, yellow and orange forms of *Melinaea marsaeus, Melinaea isocomma, H. numata* and *Heliconius hecale*. Its pattern is curiously like those obtained by backcrossing postman-patterned *m. melpomene* × *cydno cordula* F1 hybrids to *Heliconius cydno cordula*. The latter still occurs on the same eastern flanks of the Andes a few tens of kilometres to the North, while *melpomene* is sympatric with *heurippa* near Villavicencio. Mitochondrial and some nuclear sequence data suggest a closer relationship to *cydno* than to *melpomene*. However, microsatellite data suggest that *heurippa* forms a genotypic cluster separate from *cydno* as well as from *melpomene*, even though many alleles are shared among all three species. *H. heurippa* mates assortatively in the laboratory, and males of *cydno* and *melpomene* shun *heurippa* (Mavárez *et al.* 2006) – see also similar work with *H. pachinus* versus *cydno* and *melpomene* (Kronforst *et al.* 2007). This assortative mating is shown via painted models to be due to differences in colour pattern. Thus the hybrid colour pattern phenotype is itself partially reproductively isolated. These strands of evidence suggest that *heurippa* evolved by hybridization – a case of hybrid speciation (Mavárez *et al.* 2006). Furthermore, in one locality further to the North (San Cristobal, Venezuela) the same parental races, *H. m. melpomene* and *H. cydno cordula*, are in a somewhat intermediate phase, with abundant polymorphic colour pattern variants, presumably caused by recent introgression (Mavárez *et al.* 2006). This gives an example of a population where colour patterns have yet to be purified by selection for mimicry (as is normally expected, see p. 180). Such populations are the expected precursors that could start a shift to a novel, monomorphic warning colour pattern, such as that in *heurippa*.

The suggestion that *heurippa* evolved via hybrid speciation has been contentious (a discussion of these critiques can be found online at www.mailinglists.ucl.ac.uk/ pipermail/heliconius/2006-June.txt). Meanwhile, recent studies have suggested that *H. pachinus* is not a hybrid species (Kronforst *et al.* 2007), in spite of earlier suggestions (Gilbert 2003). If hybridization does not involve chromosome doubling, as here, speciation followed by introgression may be almost

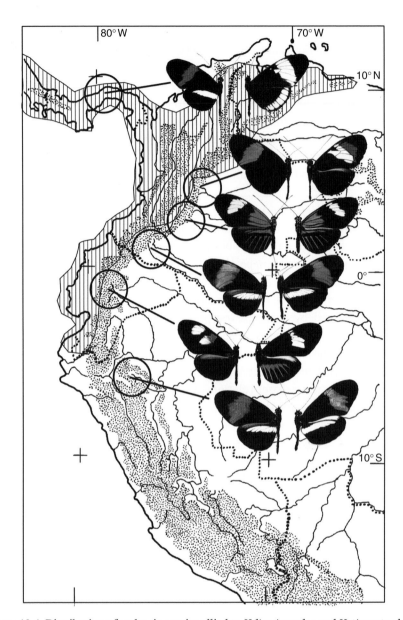

**Figure 10.4** Distribution of endemic species allied to *Heliconius cydno* and *H. timareta* along the eastern slopes of the Andes (wings to the right), compared with all the local *melpomene* races (wings to the left). *Heliconius melpomene* occurs essentially thoughout the whole area shown below about 1800 m altitude (above 1800 m shown stippled), and the mainly Western Andean distribution of *cydno* sensu stricto is shown hatched. Note, however, that some *cydno* populations are found on the eastern slopes of the Andes in northern Colombia and Venezuela just north of the distribution of *heurippa*. Geographic races of *melpomene* (lefthand wings, all of which mimic *H. erato*) are, top to bottom: *m. rosina*, *m. melpomene*, *m. malleti*, *m. bellula*, *m. plesseni* and *m. amaryllis*. *Heliconius cydno* and related species (righthand wings), top to bottom: *H. cydno chioneus* (Panama), *H. heurippa* (Villavicencio, Colombia), unnamed species (Florencia, Colombia), *H. tristero* (Mocoa, Colombia), *H. timareta* (Eastern Ecuador) and unnamed species (Upper Mayo Valley, Peru).

indistinguishable from 'true' hybrid speciation. The only clear distinction is that hybridization should be shown to be a major cause of the changes leading to the formation of the hybrid species (Mallet 2007). *H. cydno* and *H. melpomene* hybridize with introgression wherever the two overlap (Bull *et al.* 2006; Kronforst *et al.* 2006b; Mallet *et al.* 2007), and if a few colour pattern genes leaked across they could readily lead to a new stable form, like *H. (cydno) heurippa*. This is a situation where the genome will not necessarily end up an approximate 50% mosaic of the two species (Kronforst *et al.* 2007; M. Turelli & J. Coyne pers. comm.); instead, the proportion of one of the species might be very low. Backcrosses between almost any other pair of randomly picked geographic races of *cydno* and *melpomene* would fail to give the *heurippa* pattern. So this particular hybrid pattern seems to be rather convincing evidence of hybrid origin only from these particular two local forms of *melpomene* and *cydno*. And the finding that the hybrid colour pattern itself causes reproductive isolation with both parent species (Mavárez *et al.* 2006) seems a reasonable argument for calling *heurippa* a 'good' hybrid species, even though it is not known to overlap with *cydno*. Of course, there remains some doubt as to whether the nuclear genes determining the *heurippa* colour pattern were really inherited from both parents, but we will only know this with certainty when the genetic regions responsible for colour pattern changes are finally characterized and sequenced (Joron *et al.* 2006).

### H. cydno *and its relatives further south*

In Ecuador, *H. timareta* has been recognized as a separate species since 1867 (Lamas 2004). This non-mimetic form is peculiar for *Heliconius* in that it is apparently polymorphic throughout its narrow range for *melpomene*-like 'dennis' and 'nail-head ray', as well as for yellow forewing band patterns. It occurs only at rather high altitudes (for *Heliconius*) in the eastern slopes of the Andes (approx. 1000–1800 m above sea level) and co-exists with *Heliconius erato notabilis* and *Heliconius melpomene plesseni* (Fig. 10.4). The latter have co-mimetic 'twin-spot' pink and white patterns that bear no resemblance to 'dennis-ray' patterns, even though further down the Pastaza valley, these two upland races form a hybrid zone with 'dennis-ray' Amazonian races *Heliconius erato lativitta* and *Heliconius melpomene malleti*, respectively. Polymorphisms and lack of suitable co-mimics show that *timareta* is clearly non-mimetic. Most *Heliconius* are widely distributed, but, like *heurippa*, *timareta* is restricted to a small strip of submontane forest along the eastern slopes of the Andes and appears to be most closely related to *cydno*, although likelihood and other analyses of DNA sequence data do not entirely rule out *melpomene* as its closest relative. One likely possibility is that *timareta* is another form that has stabilized after hybridization between Amazonian rayed *melpomene* and an unidentified, maybe extinct form of *cydno*. Recently, a new subspecies, *Heliconius timareta timoratus*, was described (Lamas 1998). The specimen came from above 1000 m altitude in the Cordillera del Condor, on the formerly disputed border of Ecuador and Peru. The new subspecies is known from a single specimen, but in this case it is a near-perfect mimic of the local *melpomene* race *H. m. malleti*.

Another recently discovered taxon in this group is *H. tristero*. This form is found at middle elevations, >800 m altitude above Mocoa, in the Southeastern Andes of Putumayo, Colombia. The species description was based on only two males, both of which had *cydno*-like mtDNA, but were extremely similar in appearance to the local *Heliconius melpomene bellula* (Brower 1996a). *H. tristero* was therefore apparently an Eastern Andes form of *cydno* that mimicked *melpomene* and *erato* (Fig. 10.4). Perhaps understandably, several of us were sceptical that *tristero* was a separate species in its own right: could it not be an aberration due to *cydno* mtDNA introgression into *melpomene*? Most morphological characters suggested by Brower to characterize *H. tristero* appeared labile, and so only the mtDNA evidence was at all convincing. 'It is straightforward to produce wing pattern/mtDNA combinations closely resembling Brower's new taxa, … Thus, in the case of (*cydno*) *tristero* and *melpomene*, we probably have a case of *melpomene* red patterns shared by common descent from F1 fathers in an area of natural hybridization, rather than mimicry between *cydno* and *melpomene* (which is not known to exist)' (Gilbert 2003). *H. tristero* did not appear likely to be a stable form, firstly because so few specimens had been collected, so that ongoing hybridization could not be ruled out, and secondly because *cydno* and *melpomene* were unknown previously to mimic one another. As we have seen, the classical story was that *cydno* and *melpomene* had diverged as a result of a shift in mimicry ring.

Today, nuclear DNA data from Brower's original specimens (Mavárez, in preparation), as well as more recent collections from Putumayo (Arias & Huertas 2001), suggest that *tristero* does constitute a separate species from *melpomene*. As still more new *cydno*-like lineages are discovered that are also mimics of other *melpomene* races in the Eastern Andes, Brower's early insight is increasingly supported.

The first such additional example discovered was in the upper Mayo valley near the Río Serranoyacu, San Martín, Peru (Fig. 10.4). Mathieu Joron, Gerardo Lamas and I had long recognized what we thought of as a distinct 'upland' postman-patterned form of *melpomene* from above 1000 m altitude of the upper Río Mayo valley. This form occurred adjacent to and sometimes overlapping with a well-studied, but generally lower elevation form, the postman-patterned *Heliconius melpomene amaryllis* (Mallet et al. 1990). Although the overriding impression is of great similarity of the two forms, the 'upland' form is characterized by larger size, narrower and more whitish hindwing bar, more sinuous and narrower forewing band, and strongly expressed basal red spots on the undersides of fore- and hind-wings. All these colour pattern characteristics tend to enhance the otherwise *amaryllis*-like 'postman' pattern's weak mimetic alliance with another abundant upland form, *Heliconius telesiphe*, which has a twin-barred 'postman' pattern and occurs in Ecuador, Peru and Bolivia (Vane-Wright et al.

1975), and often replaces *erato* ecologically in steep submontane forests above about 800 m altitude (Benson 1978; Mallet 1993). In the 1980s, I ran allozyme electrophoresis gels on 'upland *melpomene*', and found no major differences from *H. m. amaryllis*. However, we now know that polymorphic nuclear genetic markers are frequently shared between *cydno, melpomene* and other relatives, in part because of introgression between all the species (Beltrán 1999; Bull *et al.* 2006; Kronforst *et al.* 2006b). Recently, Jesús Mavárez *et al.* (in preparation) have shown that mtDNA and some nuclear sequences of these 'upland *melpomene*' match *timareta* more closely than those of local *H. m. amaryllis*, and that morphology differences correlate perfectly with DNA sequence differences. Again we have a form with *cydno*-like DNA profile and a colour pattern whose most notable feature is near-identity with that of a local race of *melpomene*.

A third cryptic taxon is now being investigated in Colombia in Mauricio Linares' laboratory (in preparation). In upland regions near Florencia, Colombia, a 'dennis-ray' *Heliconius* with *cydno*-like mtDNA again co-occurs with the local *melpomene* form, the Amazonian 'dennis-ray' *H. m. malleti* (Fig. 10.4). The Linares group have made remarkably complete genetic, morphological, ecological and behavioural studies of these cryptic species. They have demonstrated that host-plant usage and assortative mating correlate closely with nuclear and mitochondrial DNA markers, and with slight diagnostic colour pattern differences between the two mimetic forms (Giraldo 2005). Here again is a cryptic species that shares a mimetic colour pattern with the local form of *melpomene*.

## Introgression and hybridization as a source of mimetic adaptations

Much biological work remains to be done on these new taxa. However, recent discoveries of multiple *cydno*-like or *timareta*-like *Heliconius* taxa are no longer in doubt (Fig. 10.4). These all overlap in extensive sympatry with local *melpomene*. In some cases (*heurippa, timareta*), the forms are non-mimetic, although their colour patterns strongly suggest evidence of past hybridization with *melpomene* (and in the case of *heurippa*, with *H. cydno cordula* as well). In other cases, the patterns are nearly identical to the local *melpomene*, and are best interpreted as co-mimetic with that species and with *erato*. It is striking that these include both Amazonian 'dennis-ray' (e.g. *timareta*) and also Andean 'postman' patterns (e.g. *tristero*) found in *melpomene* (Fig. 10.4) as well as in the latter's usual co-mimic *erato*. It would now not be surprising to find multiple other *cydno*-like *melpomene*-mimicking cryptic taxa at similar elevations in many other East Andean valleys in Ecuador, Peru and Bolivia.

The detailed morphology of these *cydno*-like taxa (Fig. 10.4), together with knowledge of the inheritance of colour pattern in *cydno, melpomene* and their hybrids (Naisbit *et al.* 2003), suggests that the object of mimicry itself, *melpomene*, is a likely source of the mimicry adaptation via introgression. Alternatively, mimicry-determining genes might be extremely labile, allowing repeated, homoplasious re-evolution of nearly identical mimetic patterns, including two re-inventions of nailhead rays, by *H. timareta* and the related taxon from Florencia. However, such homoplasy fails to explain why even the *non*-mimetic forms *heurippa* and *timareta* have colour patterns derivable from sympatric or closely adjacent forms of *melpomene*. In the case of *heurippa*, the pattern is clearly derived only from local *cydno* as well. These hypotheses will be testable as soon as genes for colour pattern mimicry are fully characterized and sequenced (Joron *et al.* 2006; Kronforst *et al.* 2006a). For the moment, parsimony and existing evidence suggest that *timareta*-like and *cydno*-like Andean forms have hybrid origins with local *melpomene*.

## Generality of adaptive introgression

It could be argued that mimicry is a freak adaptation for which hybridization is particularly likely to be adaptive. Relatively few loci need to be introgressed to encode major switches in mimicry (Naisbit *et al.* 2003), and the colour patterns needed are exactly those already found in other local *Heliconius* species. But I don't believe that mimicry is unusual in this respect. Adaptive radiations often occur when a dispersal event or a novel general adaptation allows a species to colonize an enhanced range of pre-existing specialized ecological opportunities. In many cases, the same range of niches may have been colonized already by the same lineage in an earlier radiation, so that hybridization between the earlier lineage and the new species could enhance the possibility of re-acquiring many of the same species-specific niche-adaptations to allow a new radiation. There are possible examples of such traits in *Heliconius* that have nothing to do with mimicry. For instance, the larvae of most heliconiine species are oligophagous within sub-groups of the plant genus *Passiflora* (Passifloraceae). However, distinct butterfly lineages from 'basal genera', *Eueides* and *Heliconius* (Fig. 10.1) have repeatedly recolonized Passifloraceae of the same large subgenera *Granadilla*, *Plectostemma*, *Distephana* and *Astrophea* (Benson *et al.* 1976). Suppose a member of a radiating lineage acquired a new key adaptation, say adult pollen-feeding, or the extraordinary pupal mating habit (Beltrán *et al.* 2007). Then hybridization and recombination with members of the earlier radiation would permit a new radiation having both the key novel adaptation as well as the earlier-developed array of species-specific host-plant chemical adaptations.

Beak shape in Darwin's finches is another characteristic which affects not only trophic niche (the range of seeds that can be handled in competition with other species), but also song characteristics that partially determine mating behaviour (Grant & Grant 1998; Podos 2001). High variance in beak shape, in part due to occasional hybridization, will enhance evolutionary flexibility (Grant 1993; Grant & Grant 1998, 2008; Podos 2001) and at the same time allow repeated associated evolution of reproductive isolation via its effect on song characteristics, as is the case with *Heliconius* mimicry. Similar cases are also evident in cichlid fish, where ecologically relevant trophic or spawning habit traits, such as water depth, may affect colouration perception during courtship, and so lead to changes in optimal male sexual colouration and reproductive isolation (Seehausen 2004).

## Conclusion

Hybridization and introgression could thus play a much greater role in adaptive radiation and speciation than traditionally assumed likely, in animals as well as in plants (Arnold 1997; Mallet 2005, 2007; Grant & Grant 2008). Species in many rapidly radiating groups are incompletely reproductively isolated, even though this hardly affects their ability to remain distinct when in sympatry with congeners. Leaky reproductive barriers are now generally acknowledged to be common, given extensive evidence from sequencing and genotyping studies (Coyne & Orr 2004; Seehausen 2004; Mallet 2005; Grant & Grant 2008); meanwhile, genes useful in adaptation should show enhanced success of transfer compared with introgression of the anonymous molecular markers normally documented. It no longer seems unlikely that rapidly diversifying lineages owe some of their evolutionary flexibility and success to genetic variance shared via occasional hybridization, providing introgression with which they can build new and useful combinations of adaptations. It is still unknown just how important hybridization is in influencing adaptive radiation, but studies on tractable groups such as *Heliconius* butterflies are likely to provide some of the answers.

## Acknowledgements

I am very grateful for discussions of many of the many exciting new discoveries on these taxa among members of the *Heliconius* Consortium. Here, I am particularly grateful for unpublished information from Kanchon Dasmahapatra on the *H. demeter* cryptic species work; from Natalia Giraldo, Adriana Tapia and Mauricio Linares on the work in Florencia, Colombia; from Jesús Mavárez and Mathieu Joron on the work in the Upper Río Mayo Valley of Peru; and from Blanca Huertas about *H. tristero* in Churumbelos, Colombia. I am also grateful to Mathieu Joron, Mauricio Linares, Jesús Mavárez, Jon Bridle and two anonymous reviewers for critically reading the manuscript. This work was supported by NERC and DEFRA Darwin Initiative.

## References

Arias, J. J. and Huertas, B. (2001) Mariposas diurnas de la Serranía de los Churumbelos, Cauca. Distribución altitudinal y diversidad de especies (Lepidoptera: Rhopalocera: Papilionoidea). *Revista Colombiana de Entomologia* **27**, 169–176.

Arnold, M. L. (1997) *Natural Hybridization and Evolution*. Oxford University Press, Oxford.

Bates, H. W. (1862) Contributions to an insect fauna of the Amazon valley. Lepidoptera: Heliconidae. *Transactions of the Linnean Society of London* **23**, 495–566.

Beltrán, M. (1999) Evidencia genética (alozimas) para evaluar el posible orígen híbrido de *Heliconius heurippa* (Lepidoptera: Nymphalidae). Universidad de los Andes, Bogotá.

Beltrán, M., Jiggins, C. D., Brower, A. V. Z., Bermingham, E. and Mallet, J. (2007) Do pollen feeding and pupal-mating have a single origin in *Heliconius*? Inferences from multilocus sequence data. *Biological Journal of the Linnean Society* **92**, 221–239.

Benson, W. W. (1978) Resource partitioning in passion vine butterflies. *Evolution* **32**, 493–518.

Benson, W. W., Brown, K. S. and Gilbert, L. E. (1976) Coevolution of plants and herbivores: passion flower butterflies. *Evolution* **29**, 659–680.

Brower, A. V. Z. (1996a) A new mimetic species of *Heliconius* (Lepidoptera: Nymphalidae), from southeastern Colombia, revealed by cladistic analysis of mitochondrial DNA sequences. *Zoological Journal of the Linnean Society* **116**, 317–332.

Brower, A. V. Z. (1996b) Parallel race formation and the evolution of mimicry in *Heliconius* butterflies: a phylogenetic hypothesis from mitochondrial DNA sequences. *Evolution* **50**, 195–221.

Brown, K. S. (1981) The biology of *Heliconius* and related genera. *Annual Review of Entomology* **26**, 427–456.

Bull, V., Beltrán, M., Jiggins, C. D., *et al.* (2006) Polyphyly and gene flow between non-sibling *Heliconius* species. *BMC Biology* **4**, 11.

Coyne, J. A. and Orr, H. A. (2004) *Speciation.* Sinauer Associates, Sunderland, Mass.

Dasmahapatra, K. K., Silva, A., Chung, J.-W. and Mallet, J. (2007) Genetic analysis of a wild-caught hybrid between non-sister *Heliconius* butterfly species. *Biology Letters* **3**, 660–663.

Eltringham, H. (1917) On specific and mimetic relationships in the genus *Heliconius*. *Transactions of the Entomological Society of London* **1916**, 101–148.

Emsley, M. G. (1964) The geographical distribution of the color-pattern components of *Heliconius erato* and *Heliconius melpomene* with genetical evidence for the systematic relationship between the two species. *Zoologica; Scientific Contributions of the New York Zoological Society* **49**, 245–286.

Estrada, C. and Jiggins, C. D. (2002) Patterns of pollen feeding and habitat preference among *Heliconius* species. *Ecological Entomology* **27**, 448–456.

Fisher, R. A. (1930) *The Genetical Theory of Natural Selection.* Clarendon Press, Oxford.

Flanagan, N. S., Tobler, A., Davison, A., *et al.* (2004) The historical demography of Müllerian mimicry in the neotropical *Heliconius* butterflies. *Proceedings of the National Academy of Sciences of the United States of America* **101**, 9704–9709.

Futuyma, D. J. (1998) *Evolutionary Biology*, 3rd edn. Sinauer, Sunderland, Mass.

Gilbert, L. E. (2003) Adaptive novelty through introgression in *Heliconius* wing patterns: evidence for a shared genetic 'toolbox' from synthetic hybrid zones and a theory of diversification. In: *Ecology and Evolution Taking Flight: Butterflies as Model Systems* (ed. C. L. Boggs), pp. 281–318. University of Chicago Press, Chicago.

Giraldo, N. (2005) Posible convergencia mimética entre las especies *Heliconius melpomene* y *H. cydno* (Lepidoptera: Nymphalidae) en Florencia (Caquetá): aproximación morfológica y genética. Universidad de Los Andes, Bogotá.

Grant, P. R. (1993) Hybridization of Darwin's finches on Isla Daphne Major, Galápagos. *Philosophical Transactions of the Royal Societies of London B* **340**, 127–139.

Grant, B. R. and Grant, P. R. (1998) Hybridization and speciation in Darwin's finches. The role of sexual imprinting on a culturally transmitted trait. In: *Endless Forms. Species and Speciation* (ed. D. J. Howard), pp. 404–422. Oxford University Press, New York.

Grant, P. R. and Grant, B. R. (2008) *How and Why Species Multiply. The Radiation of Darwin's Finches.* Princeton University Press, Princeton, NJ.

Jiggins, C. D., Emelianov, I. and Mallet, J. (2005) Assortative mating and speciation as pleiotropic effects of ecological adaptation: examples in moths and butterflies. In: *Insect Evolutionary Ecology* (ed. M. Fellowes), pp. 451–473. Royal Entomological Society, London.

Jiggins, C. D., Mallarino, R., Willmott, K. R. and Bermingham, E. (2006) The phylogenetic pattern of speciation and wing pattern change in Neotropical *Ithomia* butterflies (Lepidoptera: Nymphalidae). *Evolution* **60**, 1454–1466.

Jiggins, C. D., McMillan, W. O., King, P. and Mallet, J. (1997a) The maintenance of species differences across a *Heliconius* hybrid zone. *Heredity* **79**, 495–505.

Jiggins, C. D., McMillan, W. O. and Mallet, J. L. B. (1997b) Host plant adaptation has not played a role in the recent speciation of *Heliconius himera* and *Heliconius erato* (Lepidoptera: Nymphalidae). *Ecological Entomology* **22**, 361–365.

Jiggins, C. D., Naisbit, R. E., Coe, R. L. and Mallet, J. (2001) Reproductive isolation caused by colour pattern mimicry. *Nature* **411**, 302–305.

Joron, M., Papa, R., Beltrán, M., *et al.* (2006) A conserved supergene locus controls colour pattern diversity in *Heliconius* butterflies. *PLoS Biology* **4**, e303.

Kapan, D. D. (2001) Three-butterfly system provides a field test of Müllerian mimicry. *Nature* **409**, 338–340.

Kronforst, M. R., Kapan, D. D. and Gilbert, L. E. (2006a) Parallel genetic architecture of parallel adaptive radiations in mimetic *Heliconius* butterflies. *Genetics* **174**, 535–539.

Kronforst, M. R., Salazar, C., Linares, M. and Gilbert, L. E. (2007) No genomic mosaicism in a putative hybrid butterfly species. *Proceedings of the Royal Society B* **274**, 1255–1264.

Kronforst, M. R., Young, L. G., Blume, L. M. and Gilbert, L. E. (2006b) Multilocus analysis of admixture and introgression among hybridizing *Heliconius* butterflies. *Evolution* **60**, 1254–1268.

Kronforst, M. R., Young, L. G., Kapan, D. D., *et al.* (2006c) Linkage of butterfly mate preference and wing color preference cue at the genomic location of *wingless*. *Proceedings of the National Academy of Sciences of the United States of America* **103**, 6575–6580.

Lamas, G. (1998) Comentarios taxonómicos y nomenclaturales sobre Heliconiini neotropicales, con designación de lectotipos y descripción de cuatro subespecies nuevas (Lepidoptera: Nymphalidae: Heliconiinae). *Revista Peruana de Entomologia* **40** (1997), 111–125.

Lamas, G. (2004) Heliconiinae. In: *Hesperioidea – Papilionoidea* (ed. G. Lamas), pp. 261–274. Association for Tropical Lepidoptera. Scientific Publishers, Gainesville, FL.

Linares, M. (1989) Adaptive microevolution through hybridization and biotic destruction in the neotropics. University of Texas at Austin.

Linares, M. (1996) The genetics of the mimetic coloration in the butterfly *Heliconius cydno weymeri*. *Journal of Heredity* **87**, 142–149.

Linares, M. (1997) The ghost of mimicry past: laboratory reconstitution of an extinct butterfly 'race'. *Heredity* **78**, 628–635.

Mallet, J. (1993) Speciation, raciation, and color pattern evolution in *Heliconius* butterflies: evidence from hybrid zones. In: *Hybrid Zones and the Evolutionary Process* (ed. R. G. Harrison), pp. 226–260. Oxford University Press, New York.

Mallet, J. (2001a) Causes and consequences of a lack of coevolution in Müllerian mimicry. *Evolutionary Ecology* **13**, 777–806.

Mallet, J. (2001b) The speciation revolution. *Journal of Evolutionary Biology* **14**, 887–888.

Mallet, J. (2005) Hybridization as an invasion of the genome. *Trends in Ecology & Evolution* **20**, 229–237.

Mallet, J. (2006) What has *Drosophila* genetics revealed about speciation? *Trends in Ecology & Evolution* **21**, 386–393.

Mallet, J. (2007) Hybrid speciation. *Nature (London)* **446**, 279–283.

Mallet, J., Barton, N., Lamas, G., *et al.* (1990) Estimates of selection and gene flow from measures of cline width and linkage disequilibrium in *Heliconius* hybrid zones. *Genetics* **124**, 921–936.

Mallet, J., Beltrán, M., Neukirchen, W. and Linares, M. (2007) Natural hybridization in heliconiine butterflies: the species boundary as a continuum. *BMC Evolutionary Biology* **7**, 28.

Mallet, J. and Gilbert, L. E. (1995) Why are there so many mimicry rings? Correlations between habitat, behaviour and mimicry in *Heliconius* butterflies. *Biological Journal of the Linnean Society* **55**, 159–180.

Mallet, J., McMillan, W. O. and Jiggins, C. D. (1998a) Estimating the mating behavior of a pair of hybridizing *Heliconius* species in the wild. *Evolution* **52**, 503–510.

Mallet, J., McMillan, W. O. and Jiggins, C. D. (1998b) Mimicry and warning color at the boundary between races and species. In: *Endless Forms: Species and Speciation* (ed. D. J. Howard), pp. 390–403. Oxford University Press, New York.

Mavárez, J., Salazar, C., Bermingham, E., *et al.* (2006) Speciation by hybridization in *Heliconius* butterflies. *Nature* **441**, 868–871.

McMillan, W. O., Jiggins, C. D. and Mallet, J. (1997) What initiates speciation in passion-vine butterflies? *Proceedings of the National Academy of Sciences of the United States of America* **94**, 8628–8633.

Müller, F. (1879) *Ituna* and *Thyridia*; a remarkable case of mimicry in butterflies. *Transactions of the Entomological Society of London* **1879**, xx–xxix.

Naisbit, R. E., Jiggins, C. D., Linares, M. and Mallet, J. (2002) Hybrid sterility, Haldane's rule, and speciation in *Heliconius cydno* and *H. melpomene*. *Genetics* **161**, 1517–1526.

Naisbit, R. E., Jiggins, C. D. and Mallet, J. (2001) Disruptive sexual selection against hybrids contributes to speciation between Heliconius cydno and H. melpomene. *Proceedings of the Royal Society London B* **268**, 1849–1854.

Naisbit, R. E., Jiggins, C. D. and Mallet, J. (2003) Mimicry: developmental genes that contribute to speciation. *Evolution & Development* **5**, 269–280.

Podos, J. (2001) Correlated evolution of morphology and vocal signal structure in Darwin's finches. *Nature* **409**, 185–188.

Salazar, C. A., Jiggins, C. D., Arias, C. F., *et al.* (2005) Hybrid incompatibility is consistent with a hybrid origin of *Heliconius heurippa* Hewitson from its close relatives, *Heliconius cydno* Doubleday and *Heliconius melpomene* Linnaeus. *Journal of Evolutionary Biology* **18**, 247–256.

Schluter, D. (2000) *The Ecology of Adaptive Radiation*. Oxford University Press, New York.

Seehausen, O. (2004) Hybridization and adaptive radiation. *Trends in Ecology & Evolution* **19**, 198–207.

Sheppard, P. M., Turner, J. R. G., Brown, K. S., Benson, W. W. and Singer, M. C. (1985) Genetics and the evolution of muellerian mimicry in *Heliconius* butterflies. *Philosophical Transactions of the Royal Society of London B* **308**, 433–613.

Turner, J. R. G. (1976) Adaptive radiation and convergence in subdivisions of the butterfly genus *Heliconius* (Lepidoptera: Nymphalidae). *Zoological Journal of the Linnean Society* **58**, 297–308.

Vane-Wright, R. I., Ackery, P. R. and Smiles, R. L. (1975) The distribution, polymorphism and mimicry of *Heliconius telesiphe* (Doubleday) and the species of *Podotricha* Michener (Lepidoptera: Heliconiinae). *Transactions of the Royal Entomological Society of London* **126**, 611–636.

CHAPTER ELEVEN

# Investigating ecological speciation

DANIEL J. FUNK

## Ecological speciation

Across the years, biological thought on the causal associations between ecological factors and species formation has evolved from an initially implicit to an increasingly explicit level of recognition. Darwin (1859) recognized that adaptation to divergent environments via natural selection promoted the formation of new 'kinds'. Shortly thereafter, Benjamin Walsh (1864) offered prescient insights on the relationship between host-plant-associated divergence, interbreeding and species status in herbivorous insects (see also Bates (1862) for related inferences). Dobzhansky (1937) and Mayr (1942) later more explicitly invoked the reproductive isolation between populations as the defining characteristic of 'biological species', the concept adopted for this chapter. Simpson (1944, 1953) noted the association of ecological shifts with increased species diversity in 'adaptive radiations'. However, the questions of why and how why access to novel resources should promote the reproductive isolation required for increased speciation were long treated as a black box. Other workers of the synthesis provided verbal models that – with varying degrees of explicitness – illuminated this box. These models described how divergent adaptation might be expected to incidentally yield reproductive isolation between populations from different environments as a byproduct (Mayr 1942, 1947; Muller 1942; Dobzhansky 1951). This might occur via the pleiotropic effects of selected loci, or the direct effects of loci in linkage disequilibrium with them, on reproductive barriers. Decades later, some experiments on *Drosophila* evaluated these models (Kilias *et al.* 1980; Dodd 1989; Rice & Hostert 1993), while initial studies on sympatric speciation provided glimpses into the ecological basis of reproductive barriers (Bush 1969; Feder 1998). However, it was not until controlled, comparative, experimental studies of natural populations were introduced that the term 'ecological speciation' gained currency and an explicit, exciting and geographically neutral hypothesis on speciation biology was born (Funk 1996, 1998; Schluter 1996, 2000, 2001; Orr & Smith 1998; Funk *et al.* 2002; Rundle & Nosil 2005; Nosil 2007). Here, I follow Schluter's (2001) widely used definition: 'The ecological hypothesis of speciation is that reproductive

*Speciation and Patterns of Diversity*, ed. Roger K. Butlin, Jon R. Bridle and Dolph Schluter. Published by Cambridge University Press. © British Ecological Society 2009.

isolation evolves ultimately as a consequence of divergent natural selection on traits between environments.' Despite this definition, the existence of influential models of sympatric speciation with an essential ecological component (reviewed by Berlocher & Feder 2002; Coyne & Orr 2004) has resulted in a persisting conflation of this term with ecological speciation (Dieckmann *et al.* 2004).

The author's special interest in ecological contributions to speciation was motivated by evolutionary ecological and phylogenetic findings that had accumulated by the late 1980s. By this time, it was becoming apparent that herbivorous insect species (the author's favoured study systems) were often not ecologically homogeneous entities. Instead, evidence indicated that populations within a species often varied in the resources they used (Fox & Morrow 1981; Futuyma & Peterson 1985; Thompson 1994, 2005), providing opportunities for divergent adaptation. These observations dovetailed nicely with those of a seminal phylogenetic study by Mitter *et al.* (1988). That study introduced many biologists to a comparative approach based on 'sister-group comparisons', in which pairs of lineages that share their most recent common ancestor, and thus necessarily of equal age, are compared. Mitter *et al.* used this approach to statistically demonstrate that disparate herbivorous insect lineages repeatedly tended to be more species-rich than their non-herbivorous sister lineages. This finding was consistent with the popular notion that herbivores had a special capacity to adopt new and unexploited (host plant) resources that promoted their tremendous species-level diversity through associated adaptive radiation (Ehrlich & Raven 1964; Strong *et al.* 1984). Together, these two sorts of findings bracketed the black box, thus making clear what was inside and what needed to be empirically studied: the mechanisms by which ecological divergence promoted reproductive isolation and speciation. They thus pointed out a ground plan for future work that could connect the dots between their respective ecological and phylogenetic findings.

## Aspects of the study of ecological speciation

Connecting these dots and fully exploring the 'ecological speciation space' contained by the black box requires a variety of approaches to address questions at various biological levels. The present section provides a hierarchical taxonomy of approaches to the study of ecological speciation in order to illustrate these questions and how they relate to one another. To this end, relevant issues have been divided into four different 'aspects'. Each aspect includes multiple elements, each of which can be further studied via the approaches of the subsequent aspects. These are summarized in Table 11.1 and their relationships, in part, diagrammatically illustrated in Fig. 11.1. The primary goal of this chapter is to facilitate the development of robust and integrated research programs on ecological speciation via the description and illustration of these four

**Table 11.1** *Major aspects of the holistic investigation of ecological speciation and the assessment of its general importance*

| Traits of interest | Levels[a] of investigation | Levels of confidence in generality | Levels of taxonomic generality |
|---|---|---|---|
| $T_1{}^b$ = ecological adaptations | $I_1$ = explanation for trait<br>$I_{1a}$ = proximate explanation<br>$I_{1b}$ = ultimate explanation | $C_1$ = individual study $(N = 1)$ | $G_1$ = within lower level taxon |
| $T_2$ = reproductive barriers | $I_2$ = population divergence in trait | $C_2$ = tally of supportive studies | $G_2$ = among lower taxa, within higher level taxon |
| $T_3$ = genetic variation | $I_3$ = association between levels of divergence between traits | $C_3$ = controlled comparative studies<br>$C_{3a}$ = dichotomous approach<br>$C_{3b}$ = regression-based approach | $G_3$ = across higher level taxa |

[a] Higher 'levels' of complexity, confidence and taxonomic generality, respectively, are found as one proceeds down a column.

[b] Alphanumeric codes are those used as shorthand descriptors of the kind of analysis in the figure legends.

*Note*: Underlined words are those used in the summary descriptions at the beginning of each illustration in the text.

aspects of study and associated issues. The present chapter presents one author's views on how this goal might be achieved.

The **first aspect** of ecological speciation study, and the most fundamental, represents the evaluation of each of the three basic biological traits that play a role in this process. Here, I refer to the ecological traits that may reflect adaptive evolution, the details of reproductive biology that underlie potential reproductive barriers, and patterns of molecular genetic variation that can provide insights on such issues as gene flow and selection as these relate to a population's progress towards speciation. The **second aspect** treats the different levels at which questions about these traits can be investigated. The *first level* of the second aspect seeks 'proximate' or mechanistic explanations of how, and 'ultimate' or evolutionary/adaptive explanations of why, a given ecological trait functions and exists in a given population. The *second level* treats population differentiation in these traits, namely, ecologically adaptive population divergence, reproductive isolation, and molecular genetic differentiation (hereafter,

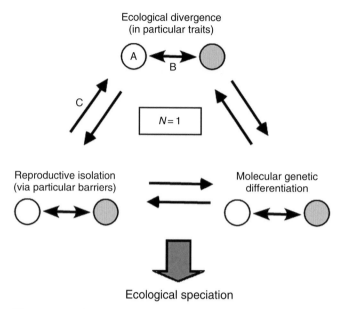

**Figure 11.1** Diagrammatic illustration of traits of interest and levels of investigation relevant to a holistic understanding of ecological speciation in a given study system (see Table 11.1). Each pair of populations that differ in the trait of interest are depicted as a pair of white and gray circles (e.g. 'A'). Differentiation between these populations in that trait is indicated by a two-way arrow (e.g. 'B'). One-way arrows indicate potentially causal associations between different aspects of population divergence (e.g. 'C'). '$N=1$' emphasizes that this diagram refers to potential studies of a single study system. By contrast, evaluating the generality of observations from such studies requires the application of comparative approaches and the evaluation of multiple taxa.

ED, RI, and GD), respectively. The *third level* investigates statistical associations between different elements of population differentiation in an attempt to evaluate possible causal relationships between them.

An essential goal of evolutionary biology is determining the relative evolutionary frequency (i.e. generality) of phenomena of interest. The last two aspects concern the evaluation of two kinds of generality. The **third aspect** concerns the level of statistically justified *confidence* one can have in the evolutionary generality of second aspect findings. The *first level* treats evidence deriving from a single study or system (so that $N=1$), for example, evidence deriving from a model system approach. The *second level* is based on tallies of evidence from multiple accumulating studies/systems in which a phenomenon has been demonstrated and is the basis for many review papers seeking adequate evidence to support a claim. At the *third level*, comparative methods are adopted that employ random taxon selection, multiple independent comparisons of populations/taxa belonging to pre-defined biological classes, and a controlled experimental design that allows for the statistical isolation and quantitative evaluation of the factor of interest against a null expectation. Recent approaches for the comparative, experimental analysis of ecological speciation are described in a separate section below. The final and **fourth aspect** of ecological speciation studies described here addresses the *taxonomic* generality at which the phenomenon is expressed. That is, such studies evaluate the level of the phylogenetic hierarchy and the diversity of disparate clades at which the phenomenon occurs. The different levels of this aspect represent increasing macroevolutionary generality.

## Comparative approaches for the study of ecological speciation

The epistemological power of too-rarely used comparative approaches (Coyne & Orr 2004) is a special focus of this chapter. As described above, the empirical study of ecological speciation was motivated by early verbal models describing the potential consequences of adaptive ecological divergence for the evolution of RI, and evaluating the relationship between these two elements of evolutionary differentiation has remained a primary focus of ecological speciation investigations (Schluter 2001). It was with the goal of rigorously isolating these contributions of ED to RI that the author developed a population-level comparative approach in the early 1990s that in various respects represents a microevolutionary analog to the Mitter *et al.* (1988) sister-group method (Funk 1996, 1998). This 'dichotomous approach' compares levels of RI in pairs of populations of two types. One type includes population pairs that are ecologically similar, for example those using the same habitat. The second type includes populations that are ecologically divergent, for example those using different habitats. Ecological speciation theory predicts that the ecologically divergent pairs should be more reproductively isolated than the ecologically similar pairs, other things being equal. This is because while both types of population pairs may have been subject to divergence due to genetic drift, sexual selection, and habitat-independent sources of selection, only the different-habitat pairs will additionally have been subject to ecologically divergent selection associated with adaptation to alternative habitats. In such comparisons, the same-habitat pairs represent a sort of null hypothesis or ecological control for expected levels of RI in the absence of the factor of interest. Comparing levels of RI between the two types thus allows the contributions of this factor to RI to be isolated and evaluated. The predicted pattern of a replicated study of this sort under the ecological speciation hypothesis is illustrated in Fig. 11.2a, where the ecologically divergent populations exhibit greater RI on average. This figure also illustrates the 'other things being equal' aspect mentioned above. That is, the hypothesis is supported only if the ecologically divergent pairs did not simply diverge longer ago – and thus have greater time to diverge under habitat-independent forces – than the ecologically similar pairs. Thus, time is an

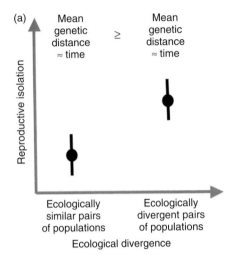

(a)

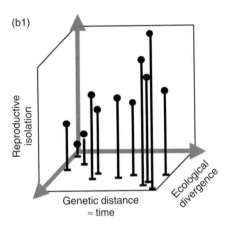

(b1)

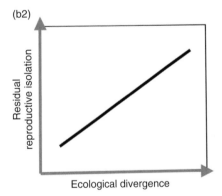

(b2)

**Figure 11.2** Comparative analyses of ecological contributions to speciation. Plots illustrate predicted patterns, for each of two approaches, in support of the hypothesis that divergent selection on ecological traits promotes the evolution of reproductive isolation. (a) The 'dichotomous' approach (based on Funk 1996; 1998). This approach compares reproductive isolation between two types of population pairs: those that have experienced similar ecological conditions and thus a small degree of divergent selection, and those that have experienced divergent ecological conditions and thus a greater degree of divergent selection. The plot also illustrates the implicit underlying assumption that ecologically different populations have not simply had a longer time to diverge. (b) The 'regression-based' approach (Funk *et al.* 2002, 2006). This approach simultaneously evaluates quantitative estimates of genetic distance (a surrogate for time since divergence), ecological divergence, and reproductive isolation. It statistically removes the contributions of time and thereby isolates and quantifies the association between ecological divergence and time-independent residual reproductive isolation. (b1) A hypothetical dataset, illustrating the distribution of paired population or species comparisons (data points) with respect to the three analysed aspects (axes) of evolutionary differentiation. (b2) A hypothetical best-fit line illustrating the predicted positive association under the ecological speciation hypothesis. Figure modified from Funk *et al.* (2006).

important cofactor in interpreting the results of this approach. Assuming a rough molecular clock, neutral genetic distance may be used as a surrogate for relative time of divergence (Coyne & Orr 1989, 2004). Despite the simplicity and utility of this approach, it has as yet been applied to few systems (Funk 1998; Jiggins *et al.* 2001; Nosil *et al.* 2002; Vines & Schluter 2006; papers summarized in Nosil 2007).

A related method is applied in studies of 'parallel speciation' (Schluter & Nagel 1995; Rundle *et al.* 2000). This approach treats replicated pairs of populations of two different ecological types that co-exist in each of multiple localities. Here, the most intriguing result is the existence of RI between types both within and between localities, but not within types between localities. If the ecological types diverged independently in each locality, these findings are most parsimoniously interpreted as evidence that reproductive isolation has evolved in parallel under divergent selection, rather than drift, in each locality.

The above approaches involve the dichotomous categorical characterization of the ecological status of population pairs. However, this discrete characterization can be viewed as a special case of a more general category of comparative approaches that aims to evaluate the association between the *degree* of ED between populations and the *degree* of to which they are reproductively isolated, both of which can also be evaluated as continuous variables. RI is commonly quantified in this manner, and various approaches may be taken to quantify ED on a continuous scale as well. This might be accomplished by experimentally quantifying ED in ecologically relevant variables or by calculating ED indices from ecological data harvested from the literature. These latter indices could be based on differences in continuous variables such as size, or on the degree of overlap across multiple categorical variables such as diet items. In this approach, the use of GD as a surrogate for time is fully incorporated in regression-based analyses. These treat population or species pairs for which data on GD, ED, and RI are all available. The effects of time are statistically removed by calculating residuals of RI on GD and then evaluating the regression of these residuals on ED (Fig. 11.2b). This approach (Funk *et al.* 2002, 2006) thus allows rigorous analyses of the time-independent association of ED with RI per se. The critical results of such analyses are partial correlation coefficients (partial $r$), parameters that provide the direction (positive or negative) of the ED-RI association as well as its strength. If phylogenetic corrections are applied to reduce historical non-independence across species pairs (Felsenstein 1985), the statistical significance of these coefficients can also be evaluated. Funk *et al.* (2006; see discussion below) applied this approach for the first time. Related applications are extremely limited. Two have applied the same method (Bolnick *et al.* 2006; Langerhans *et al.* 2007). A third demonstrates that allozyme variation is more tightly correlated with RI in *Drosophila* than is silent (and thus presumably more selectively neutral) DNA as an argument for the role of selection in speciation

(Fitzpatrick 2002). Work by Tregenza *et al.* (2000a,b) exemplifies additional studies that use history to tease apart alternative causes of RI in a comparative population context.

## Illustrating aspects of the study of ecological speciation: *Neochlamisus* and beyond

In this section, past work is described to illustrate various potential elements of the investigation of ecological speciation. Rather than confronting the reader with all 144 potential combinatorial types of studies that are implied by Table 11.1, a piecemeal approach is taken in an attempt to illustrate a variety of issues using 10 examples. The goal here is not so much to build an argument in support of the general importance of ecological speciation per se, but rather to demonstrate the insights to be gained by using various and sometimes underemployed forms of investigation for the study of this phenomenon. Some presented data are unpublished and space restrictions do not permit detailed descriptions of particular investigations. Thus, these findings should be viewed primarily as serving the goals just described. A statement that categorizes each illustrated study, following Table 11.1, will precede each description. Most of these studies derive from work on *Neochlamisus bebbianae* leaf beetles (Karren 1972).

N. bebbianae is a univoltine, eastern North American species of leaf beetle (Chrysomelidae) that uses particular species from six taxonomically disparate tree genera as host plants, which serve as the site of all life activities across all life stages. These hosts are: red maple (*Acer rubrum*, Aceraceae), certain alders (*Alnus rugosa/incana*, *Alnus serrulata*, Betulaceae), river birch (*Betula nigra*, Betulaceae), American hazel (*Corylus americana*, Corylaceae), certain oaks (*Quercus* spp., Fagaceae) and Bebbs willow (*Salix bebbiana*, Salicaceae) (Karren 1972; Funk 1998, 1999). The specialist tendencies of most *Neochlamisus* species combined with *N. bebbianae*'s use of a modest number of biologically divergent host plants seemed to offer this beetle species as a suitable subject for studies on ecological and associated evolutionary divergence, and prompted its selection as the focus of the author's dissertation research (Funk 1996). Here, I refer to those *N. bebbianae* populations associated in nature with each of these host plants as six different 'host forms'. For example, test animals collected in the field as eggs or larvae on red maple plants are considered to represent the maple host form. In the course of using *N. bebbianae* to demonstrate aspects of the study of ecological speciation, the development of this study system for such investigations will also be illustrated.

1. *Trait = ecological; level = explanation (proximate); confidence = N = 1; generality = lower*: When studying species that use multiple resources from the perspective of ecological speciation, a basic issue is the relative degree to which alternative resources are acceptable by particular populations. This issue

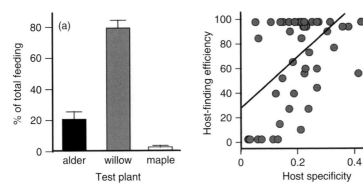

**Figure 11.3** Host plant specificity and associated host-use efficiency in the willow host form of *N. bebbianae*. (a) Mean feeding test results across individuals. A tendency to feed primarily on foliage of the native willow host plant illustrates the ecological trait of host specificity. (b) Positive association between the degree of host specificity expressed by individuals (circles) and their degree of efficiency in locating their native host in behavioural trials. This positive association suggests a potential selective advantage to specificity. Figure modified from Egan and Funk (2006); see this paper for further details. (T1, I1, C1, G1 – refer Table 11.1).

was evaluated for the willow host form by measuring the degree to which naïve individuals would feed on foliage from their native host (i.e. willow) as well as the native hosts of the alder and maple host forms (Egan & Funk 2006). These no-choice tests allowed individuals to feed ad libitum for 24 h on one of these three plants, after which feeding damage was quantified. Results (Fig. 11.3a) demonstrate that this host form does not equally accept all of the hosts cumulatively used by its species. Rather, markedly elevated feeding on willow foliage illustrates an intriguing ecological trait: a considerable degree of ecological/host specialization.

2. *Trait = ecological; level = explanation (ultimate); confidence = N=1; generality = lower*: Understanding why so many herbivorous insects exhibit such host specialization rather than exploiting a broader variety of locally available plant resources has been a major focus of evolutionary ecology, with many hypotheses proffered (Jaenike 1990). We have investigated the recent 'information processing' hypothesis for host specialization (Bernays & Funk 1999; Bernays 2001) in *N. bebbianae*. This hypothesis posits that insect specialists live in a sensorially simpler world than do generalists, who have the task of processing and making decisions about the wider variety of host-associated input provided by their greater number of local hosts. This hypothesis predicts that specialists should be more behaviourally efficient than generalists due to the lower 'distraction' that results from their selective attention to fewer host cues. We tested this hypothesis by assaying both host specificity and host-finding efficiency in individual willow host form beetles

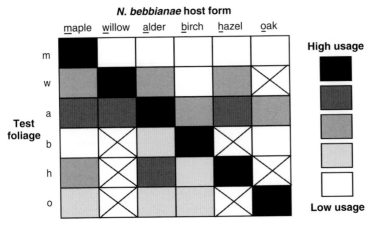

**Figure 11.4** Relative acceptance by each *N. bebbianae* host form of foliage from the native host of each host form. In relation to issues treated by this chapter, three points not yet discussed derive from a single observation: the diagonal of black boxes in this matrix. First, this indicates six cases where a given taxon exhibits a degree of apparent ecological adaptation/specificity with respect to its native host plant (T1, I1, C2, G1 – refer Table 11.1). Second, this necessarily indicates that any given pair of host forms exhibits ecological divergence in host acceptance (T1, I2, C1, G1 – refer Table 11.1). Third, it further indicates that all 21 pairwise combinations of host forms exhibit ecological divergence (T1, I2, C2, G1 – refer Table 11.1). A box with an 'X' indicates inadequate data for assessment. Results derive from Funk (1998) and Egan and Funk (2006), but primarily from unpublished data.

(Egan & Funk 2006). Specificity was estimated as the variance in feeding scores across three no-choice trials, one on each test plant (Fig. 11.3a), with higher variances indicating greater specificity (greater tendency to feed on a single plant). Host efficiency equalled the rapidity with which beetles located native host foliage in choice tests that simultaneously exposed them to all three plants. The resulting significantly positive correlation between these variables (Fig. 11.3b) supported this hypothesis. This greater efficiency may be expected to increase rates of host plant exploitation and insect fecundity, reduce time spent exposed to predators, and thus increase the fitness of specialists relative to generalists. Thus, a behaviourally based selective advantage to specialization may help account for the frequent evolution of insect host-specificity.

3.   *Trait = ecological; level = divergence; confidence = N = 1; generality = lower:* Fig. 11.4 depicts the relative degree to which each of the host forms of *N. bebbianae* can use foliage from each of the native host plants of these host forms. These patterns are based on various assays of the willingness to use and capacity to develop on alternative host foliage (Funk 1998; Egan & Funk 2006; Funk unpublished data). Consider the data for the maple and willow host forms.

**Table 11.2** *Premating isolation between* N. bebbianae *host forms*[a]

| Host forms tested | Sexual isolation $= I_{PSI}$[b] | Habitat isolation[c] |
|---|---|---|
| alder & hazel | $0.07 \pm 0.09$[NS] | 68.8 |
| alder & maple | $0.26 \pm 0.11$* | 90.7 |
| alder & willow | $-0.01 \pm 0.09$[NS] | 62.5 |
| hazel & maple | $0.50 \pm 0.08$*** | 87.4 |
| hazel & willow | $0.37 \pm 0.07$*** | 67.3 |
| maple & willow | $0.21 \pm 0.08$** | 76.3 |

[a] Data from unpublished experiment (Egan & Funk, in preparation).
[b] Index of sexual isolation and statistical tests from the JMATING program of Carvajal-Rodriguez and Rolan-Alvarez (2006).
[c] Mean host fidelity of tested host forms (50 = no host fidelity; 100 = complete host fidelity = no resting on alternative host; see text for details).
*Note*: These data are of type T2, I3, C2, G1. NS = not significant, * = $P < 0.05$, ** $P < 0.01$, *** $P < 0.001$.

Both most readily use foliage from their native host plant, using other plants to a lesser degree. This observation demonstrates that *N. bebbianae* is not an ecologically homogeneous entity, as is often assumed of insect species that employ multiple hosts (Fox & Morrow 1981). Since each host form specializes on a different plant, they exhibit ED in host-associated traits, thus demonstrating a prerequisite of ecological speciation.

4. *Trait = ecological; investigation = divergence; confidence = tally; generality = lower*: An examination of the complete plot in Fig. 11.4 demonstrates that each of the six host forms exhibits highest usage, and thus a degree of specialization, on its native host. Further, this same pattern necessarily indicates that every possible pair of host forms exhibits ED. Thus, these patterns appear to be general ecological characteristics of this species.

5. *Trait = ecological/reproductive; investigation = association; confidence = tally; generality = lower*: Given an observation of ED, a student of ecological speciation might naturally ask whether it is associated with RI. Table 11.2 presents results from a study of two forms of premating RI between all possible pairs of four *N. bebbianae* host forms (Funk & Egan, in preparation). First, sexual isolation was evaluated using no-choice mating trials that paired one male with one female in individual Petri dishes for 2 h to determine if they would mate. For any given pair of host forms, all four possible combinations of host form and sex were evaluated in this manner. Positive assortative mating was observed when the proportion of pairs that mated was higher for homotypic pairings than for heterotypic ones, illustrating sexual isolation. The degree of assortativeness (0 = random mating, 1 = only

homotypic matings) and the statistical significance of sexual isolation are presented for each cross. In five of six comparisons, positive assortative mating was observed and sexual isolation was statistically significant in four of these.

Second, habitat isolation was evaluated using trials in which female beetles were individually presented with foliage of their native host plant plus the native host of another host form in a Petri dish. The amount of time a beetle spent on each test plant was scored over a 2 h period. For each pair of host forms, the mean proportion of time spent by the two host forms on their native hosts was calculated (e.g. for the maple and willow host form comparison, the proportion of time spent by the maple host form on maple and the proportion of time spent by the willow host form on willow were averaged). In all cases, this value was >50% (Table 11.2), indicating that host forms consistently spend a minority of their time on alternative plants, thus exhibiting 'host fidelity'. Since *N. bebbianae* mate on their host plants, this host fidelity could reduce mating encounters between host forms, and provides indirect evidence for habitat isolation. Moreover, direct evidence for habitat isolation was provided by the observation that sexual isolation was significantly greater in our mating trials when foliage was present in the dishes than when it was absent. In combination, these results indicate that ED is generally associated with premating RI among *N. bebbianae* host forms.

6.  *Trait = ecological/reproductive; investigation = association; confidence = tally; generality = higher*: As interest in ecological speciation has increased, so has the number of study systems in which associations between ED and RI have been investigated and demonstrated. Table 11.3 lists some of these systems, the accumulation of which increases confidence in the suspicion that this is neither an extremely rare nor a taxonomically restricted phenomenon.

7.  *Trait = ecological/reproductive; investigation = association; confidence = comparative (dichotomous); generality = lower*: Figure 11.5 presents results from the initial application of the dichotomous comparative approach (Fig. 11.2a) for the isolation of ecology's role in speciation (Funk 1996, 1998). Each subplot illustrates a separate year of analyses and depicts the relative geographic distribution of study populations and the native host plants of each. Each year, one pair of ecologically similar ('same-host') populations and two pairs of ecologically divergent ('different-host') populations were evaluated. For each comparison, habitat isolation was evaluated using host fidelity assays, immigrant inviability (cf. Nosil *et al.* 2005) was evaluated based on the relative capacity of each population to survive on the other's host, and sexual isolation was quantified as described above. Note that in all cases, different-host population pairs exhibited greater RI than same-host pairs, consistent with the hypothesis that ecological divergence is pleiotropically

**Table 11.3** *Taxa cited in review papers[a] as possible examples of ecological speciation[b]*

| Taxon | Common name | Sample reference |
|---|---|---|
| Insects | | |
|   *Melanoplus* | grasshopper | Orr 1996 |
|   *Timema cristinae* | stick insect | Nosil 2007 |
|   *Acyrthosiphon pisum* | pea aphid | Via *et al.* 2000 |
|   *Nilaparvata lugens* | brown planthopper | Butlin 1996 |
|   *Enchenopa binotata* | two-marked treehopper | Wood & Keese 1990 |
|   *Epilachna* | (herbivorous) ladybird | Katakura *et al.* 1989 |
|   *Neochlamsisus bebbianae* | leaf beetle | Funk 1998 |
|   *Galerucella nymphae* | water lilly leaf beetle | Pappers *et al.* 2002 |
|   *Rhagoletis pomonella* | apple maggot fly | Feder *et al.* 1994 |
|   *Eurosta solidaginis* | goldenrod gall fly | Craig *et al.* 1997 |
|   *Drosophila mojavensis* | (cactophilic) pomace fly | Etges 1998 |
|   *Liriomyza brassicae* | leafmining fly | Tavormina 1982 |
|   *Spodoptera frugiperda* | fall armyworm | Pashley *et al.* 1992 |
|   *Zeiraphera diniana* | larch budmoth | Emelianov *et al.* 2002 |
|   *Heliconius erato* | butterfly | Mallet & Barton 1989 |
| Other Invertebrates | | |
|   *Hyalella azteca* | amphipod | McPeek & Wellborn 1998 |
|   *Littorina* | snail | Rolan-Alvarez *et al.* 1997 |
| Vertebrates | | |
|   *Gasterosteus* | three-spine stickleback | Nagel & Schluter 1998 |
|   *Coreogonus* | lake whitefish | Lu & Bernatchez 1999 |
|   *Bombina* | firebelly toads | Kruuk & Gilchrist 1997 |
|   *Geospiza* | Darwin's ground finches | Ratcliffe and Grant 1983 |
| Plants | | |
|   *Mimulus* | monkeyflowers | Schemske & Bradshaw 1999 |
|   *Banksia* | flowering plant | Lamont *et al.* 2003 |
|   *Artemisia tridentata* | Great Basin sagebrush | Wang *et al.* 1997 |

[a] Orr & Smith 1998; Schluter 2001; Funk *et al.* 2002; Rundle & Nosil 2005.
[b] Note that this table is not meant to present a comprehensive list of relevant study systems. Similar lists have been compiled for putatively sympatrically speciating taxa, various of which would also qualify as undergoing ecological speciation (Via 2001; Berlocher & Feder 2002; Dres & Mallet 2002; Bush & Butlin 2004).
*Note*: These data are of type T1&2, I3, C2, G3.

promoting RI and ecological speciation (see Jiggins *et al.* 2005 for a related study). Recent evidence on GD from mtDNA and AFLPs (Egan *et al.* revision in review; Funk *et al.*, unpublished data) suggest the close relationship of all host forms, further supporting these interpretations.

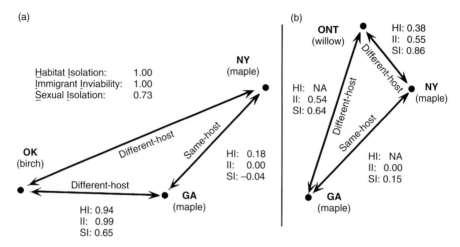

**Figure 11.5** Estimates of reproductive isolation between 'same-host' versus between 'different-host' (i.e. ecologically similar versus ecologically divergent) population pairs of *N. bebbianae* host forms. Data are presented for three reproductive barriers and two sets of geographic populations. (a) and (b) represent separate suites of experiments, each performed in a separate year. Note the consistency with which different-host populations exhibit greater reproductive isolation than same-host populations. See text for additional details. Figure modified from Funk (1998) (T1&2, I3, C3, G1 – refer Table 11.1).

8.  *Trait = ecological/genetic; investigation = association; confidence = comparative (dichotomous); generality = lower*: Recently, we have applied the dichotomous comparative approach to evaluate the degree to which host-related divergent selection is responsible for genomic differentiation between the maple and willow host forms (Egan *et al.*, 2008). In this investigation of the population genomics of ecological speciation, an AFLP-based genome scan was used to quantify genetic differentiation for hundreds of anonymous loci for 15 population comparisons (Fig. 11.6). Using simulations that determined which locus-specific $F_{ST}$ values exceeded neutral expectations, we identified putatively divergently selected 'outlier' loci for each population comparison using the program Dfdist (Beaumont & Nichols 1996; Beaumont & Balding 2004). 'Different-host-specific' outliers – those observed only in different-host population comparisons, never in same-host comparisons – were inferred to reflect host-related divergent selection. 'Same-host-specific' outliers showed the opposite pattern, being observed only in same-host comparisons, never in different-host comparisons, and were correspondingly attributed to host-independent divergent selection. While an average of 2.1% of loci per population comparison reflected host-related selection, only 0.5% reflected host-independent selection. Thus (only) by our use of this comparative approach did we obtain evidence suggesting

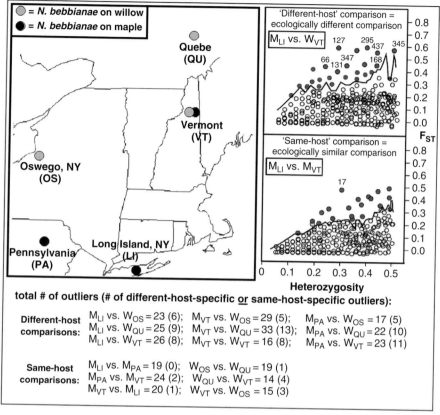

total # of outliers (# of different-host-specific or same-host-specific outliers):

**Different-host comparisons:**
$M_{LI}$ vs. $W_{OS}$ = 23 (6);   $M_{VT}$ vs. $W_{OS}$ = 29 (5);   $M_{PA}$ vs. $W_{OS}$ = 17 (5);
$M_{LI}$ vs. $W_{QU}$ = 25 (9);   $M_{VT}$ vs. $W_{QU}$ = 33 (13);   $M_{PA}$ vs. $W_{QU}$ = 22 (10);
$M_{LI}$ vs. $W_{VT}$ = 26 (8);   $M_{VT}$ vs. $W_{VT}$ = 16 (8);   $M_{PA}$ vs. $W_{VT}$ = 23 (11)

**Same-host comparisons:**
$M_{LI}$ vs. $M_{PA}$ = 19 (0);   $W_{OS}$ vs. $W_{QU}$ = 19 (1)
$M_{PA}$ vs. $M_{VT}$ = 24 (2);   $W_{QU}$ vs. $W_{VT}$ = 14 (4)
$M_{VT}$ vs. $M_{LI}$ = 20 (1);   $W_{VT}$ vs. $W_{OS}$ = 15 (3)

**Figure 11.6** AFLP-based genome scan of the maple and willow host forms of *N. bebbianae*. The map depicts the distribution of the six study populations. The two plots illustrate the empirical distribution of $F_{ST}$ values for individual loci (dots) compared to a simulated 95th quantile (line) that identifies which loci are too strongly differentiated to be accounted for by neutral evolution. These 'outlier' loci (gray dots) with $F_{ST}$ values exceeding this line are instead presumed to be evolving under divergent selection. Simulations used to identify outliers were performed in Dfdist (Beaumont & Nichols 1996; Beaumont & Balding 2004). Some pairwise population comparisons treat populations from different host forms (='different-host comparisons') while others treat those of the same host forms (='same-host comparisons'). Data from one of each type is illustrated. Numbers identify loci that were observed to be outliers in multiple population comparisons, but only in different-host comparisons (top plot) or only in same-host comparisons (bottom plot). The distribution of outliers for each of the 15 population comparisons is summarized at the bottom of the figure. The greater prevalence of outliers in different-host comparisons is consistent with the hypothesis that host-related adaptation is primarily responsible for adaptive genomic divergence in these beetles. See text for further details. Figure modified from Egan *et al.* (revision in review) (T2&3, I3, C3, G1 – refer Table 11.1).

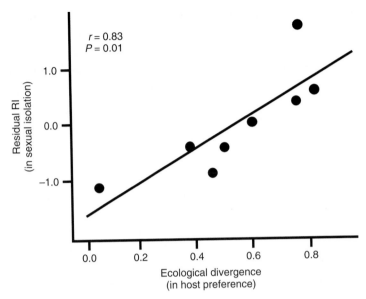

**Figure 11.7** Application of the regression-based comparative approach to data from paired comparisons of *N. bebbianae* host forms. The plot suggests a strong relationship between levels of host-associated ecological divergence and sexual isolation. Figure modified from Funk & Nosil (2008) (T1&2, I3, C3, G1 – refer Table 11.1).

that host-adaptation may play the principal role in the adaptive genomic differentiation of these host forms.

9.  *Trait = ecological/reproductive; investigation = association; confidence = comparative (regression); generality = lower*: Applying the regression-based comparative approach to data from eight host-form comparisons provides another means of evaluating the contributions of ED to the evolution of RI in *N. bebbianae* (Funk & Nosil 2008) (Fig. 11.7). In this analysis, RI is represented by sexual isolation, as quantified by Levene's isolation index after transforming it to a scale of zero (random mating) to one (completely positively assortative mating) (Ehrman & Petit 1968; Spieth & Ringo 1983). ED was estimated from the degree of divergence in host-plant preference data (see Funk & Nosil for details). GD was calculated as the mean pairwise divergence between host forms in mitochondrial DNA sequences of the cytochrome oxidase I gene (Funk 1999, unpublished data). After statistically removing contributions from GD, residual RI showed a positive, tight, and significant association with ED. As with results from all the *N. bebbianae* studies discussed above, these findings are consistent with the hypothesis that host-associated ED is a strong predictor of the evolution of RI in this species. This analysis provides perhaps the strongest evidence that these host forms represent various stages in the continuous process of ecological speciation.

10.  *Trait = ecological/reproductive; investigation = association; confidence = comparative (regression); generality = higher:* Determining just how evolutionarily important ecological speciation is requires the evaluation of the *relative frequency* with which ED drives speciation across disparate taxa. That is, it requires an estimate of the proportion of taxa in which speciation is (and the proportion in which it is not) at least partially ecologically promoted. Such issues can only be evaluated by applying controlled comparative approaches to a random selection of taxa. An opportunity to undertake such a study was provided by a recent accumulation of eight datasets of the kind first compiled by Coyne and Orr (1989) to investigate the time course of speciation in *Drosophila* (see Coyne & Orr 2004). Each of these datasets includes estimates of RI and GD for many pairwise population or species comparisons within a given taxon. Cumulatively, they represent a phylogenetically disparate array of taxa, including plants, invertebrates and vertebrates. In the first application of the regression-based comparative analysis (Funk *et al.* 2006), estimates of ED were derived from ecological data in the literature for the species treated in these data sets and added to the pre-existing estimates of RI and GD. Separate regression analyses were then conducted for each published component of RI and each calculated index of ED (based on habitat, diet, or size) for each of the eight datasets. For 7/8 datasets the mean ED-RI association across these analyses proved to be positive ($P = 0.035$, binomial test), with the sole exception being only slightly negative ($-0.02$) (Table 11.4). Furthermore a one-tailed one-sample $t$-test demonstrated that the mean of these values across datasets was significantly greater than zero ($P < 0.01$), providing strong support for the taxonomic generality of ecology's role in speciation.

In sum, the ten examples above suggest that host-associated ecological specialization in *N. bebbianae* host forms is adaptive, ubiquitous, and a general source of ED between host forms. This ED, in turn, is generally associated with RI between host forms and plays a major role in adaptive genomic differentiation. These findings put forward *N. bebbianae* as an informative study system for investigating ecological speciation, while cross-taxonomic comparative analyses suggest that the relative evolutionary frequency with which ED contributes to the evolution of RI may be high.

## Critical caveats

A miscellany of important points are worth noting in considering the observations made so far in this chapter, some of which are briefly described here. First, although the chapter has focused on evidence that supports the plausibility of the ecological speciation hypothesis, evidence contrary to its importance could certainly be found. This could, for example, take the form of the absence of

**Table 11.4** *Individual regression-based analyses of the strength of association (r) between ecological divergence and reproductive isolation[a]*

| Taxon | Ecological trait | Component of reproductive isolation | $r^b$ | Mean $r$ |
|---|---|---|---|---|
| Angiosperms | habitat | prezygotic | +0.06 | |
| Angiosperms | habitat | postzygotic | +0.43 | +0.21 |
| Angiosperms | habitat | total | +0.15 | |
| Lepidoptera | diet/habitat | postzygotic | +0.17 | +0.02 |
| Lepidoptera | diet/habitat | total | −0.14 | |
| *Drosophila* | diet/habitat | prezygotic | +0.18 | |
| *Drosophila* | diet/habitat | postzygotic | −0.27 | +0.07 |
| *Drosophila* | diet/habitat | total | +0.31 | |
| Fishes | habitat | postzygotic | +0.12 | |
| Fishes | diet | postzygotic | −0.16 | −0.02 |
| Fishes | size | postzygotic | −0.01 | |
| Darters | habitat | prezygotic | −0.03 | |
| Darters | habitat | postzygotic | −0.34 | |
| Darters | habitat | total | +0.27 | +0.11 |
| Darters | size | prezygotic | +0.53 | |
| Darters | size | postzygotic | −0.13 | |
| Darters | size | total | +0.35 | |
| Frogs | habitat | postzygotic | +0.21 | +0.21 |
| Birds | habitat | postzygotic | +0.10 | |
| Birds | diet | postzygotic | +0.13 | +0.12 |
| Birds | size | postzygotic | +0.14 | |
| Doves | habitat | postzygotic | +0.16 | |
| Doves | diet | postzygotic | +0.39 | +0.21 |
| Doves | size | postzygotic | +0.09 | |

[a] Table is modified from Funk *et al.* (2006).
[b] r-values indicate the direction and strength of the association between ecological divergence and reproductive isolation after time (genetic distance) has been statistically removed.
*Note*: These data are of type T1&2, I3, C3, G3.

positive associations between ED and RI in the sorts of comparative studies emphasized here. Moreover, this chapter does not claim that ecology is the only factor promoting speciation, or necessarily the most important one. Sexual selection, genetic drift, etc. also undoubtedly play important roles and future comparative studies should try to tease apart the relative contributions of these multiple factors.

Second, recent comparative methods have been emphasized here due to their novelty and capacity to rigorously address particular questions that other approaches cannot. However, it should be further emphasized that all levels of analysis make their own critical contributions. For example, detailed mechanistic studies of model systems (i.e. level = explanation (proximate); confidence = $N = 1$) are absolutely necessary to understand whether and how particular biological phenomena can occur and function. What is important is that findings are appropriately interpreted and generalized according to the level of study and the nature of evidence evaluated.

Third, a related topic is interpreting results as a function of the level at which study system selection is random. Scientists often and understandably select their study systems non-randomly, choosing them because they offer suitable opportunities for learning about a particular phenomenon. For example, the author first selected *N. bebbianae* as a study system specifically because its disparate host associations made it a plausible place to look for possible ED and associated RI. For this reason, while interpretations made here about generalities specific to *N. bebbianae* are legitimate, its non-random selection as a system to study ecological speciation compromises its potential contributions to our understanding of the relative frequency of this phenomenon in nature. By contrast, the eight taxa analysed in the regression-based comparative analysis by Funk *et al.* (2006) were randomly selected with respect to the factor of interest. This is because they had been compiled by different authors (Coyne & Orr 2004) for different and ecologically independent reasons (i.e. as a list of taxa for which the time course of speciation had already been studied).

Fourth, it is nonetheless important to acknowledge the limitations of any kind of approach. In the case of all comparative approaches that make use of genetic distance as a surrogate for time (Coyne and Orr 1989, Funk *et al.* 2006, Mallet *et al.* 2007), inferences are only as good as the molecular clock is reliable. For example, in the case of the Funk *et al.* (2006) study, a highly inconstant and noisy molecular clock could, in principle, negate the method's ability to factor out time using residuals, with the result that the positive ED–RI association could partly reflect the simple increase in both of these factors with time. However, datasets estimating GD from allozymes, mtDNA sequences, nuclear sequences, and DNA–DNA hybridization alike yielded positive associations in this study; hopefully there is some signal of time in at least some of these.

Fifth, for each of the three pairwise associations pointed out here (Fig. 11.1), the causal arrow may point in either direction, hence the depiction of six separate arrows in the figure. Thus, a positive EI–RI association could reflect a RI→ED rather than the ED→RI pattern of causality predicted by ecological speciation theory. This alternative pattern could, for example, arise due to the combined contributions of competitive exclusion and gene flow. On the one hand, the competitive exclusion principle predicts that strongly RI populations that are ecologically similar cannot stably co-exist while ED ones can. This will yield a tendency for strongly RI populations to exhibit considerable ED. On the other hand, weakly RI populations should experience gene flow that homogenizes their ecological traits. This will yield a tendency for weakly RI populations to exhibit little ED. In sum, this scenario would yield a positive ED–RI association that is caused by the effects of varying levels of RI on the capacity for populations exhibiting particular levels of RI to persist. In this instance, however, there is a simple way to determine the true direction of the causal arrow. Since neither competition nor gene flow can exist between allopatric populations, the ED–RI association should only exist for sympatric populations if RI→ED is due to competitive exclusion. However, preliminary analyses of two datasets for which geographic data are available show contrarily that the association holds up in allopatric scenarios as well, lending support to the original interpretation. A full analysis of this issue is underway.

Sixth, a full understanding of the evolutionary dynamics of ecological speciation requires an analysis of all six arrows, i.e. all possible associations among ED, RI, and GD. Such studies would, for example, additionally include analyses of ED←→GD to evaluate how the balance between the strength of selection and the amount of gene flow determines population divergence (Hendry & Taylor 2004; Nosil & Crespi 2004). Study systems in which all these associations have been investigated are wanting.

## Conclusions and future directions

Based on studies by other workers, the *Neochlamisus* work described here, and the results of the regression-based cross-taxon analysis (Funk *et al.* 2006), the ecological speciation hypothesis appears to describe a quite plausible and perhaps quite macroevolutionarily general contributor to speciation. The time thus seems ripe for focused attention on how we might maximize the effectiveness with which we further investigate the details and importance of this process. Hopefully, this chapter's review of the various elements of the study of ecological speciation has provided some food for thought in this regard and will help motivate the development of purposefully integrated and informative research programs to this end. Specific suggestions include: (a) broadening the taxonomic variety of study systems investigated; (b) expanding the comprehensiveness with which individual systems are studied; (c) evaluating the complex feedback loops among the elements of evolutionary divergence that drive the dynamics of this process; and (d) designing research programs and accumulating datasets with an eye towards rigorous and informative comparative analyses.

## References

Bates, H. W. (1862) Contributions to an insect fauna of the Amazon Valley. Lepidoptera: Heliconidae. *Transactions of the Entomological Society* **23**, 3, 495–566.

Beaumont, M. A. and Balding, D. J. (2004) Identifying adaptive genetic divergence among populations from genome scans. *Molecular Ecology* **13**, 969–980.

Beaumont, M. A. and Nichols, R. A. (1996) Evaluating loci for use in the genetic analysis of population structure. *Proceedings of the Royal Society of London, Series B* **263**, 1619–1626.

Berlocher, S. H. and Feder, J. L. (2002) Sympatric speciation in phytophagous insects: moving beyond controversy? *Annual Review of Entomology* **47**, 773–815.

Bernays, E. A. (2001) Neural limitations in phytophagous insects: implications for diet breadth and evolution of host affiliations. *Annual Review of Entomology* **46**, 703–727.

Bernays, E. A. and Funk, D. J. (1999) Specialists make faster decisions than generalists: experiments with aphids. *Proceedings of the Royal Society of London B* **266**, 151–156.

Bolnick, D. I., Near, T. J. and Wainwright, P. C. (2006) Body size divergence accelerates reproductive isolation during speciation. *Evolutionary Ecology Research* **8**, 903–913.

Bush, G. L. (1969) Sympatric host race formation and speciation in frugivorous flies of the genus *Rhagoletis* (Diptera, Tephritidae). *Evolution* **23**, 237–251.

Bush, G. L. and Butlin, R. K. (2004) Sympatric speciation in insects: an overview. In: *Adaptive Speciation* (ed. U. Dieckmann, M. Doebeli, J. A. J. Metz and D. Tautz), pp. 229–248. Cambridge University Press, Cambridge.

Butlin, R. K. (1996) Co-ordination of the sexual signaling system and the genetic basis of differentiation between populations in the brown planthopper, *Nilaparvata lugens*. *Heredity* **77**, 369–377.

Carvajal-Rodriguez, A. and Rolan-Alvarez, E. (2006) JMATING: a software for detailed analysis of sexual selection and sexual isolation effects from mating frequency data. *BMC Evolutionary Biology* **6**, 40.

Coyne, J. A. and Orr, H. A. (1989) Patterns of speciation in *Drosophila*. *Evolution* **43**, 362–381.

Coyne, J. A. and Orr, H. A. (2004) *Speciation*. Sinauer Associates, Sunderland, MA.

Craig, T. P., Horner, J. D. and Itami, J. K. (1997) Hybridization studies on the host races of *Eurosta solidaginis*: implications for sympatric speciation. *Evolution* **51**, 1552–1560.

Darwin, C. (1859) *On the Origin of Species by Means of Natural Selection or the Preservation of Favored Races in the Struggle for Life*. J. Murray, London.

Dieckmann, U., Metz, J. A. J., Doebeli, M. and Tautz, D. (2004) Introduction. In: *Adaptive Speciation* (ed. U. Dieckmann, M. Doebeli, J. A. J. Metz and D. Tautz), pp. 1–17. Cambridge University Press, Cambridge.

Dobzhansky, T. (1937) Genetic nature of species differences. *The American Naturalist* **71**: 404–420.

Dobzhansky, T. (1951) *Genetics and the Origin of Species*, 3rd ed. Columbia University Press, New York.

Dodd, D. M. B. (1989) Reproductive isolation as a consequence of adaptive divergence in *Drosophila pseudoobscura*. *Evolution* **43**, 1308–1311.

Dres, M. and Mallet, J. (2002) Host races in plant-feeding insects and their importance in sympatric speciation. *Philosophical Transactions of the Royal Society of London B* **357**, 471–492.

Egan, S. P. and Funk, D. J. (2006) Individual advantages to ecological specialization: insights on cognitive constraints from three conspecific taxa. *Proceedings of the Royal Society of London, Series B* **273**, 843–848.

Egan, S. P., Nosil, P. and Funk, D. J. (2008) Selection and genomic differentiation during ecological speciation: isolating the contributions of host association via a comparative genome scan of *Neochlamisus bebbianae* leaf beetles. *Evolution* **62**, 1162–1181.

Ehrlich, P. R. and Raven, P. H. (1964) Butterflies and plants: a study in coevolution. *Evolution* **18**, 586–608.

Ehrman, L. and Petit, C. (1968) Genotype frequency and mating success in the *willistoni* species group of *Drosophila*. *Evolution* **22**, 649–658.

Emelianov, I., Dres, M., Baltensweiler, W. and Mallet, J. (2002) Host-induced assortative mating in host races of the larch budmoth. *Evolution* **55**, 2002–2010.

Etges, W. J. (1998) Premating isolation is determined by larval rearing substrates in cactophilic *Drosophila mojavensis*: IV. Correlated responses in behavioral isolation to artificial selection on a life-history trait. *The American Naturalist* **152**, 1129–1144.

Feder, J. L., Opp, S. B., Wlazlo, B., *et al.* (1994) Host fidelity is an effective premating barrier between sympatric races of the apple maggot fly. *Proceedings of the National Academy of Science of the United States of America* **91**, 7990–7994.

Feder, J. L. (1998) The apple maggot fly, *Rhagoletis pomonella*: flies in the face of conventional wisdom about speciation? In: *Endless Forms: Species and Speciation* (ed. D. J. Howard and S. H. Berlocher), pp. 130–144. Oxford University Press, Oxford.

Felsenstein, J. (1985) Phylogenies and the comparative method. *The American Naturalist* **125**, 1–15.

Fitzpatrick, B. M. (2002) Molecular correlates of reproductive isolation. *Evolution* **56**, 191–198.

Fox, L. R. and Morrow, P. A. (1981) Specialization: species property or local phenomenon? *Science* **211**, 887–893.

Funk, D. J. (1996) The evolution of reproductive isolation in *Neochlamisus* leaf beetles: a role for selection. Ph.D. Dissertation, State University of New York, Stony Brook, NY.

Funk, D. J. (1998) Isolating a role for natural selection in speciation: host adaptation and sexual isolation in *Neochlamisus bebbianae* leaf beetles. *Evolution* **52**, 1744–1759.

Funk, D. J. (1999) Molecular systematics of COI and 16S in *Neochlamisus* leaf beetles and the importance of sampling. *Molecular Biology and Evolution* **16**, 67–82.

Funk, D. J., Filchak, K. E. and Feder, J. L. (2002) Herbivorous insects: model systems for the comparative study of speciation ecology. *Genetica* **116**, 251–267.

Funk, D. J. and Nosil, P. (2008) Comparative analyses and the study of ecological speciation in herbivorous insects. In: *Specialization, Speciation, and Radiation: The Evolutionary Biology of Herbivorous Insects* (ed. K. Tilmon). University of California Press, Berkeley, CA.

Funk, D. J., Nosil, P. and Etges, W. J. (2006). Ecological divergence exhibits consistently positive associations with reproductive isolation across disparate taxa. *Proceedings of the National Academy of Science of the United States of America* **103**, 3209–3213.

Futuyma, D. J. and Peterson, S. C. (1985) Genetic variation in the use of resources by insect. *Annual Review of Entomology* **30**, 217–238.

Hendry, A. P. and Taylor, E. B. (2004) How much of the variation in adaptive divergence can be explained by gene flow? An evaluation using lake-stream stickleback pairs. *Evolution* **58**, 2319–2331.

Jaenike, J. (1990) Host specialization in phytophagous insects. *Annual Review of Ecology and Systematics* **21**, 243–273.

Jiggins, C. D., Naisbit, R. E., Coe, R. L. and Mallet, J. (2001) Reproductive isolation caused by colour pattern mimicry. *Nature* **411**, 302–305.

Jiggins, C. D., Emelianov, I. and Mallet, J. (2005) Pleiotropy promotes speciation: examples from phytophagous moths and mimetic butterflies. In: *Insect Evolutionary Ecology* (ed. M. Fellowes, G. Holloway and J. Rolff), pp. 451–473. Royal Entomological Society, London.

Karren, J. B. (1972) A revision of the subfamily Chlamisinae north of Mexico (Coleoptera: Chrysomelidae). *University of Kansas Science Bulletin* **49**, 875–988.

Katakura, H., Shioi, M. and Kira, Y. (1989) Reproductive isolation by host specificity in a pair of phytophagous ladybird beetles. *Evolution* **43**, 1045–1053.

Kilias, G., Alahiotis, S. N. and Pelecanos, M. (1980) A multifactorial genetic investigation of speciation theory using *Drosophila melanogaster*. *Evolution* **34**, 730–737.

Kruuk, L. E. B. and Gilchrist, J. S. (1997) Mechanisms maintaining species differentiation: predator mediated selection in a *Bombina* hybrid zone. *Proceedings of the Royal Society of London B* **264**, 105–110.

Lamont, B. B., He, T., Enright, N. J., Krauss, S. L. and Miller, B. P. (2003) Anthropogenic disturbance promotes hybridization between *Banksia* species by altering their biology. *Journal of Evolutionary Biology* **16**, 551–557.

Langerhans, R. B., Gifford, M. E. and Joseph, E. O. (2007) Ecological speciation in *Gambusia* fishes. *Evolution* **61**, 2056–2074.

Lu, G. and Bernatchez, L. (1999) Correlated trophic specialization and genetic divergence in sympatric lake whitefish ecotypes (*Coreogonus clupeaformis*): support for the ecological speciation hypothesis. *Evolution* **53**, 1491–1505.

Mallet, J. and Barton, N. H. (1989) Strong natural selection in a warning-color hybrid zone. *Evolution* **43**, 421–431.

Mallet, J., Beltran, M., Neukirchen, W. and Linares, M. (2007) Natural hybridization in heliconiine butterflies: the species boundary as a continuum. *BMC Evolutionary Biology* **7**, 28.

Mayr, E. (1942) *Systematics and the Origin of Species*. Columbia University Press, New York.

Mayr, E. (1947) Ecological factors in speciation. *Evolution* **1**, 263–88.

McPeek, M. A. and Wellborn, G. A. (1998) Genetic variation and reproductive isolation among phenotypically divergent amphipod populations. *Limnology and Oceanography* **43**, 1162–1169.

Mitter, C., Farrell, B. and Wiegmann, B. (1988) The phylogenetic study of adaptive zones: has phytophagy promoted insect diversification? *The American Naturalist* **132**, 107–128.

Muller, H. J. (1942) Isolating mechanisms, evolution and temperature. *Biological Symposium* **6**, 71–125.

Nagel, L. and Schluter, D. (1998) Body size, natural selection, and speciation in sticklebacks. *Evolution* **52**, 209–218.

Nosil, P. (2007) Divergent host plant adaptation and reproductive isolation between ecotypes of *Timema cristinae* walking sticks. *The American Naturalist* **169**, 151–162.

Nosil, P. and Crespi, B. J. (2004) Does gene flow constrain adaptive divergence or vice versa? A test using ecomorphology and sexual isolation in *Timema cristinae* walking sticks. *Evolution* **58**, 102–112.

Nosil, P., Crespi, B. J. and Sandoval, C. P. (2002) Host-plant adaptation drives the parallel evolution of reproductive isolation. *Nature* **417**, 441–443.

Nosil, P., Vines, T. H. and Funk, D. J. (2005) Perspective: reproductive isolation caused by natural selection against immigrants from divergent habitats. *Evolution* **59**, 705–719.

Orr, M. R. (1996) Life-history adaptation and reproductive isolation in a grasshopper hybrid zone. *Evolution* **50**, 704–716.

Orr, M. R. and Smith, T. B. (1998) Ecology and speciation. *Trends in Ecology & Evolution* **13**, 502–06.

Pappers, S. M., Van der Velde, G., Ouborg, N. J. and Van Groenendael, J. M. (2002) Genetically based polymorphisms in morphology and life history associated with putative host races of the water lily leaf beetle, *Galerucella nymphae*. *Evolution* **56**, 1610–1621.

Pashley, D. P., Hammond, A. M. and Hardy, T. N. (1992) Reproductive isolating mechanisms in fall armyworm host strains (Lepidoptera, Noctuidae). *Annals of the Entomological Society of America* **85**, 400–405.

Ratcliffe, L. M. and Grant, P. R. (1983) Species recognition in Darwin's finches (*Geospiza*,

Gould). I. Discrimination by morphological cues. *Animal Behaviour* **31**, 1139–1153.

Rice, W. R. and Hostert. E. E. (1993) Laboratory experiments on speciation: what have we learned in 40 years? *Evolution* **47**, 1637–1653.

Rolan-Alvarez, E., Johannesson, K. and Erlandsson, J. (1997) The maintenance of a cline in the marine snail *Littorina saxatilis*: the role of host site advantage and hybrid fitness. *Evolution* **51**, 1838–1847.

Rundle, H. D., Nagel, L., Boughman, J. W. and Schluter, D. (2000) Natural selection and parallel speciation in sympatric sticklebacks. *Science* **287**, 306–308.

Rundle, H. D. and Nosil, P. (2005) Ecological speciation. *Ecology Letters* **8**, 336–352.

Schemske, D. W. and Bradshaw, H. D., Jr. (1999) Pollination preference and the evolution of floral traits in monkey flowers (*Mimulus*). *Proceedings of the National Academy of Sciences of the United States of America* **96**, 11910–11915.

Schluter, D. (1996) Ecological speciation in postglacial fishes. *Philosophical Transactions of the Royal Society of London, Series B* **351**, 807–814.

Schluter, D. (2000) *The Ecology of Adaptive Radiation*. Oxford University Press, Oxford.

Schluter, D. (2001) Ecology and the origin of species. *Trends in Ecology & Evolution* **16**, 372–380.

Schluter D. and Nagel, L. M. (1995) Parallel speciation by natural selection. *The American Naturalist* **146**, 292–301.

Simpson, G. G. (1944) *Tempo and Mode in Evolution*. Columbia University Press, New York.

Simpson, G. G. (1953) *The Major Features of Evolution*. Columbia University Press, New York.

Spieth, H. T. and Ringo, J. M. (1983) Mating behavior and sexual isolation in *Drosophila*. In: *The Genetics and Biology of Drosophila. Vol. 3c* (ed. M. Ashburner, H. L. Carson and J. N. Thompson Jr.), pp. 223–285. Academic Press, New York.

Strong, D. R., Lawton, J. H. and Southwood, R. (1984) *Insects on Plants: Community Patterns and Mechanisms*. Blackwell Scientific Publications, London.

Tavormina, S. J. (1982) Sympatric genetic divergence in the leaf-mining insect *Liriomyza brassicae* (Diptera: Agromyzidae). *Evolution* **36**, 523–534.

Thompson, J. N. (1994) *The Coevolutionary Process*. University of Chicago Press, Chicago.

Thompson, J. N. (2005) *The Geographic Mosaic of Coevolution*. University of Chicago Press, Chicago.

Tregenza, T., Pritchard, V. L. and Butlin, R. K. (2000a) Patterns of trait divergence between populations of the meadow grasshopper, *Chorthippus parallelus*. *Evolution* **54**, 574–585.

Tregenza, T., Pritchard, V. L. and Butlin, R. K. (2000b) The origins of premating reproductive isolation: testing hypotheses in the grasshopper *Chorthippus parallelus*. *Evolution* **54**, 1687–1698.

Via, S., Bouck, A. C. and Skillman, S. (2000) Reproductive isolation between divergent races of pea aphids on two hosts. II. Selection against migrants and hybrids in the parental environments. *Evolution* **54**, 1626–1637.

Via, S. (2001) Sympatric speciation in animals: the ugly duckling grows up. *Trends in Ecology & Evolution* **16**, 381–390.

Vines, T. H. and Schluter, D. (2006) Strong assortative mating between allopatric sticklebacks as a by-product of adaptation to different environments. *Proceedings of the Royal Society of London, Series. B* **273**, 911–916.

Wang, H., McArthur, E. D., Sanderson, S. C., Graham, J. H. and Freeman, D. C. (1997) Narrow hybrid zone between two subspecies of big sagebrush (*Artemisia tridentata*: Asteraceae). IV. reciprocal transplant experiments. *Evolution* **51**, 95–102.

Walsh, B. J. (1864) On phytophagous varieties and phytophagous species. *Proceedings of the Entomological Society of Philadelphia* **3**, 403–430.

Wood, T. K. and Keese, M. C. (1990) Host-plant-induced assortative mating in *Enchenopa* treehoppers. *Evolution* **44**, 619–628.

CHAPTER TWELVE

# Biotic interactions and speciation in the tropics

## DOUGLAS W. SCHEMSKE

Tropical environments provide more evolutionary challenges than do the environments of temperate and cold lands. Furthermore, the challenges of the latter arise largely from physical agencies, to which organisms respond by relatively simple physiological modifications.... The challenges of tropical environments stem chiefly from the intricate mutual relationships among inhabitants.

Dobzhansky (1950, p. 221)

In virtually all groups of organisms, species richness increases from polar to equatorial regions (Wallace 1878; Dobzhansky 1950; Rosenzweig 1995; Brown & Lomolino 1998; Hillebrand 2004). This pattern is observed for extinct and living species, plants and animals, and in terrestrial and marine environments (Table 15.1; Brown & Lomolino 1998). The contrast in species richness between temperate and tropical communities is often substantial. For example, the breeding bird diversity in North America varies from <100 species in high latitude areas of the boreal zone, to 300 species in Central Mexico and to 600 species in equatorial regions of the New World tropics (MacArthur 1969; Hawkins *et al.* 2006). For trees, the latitudinal differences are astonishing. There are 620 tree species in all of North America (Currie & Paquin 1987), as compared to 1017 species on just 15 hectares in Yasuni National Park, Ecuador (Pitman *et al.* 2002), and an estimated 22 500 species in the New World tropics (Fine & Ree 2006).

The cause of the latitudinal biodiversity gradient has been the subject of intense interest from biologists ever since Wallace (1878) first drew attention to this striking biogeographic pattern (Dobzhansky 1950; Fischer 1960; Pianka 1966; MacArthur 1969; Gaston 2000; Willig *et al.* 2003, Leigh *et al.* 2004). As reviewed recently by Mittelbach *et al.* (2007), there are three primary categories of explanations. First, historical hypotheses propose that tropical climates are older and were once more widespread, and that tropical regions escaped large-scale glaciation events that may have caused extinction in northern latitudes. Thus, the higher diversity in the tropics is a consequence of greater effective time for diversification (Fig. 12.1a). Second, ecological hypotheses suggest that factors such as higher productivity and higher mean temperatures lead to larger population sizes and a lower rate of extinction in the tropics (Fig. 12.1b). Finally, evolutionary hypotheses suggest that

*Speciation and Patterns of Diversity*, ed. Roger K. Butlin, Jon R. Bridle and Dolph Schluter. Published by Cambridge University Press. © British Ecological Society 2009.

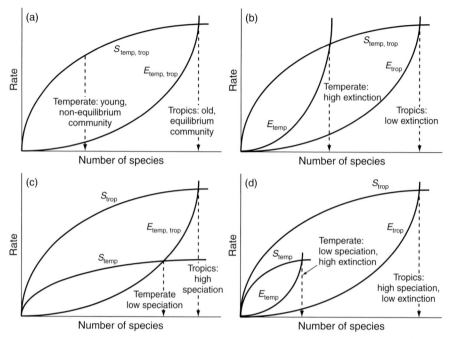

**Figure 12.1** Alternative hypotheses for the evolution of high tropical biodiversity, adapted from MacArthur (1969). Rates of extinction ($E$) and speciation ($S$) are shown as a function of the number of species in a community, and the vertical dashed lines give the expected diversity. If the speciation and extinction functions intersect, the expected diversity represent an equilibrium. (a) Historical hypotheses propose that speciation and extinction rates do not differ between regions, but that tropical communities are older, and have had more time to diversify, while temperate communities have not reached their equilibrium diversity. (b) Ecological hypotheses propose that speciation rates do not differ between regions, but that tropical communities have lower extinction rates. (c) Evolutionary hypotheses propose that extinction rates do not differ between regions, but that tropical communities have higher speciation rates. (d) MacArthur's (Fig. 4, 1969) model for the latitudinal diversity gradient proposes that extinction rates are lower and speciation rates are higher in the tropics.

latitudinal differences in species richness are due to higher speciation in tropical regions because of higher mutation rates and/or a greater role of biotic interactions (Fig. 12.1c). A complete explanation for the gradient will probably require elements from historical, ecological and evolutionary hypotheses (Mittelbach *et al.* 2007), as illustrated by MacArthur's (1969) composite model (Fig. 12.1d).

Evolutionary hypotheses have received far less attention than historical or ecological hypotheses, owing perhaps to the difficulty in comparing evolutionary rates and in establishing modes and rates of speciation (Mittelbach *et al.* 2007). Yet, understanding the evolutionary mechanisms that have contributed to the origins of diversity is critical to identifying the cause of the latitudinal gradient. As suggested by Dobzhansky (1950, p. 209): 'Since the animals and plants which exist in the world are products of the evolutionary development of living matter, any differences between tropical and temperate organisms must be the outcome of differences in evolutionary patterns' (Dobzhansky 1950, p. 209). This point was later underscored by Rohde (1992, p. 516): 'with regard to increased species richness in the tropics, the main problem is ... to answer the question of the origin of the greater numbers of competitors, predators, or organisms ... responsible for creating complex habitats for other species.'

## Climate, history and topography

Climatic factors such as temperature and water are often strongly associated with species richness and with latitude (Currie et al. 2004). For example, Currie (1991) found in North America that the correlation between annual mean temperature and species richness for trees, birds, mammals, amphibians and reptiles varied from 0.61 to 0.94 (Mean = 0.68, Table 2). For vertebrates, the best predictor of species richness was Potential evapotranspiration (PET), a measure of energy correlated with latitude that represents the amount of water that evaporates from a saturated surface (Currie 1991). Kreft and Jetz (2007) found that PET was the most important predictor of global plant species richness in a combined model including six explanatory variables. These results mirror those from studies conducted on a broad diversity of organisms indicating that geographic patterns of species richness are highly correlated with temperature and other related climatic variables (Adams & Woodward 1989; Hawkins et al. 2003; Francis & Currie 2003; Currie et al. 2004). Furthermore, the patterns of association between richness and climatic factors in different geographic regions suggest the action of similar causal mechanisms (Currie et al. 2004).

Francis and Currie (2003) concluded that history and topography were not as important as climate in explaining patterns of species richness, yet these two factors do contribute significantly to community diversity. For example, Fine and Ree (2006) proposed that time-integrated biome area was a primary driver of the latitudinal gradient in tree species richness, and Farrell and Mitter (1994) concluded that the high diversity of insects in the tropics is due to the greater age of tropical communities. The observation of significant regional effects in global comparisons of species richness also supports a role for historical factors (Latham & Ricklefs 1993; Ricklefs & Latham 1999; Kreft & Jetz 2007). The tropical origin of many families, coupled with dispersal limitation and/or constraints on adaptation to a temperate environment, may suggest that historical factors contribute to the observed latitudinal biodiversity gradient (Wiens & Donoghue 2004; Ricklefs 2006). Nevertheless, Kreft and Jetz (2007) found that contemporary environmental factors were highly predictive of latitudinal patterns in plant diversity, with no correlation between residuals in their global model and latitude as might be expected if historical factors were also important.

While topographic heterogeneity clearly contributes to global diversity patterns, for example, in plants (Kreft & Jetz 2007) and birds (Davies *et al.* 2007), the diversity of landforms is no greater in tropical than in temperate regions (Fischer 1960), and spatial heterogeneity at the landscape scale cannot explain higher diversity within habitats (Pianka 1966). Thus, while historical factors and topography are clearly associated with global richness, their role in producing the latitudinal diversity gradient remains uncertain. This suggests that we need to explore hypotheses with a direct link to environmental factors such as temperature and PET that are strongly correlated with the latitudinal gradient.

## Evolution and the latitudinal gradient

Although environmental factors are often correlated with latitudinal patterns of species richness, a key issue is to determine how these factors contribute to the origin of species. Mittelbach *et al.* (2007) discuss seven evolutionary mechanisms that have been proposed to increase speciation rates in tropical regions. Two of these, the evolutionary speed hypothesis and the biotic interactions hypothesis, suggest that higher speciation rates in tropical regions are related to temperature. As discussed above, the strong correlation between temperature and global species richness patterns suggests that temperature is a key driver of the latitudinal biodiversity gradient, and we might expect that the primary causal factor(s) underlying the biodiversity gradient will show a geographic pattern similar to the gradient itself.

The evolutionary speed hypothesis suggests that the kinetic effects of temperature on mutation rates and generation time lead to faster rates of genetic divergence among populations (Rohde 1992; Allen *et al.* 2002). While there is evidence for higher rates of molecular evolution in some tropical organisms (Wright *et al.* 2003; Davies *et al.* 2004; Allen *et al.* 2006), it is not clear how this molecular mechanism will cause speciation. Davies *et al.* (2004) found a positive effect of temperature on rates of molecular evolution in 86 sister-family comparisons of plants, but no effect of molecular evolution on species richness.

The biotic interactions hypothesis suggests that natural selection in temperate regions is governed primarily by abiotic factors, particularly low temperature, while in tropical regions, a greater role of biotic interactions may increase the opportunity for evolutionary novelty (Dobzhansky 1950) and rapid diversification (Schemske 2002). The importance of strong biotic interactions is a common theme in discussions of the maintenance of high tropical diversity (Janzen 1970; Connell 1971; Leigh *et al.* 2004; Burslem *et al.* 2005), but their role in speciation has received far less attention. Wallace (1878) was probably the first to highlight the role of biotic interactions in the origin of high tropical diversity, suggesting that 'equatorial lands must always have remained thronged with life; and have been unintermittingly subject to those complex influences of organism upon organism, which seem the main agents in developing the greatest variety of forms and filling up every vacant place in nature' (1878, p. 122). By contrast, he viewed the lower diversity of temperate regions as a result of extinctions caused by successive glacial periods. Hence, the greater effective time for diversification in the tropics facilitated the evolution of high species richness: 'In the one, evolution has had a fair chance; in the other it has had countless difficulties thrown in its way' (Wallace 1878, p. 123). It is now clear, however, that the latitudinal biodiversity gradient existed long before the Pleistocene glaciations (Mittelbach *et al.* 2007), thus time since glaciation cannot be the primary explanatory factor.

Dobzhansky (1950) and Fischer (1960) both suggested that density independent mortality caused by a harsh climate plays a greater role in temperate than tropical environments, and that biotic interactions constitute the primary selective pressures in the tropics. Furthermore, they proposed that this latitudinal difference in the nature of selection could influence evolutionary diversification and ultimately species richness. In his discussion of species diversity and coexistence in the tropics, MacArthur (1969, p. 26) concluded that 'The simplest – in fact the only – hypothesis I know which can account for all of these properties of tropical communities is that species interactions are important…' (p. 26). Finally, expanding on ideas proposed by MacArthur (1972), Kaufman (1995) suggested that high latitude range boundaries are limited by abiotic factors while low latitude range boundaries are limited by biotic factors, and that these geographic differences in limiting factors produce the latitudinal diversity gradient.

None of the earlier proponents of the biotic interactions hypothesis offered specific mechanisms linking biotic interactions to the evolution of the latitudinal gradient. In an earlier paper, I suggested that coevolution would promote rapid diversification in regions where biotic interactions play a primary role in determining the fitness of species populations (Schemske 2002). In this scenario, the phenotypic optimum in tropical regions is constantly changing because of reciprocal selection and coevolution. In contrast, where abiotic interactions play the primary role, the phenotypic optimum is more static since abiotic factors do not coevolve. This argument assumes that speciation is a byproduct of ecological adaptation, either directly, for example, habitat shifts or differences in mating preferences, or indirectly, due to the fixation of adaptive genes that contribute to hybrid incompatibilities.

This is not to say that biotic interactions are unimportant in extra-tropical regions. Competition, predation, parasitism and mutualism are clearly significant factors in the evolution of virtually all organisms, and there are countless examples of adaptation to these biotic challenges in all geographic regions. Climatic stresses might actually magnify some biotic interactions. As suggested by Darwin (1859, p. 121) 'in so far as climate chiefly acts to reduce food, it brings on the most severe struggle between individuals'.

Here I expand on the ideas first presented in Schemske (2002), and explore the proposed causal relationship between latitude, biotic interactions and the evolution of the latitudinal biodiversity gradient. To this end, I address the following questions: Are biotic interactions stronger in the tropics? What are the evolutionary implications of strong biotic interactions? How do biotic interactions facilitate speciation? Do strong biotic interactions lead to greater species richness? Moreover, are speciation rates higher in the tropics? I then summarize the assumptions and predictions of the biotic interactions hypothesis, and discuss future research directions. I acknowledge at the outset that many of these ideas are speculative and have not yet been subject to empirical or theoretical tests. Nevertheless, given the extraordinary history and diversity of ideas on the cause of the latitudinal biodiversity gradient, a discussion of biotic interactions and coevolution as potential drivers of tropical diversification is warranted.

## Are biotic interactions stronger in the tropics?

Despite the early statements of Dobzhansky (1950) and Fischer (1960) that biotic interactions are more important in the tropics, there is no comprehensive treatment of the subject (Mittelbach *et al.* 2007). Latitudinal variation in the role of biotic interactions is potentially manifest by both a greater strength and variety of interactions. For example, consider that the typical tropical tree requires animals for pollination and seed dispersal and often is engaged in ant–plant mutualisms for protection from herbivores. While such a diversity of biotic interactions is also observed in temperate species (e.g. *Prunus*), it is the exception. This suggests that empirical tests of the biotic interactions hypothesis will require latitudinal comparisons of both the magnitude and types of biotic interactions. A full review is beyond the scope of the present chapter, but here I briefly discuss results from a number of studies that have investigated latitudinal patterns in biotic interactions. Three categories of observations are recognized, including latitudinal patterns in (1) the frequency of different biotic interactions, (2) the traits involved in biotic interactions and (3) the strength of biotic interactions.

*Latitudinal patterns in the frequency of different biotic interactions.* Two studies did not find a latitudinal pattern. The proportion of tree species that experience density-dependent mortality does not vary with latitude (Hille Ris Lambers *et al.* 2002), although it is not known if the magnitude of density-dependence differs between regions. Parasitoid attack shows no relationship with latitude (Stireman *et al.* 2005).

In contrast, several studies have noted a pattern. Moles *et al.* (2007) found that the fraction of animal-dispersed plants increased from high to low latitude regions, and Jordano (2000) observed that >90% of tropical plant species rely on animals for seed dispersal. Symbiotic ant–plant mutualisms are restricted to the tropics (>100 plant genera and >40 ant genera; Davidson & McKey 1993). Coley and Aide (1991) found that the proportion of plant families that produce extra-floral nectaries, an adaptation for ant defense against herbivores, is higher in tropical (39%) than in temperate communities (12%).

*Latitudinal patterns in the traits involved in biotic interactions.* Wallace (1878, p. 98) was probably the first naturalist to offer evidence of latitudinal differences in traits associated with biotic interactions: 'It is in the tropics that we find most largely developed, whole groups of organisms which are unpalatable to almost all insectivorous creatures, and it is among these that some of the most gorgeous colours prevail'. Gauld *et al.* (1992) also found out that the proportion of such aposematic insects is often higher in the tropics. Leaf toughness is higher in tropical trees, suggesting greater allocation to structural defenses against herbivores (Coley & Aide 1991). Levin (1976) found a negative correlation between the percentage of alkaloid containing plant species and latitude, and, in general, tropical plants possess greater chemical defenses (Dyer & Coley 2002). Moles *et al.* (2007) found a negative relationship between seed mass and latitude, with geometric mean seed mass decreasing 320-fold from equatorial to high latitude regions. This pattern was predicted by a regression including vegetation type, growth form and dispersal syndrome and may reflect, in part, an advantage of greater investment in resources to enhance seedling survival in tropical regions. Finally, molluscan shell ornamentation in modern oceans is greater at lower latitudes, perhaps in response to higher predation pressure in the tropics (Leighton 2000).

*Latitudinal patterns in the strength of biotic interactions.* This is the most direct approach to evaluating the biotic interactions hypothesis. In a survey of multiple bird species in temperate and tropical regions, Robinson *et al.* (2000) concluded that nest predation was higher in the tropics. In the red-winged blackbird, nest predation was considerably higher in the Neotropics (Costa Rica: Orians 1973) in comparison to a temperate locality (Washington State, USA: Haigh 1968; Beletsky & Orians 1996). Cornell and Hawkins (1995) found a trend that pre-adult mortality of holometabolous herbivorous insects due to enemies is more frequent in tropical systems. Endophytic fungi have been shown to provide protection from pathogens in some plant hosts (Arnold *et al.* 2003; Herre *et al.* 2007), and the incidence of infections from endophytes increases from <1% in the arctic to >99% in the tropics (Arnold & Lutzoni 2007). Coley and Barone (1996) found significantly higher rates of herbivory of forest trees in the tropics. Novotny *et al.* (2006) found that predation on insect baits was 18-fold higher on tropical than on temperate trees. In a meta-analysis of tri-trophic interactions, Dyer and Coley (2002) found significantly greater effects of predators on prey and of herbivores on plant biomass in the tropics. In addition, the top–down effects of enemies on plants, i.e. the trophic cascade, was stronger in the tropics. In marine systems, shell crushing of gastropods was higher in tropical regions (Bertness *et al.* 1981), and drilling frequency in bivalves increased southward along the eastern coast of NA (Alexander & Dietl 2001). Jeanne (1979) measured ant predation of wasp larvae in experimental depots distributed along a latitudinal gradient extending from New Hampshire (USA) to

Brazil, and found a greater risk of predation at lower latitudes. Finally, most crops grown in the tropics have more diseases than in temperate zones (Wellman 1968).

*Summary.* These results provide some support for the view first proposed by Dobzhansky (1950) that biotic interactions are stronger in tropical regions, but the number of comparative studies is limited. A more complete description of latitudinal patterns in biotic interactions is in preparation (Schemske *et al.*, unpublished). It is important to note that the greater species richness of tropical regions may confound latitudinal comparisons of interaction strength. Further investigations are needed that compare the strength of species interactions in different geographic regions, while accounting for the geographic differences in species richness. These studies should also simultaneously estimate the relative importance of abiotic factors, as only then will it be possible to determine if the agents of selection differ between temperate and tropical regions.

## What are the evolutionary implications of strong biotic interactions?

If we assume that all adaptations are costly, and that organisms must allocate resources for different adaptive challenges from a limited resource pool, then adaptation to one selective factor must necessarily decrease the potential for adaptation to others. Such tradeoffs based on resource allocation are a key feature of adaptive evolution (Bell 1980; Stearns 1992). Abiotic factors such as low temperatures require the evolution of specific physiological, morphological and developmental traits that ensure survival under the most extreme climatic conditions (Storey & Storey 1988; Smallwood & Bowles 2002; Voituron *et al.* 2002; Broggi *et al.* 2004; Pörtner 2004; Margesin *et al.* 2007). As a result, we may expect that organisms in temperate regions must allocate proportionally more resources to climatic adaptation, while in tropical regions there is a lesser constraint. Vermeij (2005, p. 18) suggested 'many functions become less expensive as temperatures rise…', and concluded that the temperature regimes of tropical environments are more conducive to the evolution of phenotypic diversity by virtue of the lower cost of metabolic processes. Fischer (1960, p. 75) reached the same conclusion 'tropical conditions permit a wider range of physiological and structural variation than do high-latitude conditions, which seasonally approach the lower limits at which life can be maintained …'. In this regard, Deutsch *et al.* (2008) found that tropical insects live closer to their optimal temperature than do temperate insects, which may suggest that temperate insects must allocate resources to traits that can influence body temperature, and hence, overall performance.

In addition to the effects of physiologic adaptation on the allocation of resources, climate may play a direct role in determining the relative importance of abiotic and biotic interactions in different geographic regions. Highly coevolved interactions, such as specialized mutualisms, may be particularly sensitive to climatic fluctuations, which may explain their greater development in tropical regions (Price, pers. comm.). Temperate regions have experienced pronounced climate change caused by Milankovitch cycles that result in turnover in species communities (Dynesius & Jansson 2000). Similarly, higher extinction rates caused by Pleistocene glaciation in temperate regions may have disrupted the coevolution of species interactions (Schluter, pers. comm.).

Two correlates of strong biotic interactions are proposed to facilitate speciation, (1) coevolution, and (2) biotic drift, i.e. the stochastic change in community composition that occurs during geographic isolation (Turner & Mallet 1996). These are not mutually exclusive, and in fact, will often act in concert to influence the direction and strength of selection mediated by biotic interactions. Below I briefly discuss these mechanisms, and their potential role in the diversification of tropical communities.

*Coevolution.* Strong biotic interactions and reciprocal selection increase the opportunity for coevolution between interacting species to influence the direction and rate of adaptation (Schemske 2002). The optimum phenotype in populations that experience strong biotic interactions is constantly changing as species within the interaction network adapt to the shifting selective environment. In this sense, an adaptive phenotype achieves only temporary success. It is soon exploited by coevolving mutualists and antagonists, and is exposed to a new set of selective pressures. These dynamics are quite different from those expected in temperate habitats where abiotic factors are the major source of selection. Here adaptation to a climatic factor such as temperature is not met by a reciprocal response. Thus, the trajectory of adaptation in temperate regions is towards a relatively static optimum, and once achieved, there is far less opportunity for major shifts in the selective landscape. In addition, the physiological costs of cold tolerance in temperate regions may constrain adaptation to biotic factors. We might predict, therefore, that the greater role of biotic interactions and coevolution in the tropics will increase rates of adaptation and speciation in tropical regions. This is consistent with Thompson's (2005) suggestion that coevolution can increase the rate of evolution over that observed for adaptation to physical environments.

There are many examples of coevolutionary diversification in the tropics, including plant–herbivore interactions (Benson *et al.* 1975; Becerra 2003), ant–plant mutualisms (Quek *et al.* 2004), and interactions between predators and prey (Duda & Palumbi 1999). One of the best-known and most spectacular examples of coevolution and speciation in the tropics is the interaction between the nearly 750 species of figs (*Ficus*) and the wasps that pollinate them (Wiebes 1979; Machado *et al.* 2001; Weiblen & Bush 2002; Jousselin *et al.* 2003; Weiblen 2004). It is often assumed that each fig species is pollinated by a single species of wasp, although recent genetic evidence suggests the coexistence of cryptic wasp species in some figs (Molbo *et al.* 2003). Nevertheless, the interaction between figs and their pollinating wasps is highly specialized, and coevolution has likely facilitated speciation in both groups.

Coevolution does play a role in tropical speciation, but is it more important in tropical than in temperate communities? There is insufficient evidence at present to answer this question. Detailed ecological and phylogenetic studies are needed to assess the relative importance of coevolution in temperate and tropical regions.

*Biotic drift.* Where biotic interactions are strong, geographic isolation may cause stochastic changes in species composition, which in turn could alter selection on traits involved in species interactions (Schemske 2002). Turner and Mallett (1996) first coined the term 'biotic drift' to describe stochastic changes in the biotic environment. They suggested that an isolated community 'contains at any one time only a stochastic subsample of the species which might be present', and that 'the form which natural selection takes, for the constituent species, will differ from island to island or patch to patch' (Turner & Mallet 1996, p. 837). These observations were based upon studies of the evolutionary mechanisms responsible for parallel race formation in *Heliconius erato* and *H. melpomene*, two species of butterflies found in the tropical forests of South America. Sympatric populations of *Heliconius erato* and *H. melpomene* possess virtually identical colour and wing patterns, yet the two species have formed multiple distinct geographical races. Predators exert strong frequency-dependent and stabilizing selection for warning colouration, so local convergence to a common phenotype is expected. However, it is not obvious why different populations should evolve distinct patterns. Biotic drift is one explanation (Turner & Mallet 1996). Brown *et al.* (1974, p. 375) had earlier suggested that if species composition or relative abundance differed among refugia 'then the most advantageous pattern in each would have been different since it depended on the particular pattern of the most distasteful and most abundant species.'

The effects of biotic drift on the direction and magnitude of selection will depend on both the size of the isolated communities and on the degree of interconnection between species. Biotic drift in small communities with strong species interactions will have a greater effect on the focal species than it will in large communities with weak interactions (Price, pers. comm.). Potentially in support of this idea, there is greater turnover in species composition and greater morphological variation among bird communities on small islands (Price 2007).

Just as the expected change in allele frequency caused by random genetic drift is inversely proportional to population size, the stochastic changes in community composition caused by biotic drift will increase as species abundance declines. With each successive round of speciation, competition between coexisting species should reduce average population density. This, in turn, will magnify biotic drift in subsequent episodes of geographic isolation. In this way, biotic drift potentially has an autocatalytic property.

## How do biotic interactions facilitate speciation?

If biotic interactions are stronger in the tropics, we may expect that the nature of selection will differ between temperate and tropical regions. Figure 12.2

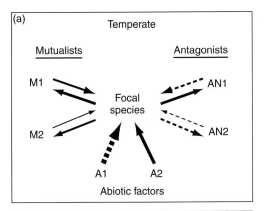

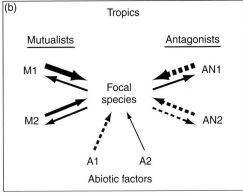

**Figure 12.2** Interaction strengths in temperate (a) and tropical communities (b) Arrows indicate the proportion of variation in fitness for a given focal species caused by biotic interactions with mutualists or antagonists, or by different abiotic factors. Positive interactions are represented by solid lines and negative interactions by dashed lines.

depicts a hypothetical scenario for the relative contributions of biotic factors (mutualisms, antagonisms) and abiotic factors (e.g. temperature, moisture) to the total variation in fitness for a given focal species. Organisms in temperate regions are expected to experience stronger selection from abiotic factors (Fig. 12.2a), while in tropical regions, selection imposed by biotic factors is greater, facilitating coevolution (Fig. 12.2b). Thus, the net trajectories and rates of adaptive evolution should differ between regions, with the evolution of traits related to species interactions playing a larger role in the tropics.

Building on this idea, Fig. 12.3 depicts the adaptive divergence of traits and the evolution of reproductive isolation following geographic isolation in temperate and tropical regions. Traits involved in abiotic adaptation are expected to experience stronger selection, and hence greater divergence in temperate than in tropical regions (Figs. 12.3a and 12.3b), while the converse is expected for characters involved in biotic interactions (Figs. 12.3c and 12.3d). In temperate regions, adaptation to contrasting abiotic environments causes an initial change in phenotype following geographic isolation, but further evolution is limited once the local optimum is achieved since abiotic factors do not coevolve (Fig. 12.3a). By contrast, in tropical regions, reciprocal selection and coevolution

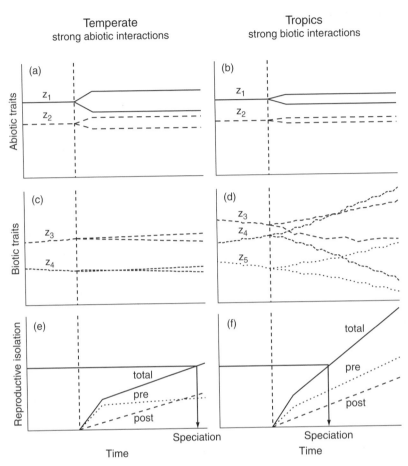

**Figure 12.3** The contribution of abiotic factors and biotic interactions to phenotypic evolution and speciation in temperate and tropical communities. The figure depicts the hypothesis that abiotic interactions are stronger in temperate regions, and that biotic interactions are stronger and more varied in the tropics. The greater importance of biotic factors in the tropics contributes to coevolution and continued phenotypic divergence following geographic isolation, and this in turn results in a higher rate of accumulation of reproductive isolation in the tropics and a shorter time to speciation. The vertical dashed line in each panel indicates the time of geographic isolation, separating the ancestral population (left side) into two isolated populations (right side). (a), (b) The evolutionary dynamics of traits ($z_i$) related to abiotic adaptation in temperate (a) and tropical (b) regions. (C, D) The evolutionary dynamics of traits ($z_i$) related to biotic adaptation in temperate (c) and tropical (d) regions. (E, F) The evolution of pre- and post-zygotic reproductive isolation and total isolation between geographically isolated populations in temperate (e) and tropical (f) regions. The horizontal lines indicate the total reproductive isolation required for speciation, and the vertical arrows indicate the time to speciation.

caused by biotic interactions results in continued phenotypic evolution in response to the shifting optimum (Fig. 12.3d). This is analogous to Thompson's suggestion: 'If gene flow between populations is sufficiently restricted, specia-tion may result as a fortuitous consequence of populations adapting to different local ecological conditions as the interactions evolve in different ways' (Thompson 1994, p. 239).

By virtue of the greater importance of biotic interactions and coevolution to adaptation in tropical regions, both pre- and post-zygotic reproductive iso-lation are expected to accumulate faster in tropical than in temperate regions (Figs. 12.3e and 12.3f). The evolution of ecological and reproductive traits driven by biotic interactions in tropical regions should contribute to rapid population divergence for components of pre-zygotic isolation. This in turn could acceler-ate the accumulation of post-zygotic barriers, both for extrinsic isolation caused by ecological factors, and for intrinsic isolation caused by Muller–Dobzhansky incompatibilities for genes involved in biotic interactions (Mittelbach et al. 2007). As a result, the rate of accumulation of total reproductive isolation will be faster in tropical than temperate regions, and time to speciation should be shorter in the tropics (Figs. 12.3e and 12.3f).

Consider two examples of how biotic factors may contribute to speciation in the tropics. First, Leigh et al. (2004) suggest that habitat shifts to escape pests are an important mechanism of speciation in tropical trees. This is consistent with experimental evidence that habitat specialization in Amazonian trees is driven largely by differential pest pressure (Fine et al. 2004). If herbivorous insects in the tropics are more specialized than their temperate counterparts, as has been observed in the New World (Dyer et al. 2007), habitat shifts triggered by selection to escape herbivores could facilitate tropical speciation. However, it should be noted that in the Old World there is apparently no latitudinal difference in the specificity of herbivores (Novotny et al. 2006). Second, it has often been sug-gested that pollinator-mediated speciation has contributed to the rapid radia-tion of many tropical plant lineages (Gentry 1982). In support of this view, Kay et al. (2005) found that hummingbird pollination evolved from bee pollination at least seven different times in Neotropical *Costus*, and that one-fifth of the speciation events in the genus involved a pollinator shift. The rate observed in *Costus* exceeds that of any temperate lineage studied to date (Kay et al. 2005). These examples illustrate possible mechanisms of speciation mediated by biotic interactions, but further studies are needed to assess their contribution to the latitudinal diversity gradient.

## Do strong biotic interactions lead to greater species richness?

The above arguments suggest that biotic interactions facilitate speciation, yet this mechanism alone may be insufficient to increase community diversity without parallel mechanisms that contribute to species coexistence. It has long been proposed that strong density-dependent mortality in tropical tree species may promote the maintenance of high species diversity, the 'Janzen-Connell effect' (Janzen 1970; Connell 1971; Leigh *et al.* 2004). However, the proportion of tree species that display density dependence does not vary with latitude, although the magnitude of effect may differ between temperate and tropical regions (Hille Ris Lambers *et al.* 2002). The higher productivity of the tropics could increase species packing (Price, pers. comm.), and is often proposed as an ecological factor that could reduce extinction rates (e.g. via the 'more individuals' hypothesis), yet the relationship between productivity and diversity is often not apparent (Currie *et al.* 2004; Leigh *et al.* 2004).

Hutchinson (1959) proposed that community diversification is enabled by species interactions, and becomes a self-inducing process, for example, increased predator efficiency in the late Precambrian facilitated the evolution of prey defenses such as hard skeletons. While his discussion was not focused on the latitudinal diversity gradient, it does support the notion that species interactions contribute to community diversity. In a similar vein, Whittaker (1969) proposed that the diversity of plant communities has no upper bound, and is subject to self-augmentation because of the expansion of the number of niche dimensions through interactions with predators and edaphic factors.

As suggested by Vermeij (2005, p. 20), 'Every species is potentially a resource on which some other species can in principle specialize or to which another species must adapt.' In this sense, diversification in the tropics may be less constrained than in temperate regions, and each new species in the tropics could provide opportunities for further community diversification. As one striking example, consider the diverse assemblage of Neotropical species that depend to varying degrees on arthropods that are flushed from the forest understory by the army ant *Eciton burchelli*. At least 50 species of birds regularly follow army ants, and many forage nowhere else (Willis & Oniki 1978). Army ant swarms support >300 species of insects (Rettenmeyer pers. comm.), including flies that parasitize fleeing arthropods (Rettenmeyer 1963; Brown & Feener 1998), ant-mimicking beetles that prey on insects in the midst of the swarm, and mimics of ant larvae that live within the colony (Hölldobler & Wilson 1990). Furthermore, in a Brazilian rainforest males of >400 species of skipper butterflies (Hesperiidae) use the droppings deposited by ant-following birds. Diurnal partitioning of this resource may provide significant reproduction isolation among butterfly species (DeVries *et al.* in press). The diversity of ant-following species is likely catalyzed by intense biotic interactions and an ever-expanding number of niche dimensions.

## Are speciation rates higher in the tropics?

Analyses of extant taxa can reliably estimate only net diversification (speciation–extinction). To estimate speciation rates requires simplifying assumptions about rates of extinction, which are difficult to evaluate. Nevertheless, there is evidence of both higher tropical diversification and speciation rates, although there are also exceptions (Mittelbach *et al.* 2007). Phylogenetic studies using sister clade comparisons indicate higher tropical diversification rates for angiosperms (Davies *et al.* 2004) and for passerine birds and swallowtail butterflies (Cardillo 1999). Paleontological studies estimate higher tropical diversification rates for marine bivalves (Crame 2002) and benthic foraminifera (Buzas *et al.* 2002), as well as higher tropical origination rates for marine invertebrates (Jablonski 1993), marine bivalves (Jablonski *et al.* 2006), and fossil foraminifera (but not nano-plankton or radiolarians) (Allen *et al.* 2006). A recent phylogenetic study using a birth–death model for birds and mammals suggested higher speciation and extinction rates in the temperate zone, yet there was no significant latitudinal gradient in net diversification (Weir & Schluter 2007). Cardillo *et al.* (2005) and Ricklefs (2006), on the other hand, found that net diversification was faster near the equator than near the poles. The factors contributing to the higher diversification or speciation rates observed in these studies are unknown. To evaluate the biotic interactions hypothesis for high tropical diversity will require many more estimates of latitudinal patterns in speciation rates, and of the ecological and evolutionary mechanisms that contribute to speciation in temperate and tropical regions.

## Synopsis and future directions

Dobzhansky (1950) proposed that biotic interactions are more important in tropical regions, and that they contribute to the latitudinal biodiversity gradient. The review of the literature presented here suggests that biotic interactions are stronger, and perhaps more diverse, in tropical regions. It is proposed that the greater importance of abiotic factors in temperate regions constrains adaptation, such that the rate of speciation is reduced in comparison to tropical regions where biotic interactions and coevolution create a shifting selective landscape. Tropical environments thus provide an example of the Red Queen hypothesis (Van Valen 1973), where dynamic species interactions and coevolution continually drive phenotypic change (Stenseth 1984). Furthermore, biotic drift may amplify divergence in the tropics by virtue of stochastic changes in species interactions. This process, coupled with an ever-expanding number of niche axes as communities diversify, may cause the species richness of tropical communities to display a self-augmenting property, i.e. 'diversity begets diversity'.

A recent Symposium volume focusing on the role of biotic interactions in the maintenance of high tropical diversity (Burslem *et al.* 2005) illustrates that this is a subject of considerable interest. Yet, much more research is needed to evaluate the role of species interactions and coevolution in the origin of high tropical diversity. Questions to be addressed in future work include: Does the strength of biotic interactions vary with latitude? Does physiological adaptation to abiotic stress, for example, low temperature, constrain the evolution of species interactions? Is speciation faster in the tropics, and if so, what are the main contributing factors? Does a greater role of species interactions and coevolution increase the number of niche dimensions, and ultimately, species richness? Furthermore, does the biotic interactions hypothesis for the origin of the latitudinal diversity gradient also apply to other ecological patterns, for example, diversity gradients correlated with altitude or precipitation. Answers to these questions will require empirical and theoretical approaches that investigate mechanisms of diversification and coexistence along latitudinal gradients.

## Acknowledgements

This paper benefited from participation in the Gradients in Biodiversity and Speciation Working Group supported by the National Center for Ecological Analysis and Synthesis (NCEAS), a Center funded by NSF (Grant #DEB-00-72909), the University of California at Santa Barbara, and the State of California. I am grateful to T. Curtis, E. A. Herre, C. M. Johnston, E. G. Leigh Jr., G. G. Mittelbach, T. D. Price, D. Schluter and J. M. Sobel for helpful comments on the manuscript, and to P. DeVries and R. R. Ricklefs for discussion.

## References

Adams, J. M. and Woodward, F. I. (1989) Patterns in tree species richness as a test of the glacial extinction hypothesis. *Nature* **339**, 699–701.

Alexander, R. R. and Dietl, G. P. (2001) Latitudinal trends in naticid predation on *Anadara ovalis* (Bruguière, 1789) and *Divalinga quadrisulcata* (Orbigny, 1842) from New Jersey to the Florida Keys. *American Malacological Bulletin* **16**, 179–194.

Allen, A. P., Brown, J. H. and Gillooly, J. F. (2002) Global biodiversity, biochemical kinetics, and the energetic-equivalence rule. *Science* **297**, 1545–1548.

Allen, A. P., Gillooly, J. F., Savage, V. M. and Brown, J. H. (2006) Kinetic effects of temperature on rates of genetic divergence and speciation. *Proceedings of the National Academy of Sciences of the United States of America* **103**, 9130–9135.

Arnold, A. E. and Lutzoni, F. (2007) Diversity and host range of foliar fungal endophytes: are tropical leaves biodiversity hotspots? *Ecology* **88**, 541–549.

Arnold, A. E., Mejía, L. C., Kyllo, D., *et al.* (2003) Fungal endophytes limit pathogen damage in a tropical tree. *Proceedings of the National Academy of Sciences of the United States of America* **100**, 15649–15654.

Becerra, J. X. (2003) Synchronous coadaptation in an ancient case of herbivory. *Proceedings of the National Academy of Sciences of the United States of America* **100**, 12807–12807.

Beletsky, L. D. and Orians, G. H. (1996) *Red-Winged Blackbirds: Decision-Making and Reproductive Success.* University of Chicago Press, Chicago.

Bell, G. (1980) The costs of reproduction and their consequences. *American Naturalist* **116**, 45–76.

Benson, W. W., Brown, K. S. and Gilbert, L. E. (1975) Coevolution of plants and herbivores: passion flower butterflies. *Evolution* **29**, 659–680.

Bertness, M. D., Garrity, S. D. and Levings, S. C. (1981) Predation pressure and gastropod foraging: a tropical-temperate comparison. *Ecology* **35**, 995–1007.

Broggi, J., Orell, M., Hohtola, E. and Nilsson, J.-A. (2004) Metabolic response to temperature variation in the great tit: an interpopulation comparison. *Journal of Animal Ecology* **73**, 967–972.

Brown, B. V. and Feener, D. H., Jr. (1998) Parasitic phorid flies (Diptera: Phoridae) associated with army ants (Hymenoptera: Formicidae: Ecitoninae, Dorylinae) and their conservation biology. *Biotropica* **30**, 482–487.

Brown, J. H. and Lomolino, M. V. (1998) *Biogeography*, 2nd edn. Sinauer Associates, Sunderland, MA.

Brown, K. S., Jr., Sheppard, P. M. and Turner, J. R. G. (1974) Quaternary refugia in tropical America: evidence from race formation in *Heliconius* butterflies. *Proceedings of the Royal Society of London Series B* **187**, 369–378.

Burslem, D., Pinard, M. and Hartely, S. (eds) (2005) *Biotic Interactions in the Tropics: Their Role in the Maintenance of Species Diversity*. Cambridge University Press, Cambridge.

Buzas, M. A., Collins, L. S. and Culver, S. J. (2002) Latitudinal differences in biodiversity caused by higher tropical rate of increase. *Proceedings of the National Academy of Sciences of the United States of America* **99**, 7841–7843.

Cardillo, M. (1999) Latitude and rates of diversification in birds and butterflies. *Proceedings of the Royal Society of London. Series B* **266**, 1221–1225.

Cardillo, M., Orme, C. D. L. and Owens, I. P. F. (2005) Testing for latitudinal bias in diversification rates: an example using New World birds. *Ecology* **86**, 2278–2287.

Coley, P. and Aide, T. M. (1991) Comparison of herbivory and plant defenses in temperate and tropical broad-leaved forests. In: *Plant-Animal Interactions: Evolutionary Ecology in Tropical and Temperate Regions* (ed. P. W. Price, T. M. Lewinsohn, G. W. Fernandes and W. W. Benson), pp. 25–49. John Wiley & Sons, New York.

Coley, P. D. and Barone, J. A. (1996) Herbivory and plant defenses in tropical forests. *Annual Review of Ecology and Systematics* **27**, 305–335.

Connell, J. H. (1971) On the role of natural enemies in preventing competitive exclusion in some marine animals and in rain forest trees. In: *Dynamics of Populations* (ed. P. J. den Boer and G. R. Gradwell), pp. 298–312. Centre for Agricultural Publication and Documentation. Wageningen, The Netherlands.

Cornell, H. V. and Hawkins, B. A. (1995) Survival patterns and mortality sources of herbivorous insects: some demographic trends. *American Naturalist* **145**, 563–593.

Crame, J. A. (2002) Evolution of taxonomic diversity gradients in the marine realm: a comparison of late Jurassic and Recent bivalve faunas. *Paleobiology* **28**, 184–207.

Currie, D. J. (1991) Energy and large scale patterns of animal and plant species richness. *American Naturalist* **137**, 27–49.

Currie, D. J. and Paquin, V. (1987) Large-scale biogeographical patterns of species richness in trees. *Nature* **329**, 326–327.

Currie, D. J., Mittelbach, G. G., Cornell, H. V., *et al.* (2004) Predictions and tests of climate-based hypotheses of broad-scale variation in taxonomic richness. *Ecology Letters* **7**, 1121–1134.

Darwin, C. (1859) *The Origin of Species by Means of Natural Selection, or the Preservation of Favoured Races in the Struggle for Life*. John Murray, London.

Davidson, D. W. and McKey, D. (1993) The evolutionary ecology of symbiotic ant-plant relationships. *Journal of Hymenoptera Research* **2**, 13–83.

Davies, R. R., Orme, C. D. L., Storch, D., *et al.* (2007) Topography, energy and the global distribution of bird species richness. *Proceedings of the Royal Society of London Series B* **274**, 1189–1197.

Davies, T. J., Savolainen, V., Chase, M. W., Moat, J. and Barraclough, T. G. (2004) Environmental energy and evolutionary rates in flowering plants. *Proceedings of the Royal Society of London Series B* **271**, 2195–2200.

Deutsch, C. A., Tewksbury, J. J., Huey, R. B., *et al.* (2008) Impacts of climate warming on terrestrial ectotherms across latitude. *Proceedings of the National Academy of Sciences of the United States of America* **105**, 6668–6672.

DeVries, P. J., Austin, G. T. and Martin, N. M. (2008) Diel activity and reproductive isolation in a diverse assemblage of Neotropical skippers (Lepidoptera: Hesperiidae). *Biological Journal of the Linnaean Society* **94**, 723–736.

Dobzhansky, T. (1950) Evolution in the tropics. *American Scientist* **38**, 209–221.

Duda, T. F., Jr. and Palumbi, S. R. (1999) Molecular genetics of ecological diversification: duplication and rapid evolution of toxin genes of the venomous gastropod *Conus*. *Proceedings of the National Academy of Sciences of the United States of America*, **96**, 6820–6823.

Dyer, L. A. and Coley, P. D. (2002) Tritrophic interactions in tropical and temperate communities. In: *Multitrophic Level Interactions* (ed. T. Tscharntke and B. Hawkins), pp. 67–88. Cambridge University Press, Cambridge.

Dyer, L. A., Singer, M. S., Lill, J. T., *et al.* (2007) Host specificity of Lepidoptera in tropical and temperate forests. *Nature* **448**, 696–699.

Dynesius, M. and Jansson, R. (2000) Evolutionary consequences of changes in species' geographical distributions driven by Milankovitch climate oscillations. *Proceedings of the National Academy of Sciences of the United States of America* **97**, 9115–9120.

Farrell, B. D. and Mitter, C. (1994) Adaptive radiation in insects and plants: time and opportunity. *American Zoologist* **34**, 57–69.

Fine, P. V. A. and Ree, R. H. (2006) Evidence for a time-integrated species-area effect on the latitudinal gradient in tree diversity. *American Naturalist* **168**, 796–804.

Fine, P. V. A., Mesones, I. and Coley, P. D. (2004) Herbivores promote habitat specialization by trees in Amazonian forests. *Science* **305**, 663–665.

Fischer, A. G. (1960) Latitudinal variations in organic diversity. *Evolution* **14**, 64–81.

Francis, A. P. and Currie, D. J. (2003) A globally consistent richness-climate relationship for angiosperms. *American Naturalist* **161**, 523–536.

Gaston, K. J. (2000) Global patterns in biodiversity. *Nature* **405**, 220–227.

Gauld, I. D., Gaston, K. J. and Janzen, D. H. (1992) Plant allelochemicals, tritrophic interactions and the anomalous diversity of tropical parasitoids: the 'nasty' host hypothesis. *Oikos* **65**, 353–357.

Gentry, A. H. (1982) Neotropical floristic diversity: phytogeographical connections between Central and South America, Pleistocene climatic fluctuations, or an accident of the Andean orogeny? *Annals of the Missouri Botanical Garden* **69**, 557–593.

Haigh, C. (1968) Sexual dimorphism, sex ratios, and polygyny in the red-winged blackbird. *Ph.D. Dissertation*, University of Washington, Seattle.

Hawkins, B. A., Diniz-Filho, J. A. F., Jaramillo, C. A. and Soeller, S. A. (2006) Post-Eocene climate change, niche conservatism, and the latitudinal diversity gradient of new World birds. *Journal of Biogeography* **33**, 770–780.

Hawkins, B. A., Field, R., Cornell, H. V., *et al.* (2003) Energy, water and broad-scale geographic patterns of species richness. *Ecology* **84**, 3105–3117.

Herre, A. E., Mejía, L. C., Kyllo, D. A., *et al.* (2007) Ecological implications of anti-pathogen effects of tropical fungal endophytes and mycorrhizae. *Ecology* **88**, 550–558.

Hillebrand, H. (2004) On the generality of the latitudinal diversity gradient. *American Naturalist* **163**, 192–211.

Hille Ris Lambers J. H., Clark, J. S. and Beckage, B. (2002) Density-dependent mortality and the latitudinal gradient in species diversity. *Nature* **417**, 732–735.

Hölldobler, B. and Wilson, E. O. (1990) *The Ants.* Belknap Press, Cambridge.

Hutchinson, G. E. (1959) Homage to Santa Rosalia or Why are there so many kinds of animals? *American Naturalist* **93**, 145–159.

Jablonski, D. (1993) The tropics as a source of evolutionary novelty through geological time. *Nature* **364**, 142–144.

Jablonski, D. and Roy, K. (2003) Geographical range and speciation in fossil and living mollusks. *Proceedings of the Royal Society of London Series B* **270**, 401–406.

Jablonski, D., Roy, K. and Valentine, J. W. (2006) Out of the tropics: evolutionary dynamics of the latitudinal diversity gradient. *Science* **314**, 102–106.

Janzen, D. H. (1970) Herbivores and the number of tree species in tropical forests. *American Naturalist* **104**: 501–528.

Jeanne, R. L. (1979) A latitudinal gradient in rates of ant predation. *Ecology* **60**, 1211–1224.

Jordano P. (2000) Fruits and frugivory. In: *Seeds: The Ecology of Regeneration in Natural Plant Communities* (ed. M. Fenner), pp. 125–166. Commonwealth Agricultural Bureau International, Wallingford.

Jousselin, E., Rasplus, J.-Y, and Kjellberg, F. (2003) Convergence and co-evolution of in a mutualism: evidence from a molecular phylogeny of *Ficus*. *Evolution* **57**, 1255–1269.

Kaufman, D. M. (1995) Diversity of new world mammals: universality of the latitudinal gradients of species and bauplans. *Journal of Mammalogy* **76**, 322–334.

Kay, K. M., Reeves, P. A., Olmstead, R. G. and Schemske, D. W. (2005) Rapid speciation and the evolution of hummingbird pollination in Neotropical *Costus* subgenus *Costus* (Costaceae): evidence from nrDNA ITS and ETS sequences. *American Journal of Botany* **92**, 1899–1910.

Kreft, H. and Jetz, W. (2007) Global patterns and determinants of vascular plant diversity. *Proceedings of the National Academy of Sciences of the United States of America* **104**, 5925–5930.

Latham, R. E. and Ricklefs, R. E. (1993) Continental comparisons of temperate-zone tree species diversity. In: *Species Diversity in Ecological Communities: Historical and Geographical Perspectives* (ed. R. E. Ricklefs and D. Schluter), pp. 294–314. University of Chicago Press, Chicago.

Leigh, E. G., Jr., Davidar, P., Dick, C. W., *et al.* (2004) Why do some tropical forests have so many species of trees? *Biotropica* **36**, 447–473.

Leighton, L. R. (2000) Possible latitudinal predation gradient in Middle Paleozoic oceans. *Geology* **27**, 47–50.

Levin, D. A. (1976) Alkaloid-bearing plants: an ecogeographic perspective. *American Naturalist* **110**, 261–284.

MacArthur, R. H. (1969) Patterns of communities in the tropics. *Biological Journal of the Linnaean Society* **1**, 19–30.

MacArthur, R. H. (1972) *Geographical ecology: patterns in the distribution of species.* Princeton University Press, Princeton, New Jersey.

Machado, C. A., Jousellin, E., Kjellberg, F., Compton, S. G. and Herre, E. A. (2001) The evolution of fig pollinating wasps: phylogenetic relationships, character evolution, and historical biogeography. *Proceedings of the Royal Society of London Series B* **268**, 685–694.

Margesin, R., Neuner, G. and Storey, K. B. (2007) Cold-loving microbes, plants and animals – fundamental and applied aspects. *Naturwissenschaften* **94**, 77–99.

Mittelbach, G. G., Schemske, D. W., Cornell, H. V., *et al.* (2007) Evolution and the latitudinal diversity gradient: speciation, extinction, and biogeography. *Ecology Letters* **10**, 315–331.

Molbo, D., Machado, C. A., Sevenster, J. G., Keller, L. and Herre, E. A. (2003) Cryptic species of fig-pollinating wasps: implications for the evolution of the fig-wasp mutualism, sex allocation, and precision of adaptation. *Proceedings of the National Academy of Sciences of the United States of America*, **100**, 5867–5872.

Moles, A. T., Ackerly, D. A., Tweddle, J. C., *et al.* (2007) Global patterns in seed size. *Global Ecology and Biogeography* **16**, 109–116.

Novotny, V., Drozd, P., Miller, S. E., *et al.* (2006) Why are there so many species of herbivorous insects in tropical forests? *Science* **313**, 1115–1118.

Orians, G. H. (1973) The Red-winged blackbird in tropical marshes. *The Condor* **75**, 28–42.

Pianka, E. R. (1966) Latitudinal gradients in species diversity: a review of concepts. *American Naturalist* **100**, 33–46.

Pitman, N. C. A., Terborgh, J. W., Silman, M. R. V., *et al.* (2002) A comparison of tree species diversity in two upper Amazonian forests. *Ecology* **83**, 3210–3224.

Pörtner, H. O. (2004) Climate variability and the energetic pathways of evolution: the origin of endothermy in mammals and birds. *Physiological and Biochemical Zoology* **77**, 959–981.

Price, T. D. (2007) *Speciation in Birds*. Roberts and Company Publishers. Greenwood Village, CO.

Quek, S.-P., Davies, S. J., Itino, I., and Pierce, N. E. (2004) Codiversification in an ant-plant mutualism: stem texture and the evolution of host use in *Crematogaster* (Formicidae: Mymicinae) inhabitants of *Macaranga* (Euphorbiaceae). *Evolution* **58**, 554–570.

Rettenmeyer, C. W. (1963) Behavioral studies of army ants. *University of Kansas Scientific Bulletin* **44**, 281–465.

Ricklefs, R. E. (2006) Global variation in the diversification rate of passerine birds. *Ecology* **87**, 2468–2478.

Ricklefs, R. and Latham, R. E. (1999) Global patterns of tree species richness in moist forests: distinguishing ecological influences and historical contingency. *Oikos* **86**, 369–373.

Robinson, W. D., Robinson, T. R., Robinson, S. K. and Brawn, J. D. (2000) Nesting success of understory forest birds in central Panama. *Journal of Avian Biology* **31**, 151–164.

Rohde, K. (1992) Latitudinal gradients in species diversity: the search for the primary cause. *Oikos* **65**, 514–527.

Rosenzweig, M. L. (1995) *Species Diversity in Space and Time*. Cambridge University Press, Cambridge.

Schemske, D. (2002) Tropical diversity: patterns and processes. In: *Ecological and Evolutionary Perspectives on the Origins of Tropical Diversity: Key Papers and Commentaries* (ed. R. Chazdon and T. Whitmore), pp. 163–173. University of Chicago Press, Chicago.

Smallwood, M. and Bowles, D. J. (2002) Plants in a cold climate. *Philosophical Transactions of the Royal Society of London, Series B* **357**, 831–847.

Stearns, S. C. (1992) *The Evolution of Life Histories*. Oxford University Press, Oxford.

Stenseth, N. C. (1984) The tropics: cradle or museum? *Oikos* **43**, 417–420.

Stireman, J. O., Dyer, L. A., Janzen, D. H., *et al.* (2005) Climatic unpredictability and parasitism of caterpillars: implications of global warming. *Proceedings of the National Academy of Sciences of the United States of America* **102**, 17384–17387.

Storey, K. B. and Storey, J. M. (1988) Freeze tolerance in animals. *Physiological Reviews* **68**, 27–84.

Thompson, J. N. (1994) *The Coevolutionary Process*. University of Chicago Press, Chicago.

Thompson, J. N. (2005) *The Geographic Mosaic of Coevolution*. The University of Chicago Press, Chicago.

Turner, J. R. G. and Mallet, J. L. B. (1996) Did forest islands drive the diversity of warningly coloured butterflies? Biotic drift and the shifting balance. *Philosophical Transactions of the Royal Society of London. Series B* **351**, 835–845.

Van Valen, L. (1973) A new evolutionary law. *Evolutionary Theory* **1**, 1–30.

Vermeij, G. J. (2005) From phenomenology to first principles: towards a theory of diversity. *Proceedings of the California Academy of Sciences* **56** (Suppl. I, No. 2), 12–23.

Voituron, Y., Mouquet, N., de Mazancourt, C. and Cloberts, J. (2002) To freeze or not to freeze? An evolutionary perspective on the cold-hardiness strategies of overwintering ectotherms. *American Naturalist* **160**, 255–270.

Wallace, A. R. (1878) *Tropical Nature and Other Essays*. Macmillan, New York.

Weiblen, G. D. (2004) Correlated evolution in fig pollination. *Systematic Biology* **53**, 128–139.

Weiblen, G. D. and Bush, G. L. (2002) Speciation in fig pollinators and parasites. *Molecular Ecology* **11**, 1573–1578.

Weir, J. T. and Schluter, D. (2007) The latitudinal gradient in recent speciation and extinction rates. *Science* **315**, 1574–1576.

Wellman, F. L. (1968) More diseases on crops in the tropics than in the temperate zone. *Ceíba* **14**, 17–28.

Whittaker, R. H. (1969) Evolution of diversity in plant communities. In Diversity and stability in ecological systems. *Brookhaven Symposia in Biology* **22**, 178–195.

Wiebes, J. T. (1979) Co-evolution of figs and their insect pollinators. *Annual Review of Ecology and Systematics* **10**, 1–12

Wiens, J. J. and Donoghue, M. J. (2004) Historical biogeography, ecology and species richness. *Trends in Ecology & Evolution* **19**, 639–644.

Willig, M. R., Kaufman, D. M. and Stevens, R. D. (2003) Latitudinal gradients of biodiversity: pattern, process, scale, and synthesis. *Annual Review of Ecology and Systematics* **34**, 273–309.

Willis, E. O. and Oniki, Y. (1978) Birds and army ants. *Annual Review of Ecology and Systematics* **9**, 243–263.

Wright, S., Gray, R. D. and Gardner, R. C. (2003) Energy and the rate of evolution: inferences from plant rDNA substitution rates in the western Pacific. *Evolution* **57**, 2893–2898.

# CHAPTER THIRTEEN

# Ecological influences on the temporal pattern of speciation

## ALBERT B. PHILLIMORE AND TREVOR D. PRICE

## Introduction

Nearly all speciation events in animals seem to involve spatial separation of populations, at least in the initial stages of divergence, and we have few compelling examples of sympatric speciation (Mayr 1947; Coyne & Orr 2004). Given the requirement of geographical isolation, following Mayr (1947), we consider three steps that limit the rate at which new species form. First, gene flow between populations must be restricted. Second, populations diverge in various traits that generate reproductive isolation. Third, populations must expand ranges. This third stage is essential, for without range expansions, newly produced species would remain geographical replacements of one another, severely limiting the total number of species that can be produced from a common ancestor.

Stage 2, that is the generation of reproductive isolation, is usually taken to be synonymous with speciation (Coyne & Orr 2004). Ecological conditions experienced by different taxa can drive and accelerate the rate of attainment of reproductive isolation ('ecological speciation', Schluter 2001; Rundle & Nosil 2005), thereby accelerating speciation. This has been the focus of much recent study and is one of the main topics considered in this volume. However, ecological influences on any of the three stages – not just the acquisition of reproductive isolation – can limit the rate at which new species form (Mayr 1947).

In this chapter, we first summarize various ecological and nonecological factors that may limit each of the three stages of speciation. Then we ask if speciation events are temporally distributed in some way, which is predicted from many ecological models of speciation. In particular, ecological factors may promote speciation more rapidly early in adaptive radiation rather than later, because range expansion is easier at this time due to fewer competitors (Mayr 1942, 1947), or because reproductive isolation evolves more quickly (Rice & Hostert 1993; Schluter 2000, 2001). Gavrilets and Vose (2005, this volume) also find a burst of speciation early in adaptive radiation in their simulations, partly because as species evolve to become specialists in one particular niche, they can

*Speciation and Patterns of Diversity*, ed. Roger K. Butlin, Jon R. Bridle and Dolph Schluter. Published by Cambridge University Press. © British Ecological Society 2009.

less easily evolve to occupy alternative niches when they become available. If any of these factors apply in adaptive radiation, speciation should slow through time. In the second part of the chapter we test this prediction, by investigating patterns of speciation through time, based on an analysis of a collection of molecular phylogenies of birds.

## Stages of speciation

In this section we briefly review ecological and nonecological influences on the three stages of speciation, as envisaged by Mayr (1942, 1947), drawing on some examples from birds by way of illustration.

Stage 1. *Populations become geographically isolated.* A geographical barrier, such as a water gap for land birds, is one common nonecological way that populations may be separated, but intervening ecologically unsuitable habitat in which it is difficult to survive and reproduce is another (Mayr 1947). Alterations in the distribution of preferred habitats in response to climate change can lead to range fragmentation, setting the stage for population divergence and speciation (Wiens 2004). For example, Keast (1961) and Cracraft (1986) showed how distributions of related bird species fit postulated climatic refugia in Australia. The ongoing debate about causes of speciation in the Amazon is focused on when and where both nonecological and ecological barriers appeared (reviewed by Moritz *et al.* 2000; Newton 2003, his chapters 10 and 11; Aleixo 2004). Nonecological factors include wide rivers and islands created by a rise in sea level, and ecological factors include both pre-Pleistocene and Pleistocene climatic refugia.

Sometimes the barriers, whether ecological or nonecological, may be absolute. Other times they may be leaky, with migrants occasionally dispersing across the barrier. In this case, ecological factors in both the recipient and the source population can cause gene flow to be reduced to zero. For example, immigrants from one population may have low fitness in the alternative environment, especially in competition with residents (Mayr 1947; Nosil *et al.* 2005). Alternatively, selection in the source population may favour reduced dispersal propensity, as in the evolution of flightlessness on islands (see below).

Stage 2. *Geographically separated populations diverge.* In the hypothesis of ecological speciation, populations occupying different environments experience divergent selection pressures, causing evolution of traits that affect reproductive isolation (Schluter 2001; Rundle & Nosil 2005; Funk *et al.* 2006). Traits may diverge in response to both natural and sexual selection pressures. For example, body size and habitat choice may evolve in response to natural selection pressures, whereas traits used in communication and mate attraction may diverge in response to shifting targets of sexual selection, for example, some environments favour transmission of songs at different frequencies than others (Slabbekoorn & Smith 2002). Such differences often affect premating isolation when populations spread into sympatry. Habitat differences automatically result in many members of one incipient species not encountering members of another, and in many taxa songs are used to recognize conspecifics. Differences may also lead to postmating isolation, because hybrids are often different from either parent in ecologically relevant traits (e.g. body size) and thus may be at a disadvantage, at least in competition with the parental species. Finally, hybrids may also suffer intrinsic loss of fitness (i.e. developmental problems leading to infertility or inviability), and this may sometimes arise as a side effect of adaptation of the parental taxa to ecological conditions (Funk *et al.* 2006). Muller (1942) noted that adaptation to different temperature regimes could lead to developmental problems for hybrids, irrespective of the temperature they experience.

Ecological speciation can be compared to alternative nonecological factors that might drive reproductive isolation. Although the role of genetic drift and founder effects has been emphasized in the literature on nonecological speciation (Gavrilets 2004; see Coyne & Orr 2004 for a critique), an alternative and potentially important mode of nonecological divergence is through the occasional production of a mutation that is favoured on the genetic background on which it arises. This includes mutations that may be favoured under sexual selection (Gulick 1890), sexual conflict (Rice 1998) and intragenomic conflict (Hurst *et al.* 1996; Burt & Trivers 2006). For example, consider a new mutation that increases the frequency at which it is itself transmitted from parent to offspring, for example, by killing sperm carrying the other nonmutated form (meiotic drive). Such a mutation may rapidly increase in a population, and different mutations may spread in different populations. Interference between such mutations could cause postmating reproductive isolation. Whilst such mechanisms may be important in speciation, empirical evidence is only just accumulating (Orr *et al.* 2004; Coyne & Orr 2004).

A major difference between the nonecological model described here and ecological speciation is that gene flow is much more homogenizing in the nonecological model. Under the nonecological model, a new mutation may be favoured everywhere in the species range, and even a trickle of gene flow between two populations greatly slows the rate at which they diverge. The favoured mutation rapidly increases in the population in which it arises and once it is fixed in that population it will be introduced into the other population with every immigrant, giving the mutation a high chance of spreading through the whole species (Barton 1979). In order for populations to become substantially divergent in this model, either a long period of complete separation is needed so that many different mutations can accumulate in each population which then interfere with each other, or else the mutation rate has to be so high that different mutations arise more or less simultaneously in different parts of the species range (Kondrashov 2003). Alternatively, in the ecological model, populations occupy different environments and alleles that are favoured in one place may be disfavoured in another place. In this scenario populations will diverge if selection is strong enough to overcome any retarding effect of gene flow.

Stage 3. *Species expand ranges enabling renewed rounds of geographical isolation.* In the absence of range expansions any successive speciation events take place within smaller and smaller areas. When area is small, the likelihood that a new barrier will arise within the area is low, and if a barrier does appear, small population sizes may lead to extinction (Rosenzweig 1995). Thus, without range expansions, opportunities for speciation decay through time.

Range expansions may result from nonecological factors, specifically the disappearance of barriers. In addition, many ecological factors affect the chances of a population getting to, and becoming established in, a new location. Range expansions are likely to be easiest when individuals colonize an environment that is generally similar to that of the ancestral environment, but has few competitors, predators, etc. (Mayr 1947). Many examples (Schluter 2000) make a clear case for the absence of competitors in enabling a population to persist in an unusual niche, eventually leading to substantial divergence from its ancestor. Striking examples come from oceanic islands. Darwin (1859, p. 391) noted that 'islands are sometimes deficient in certain classes and their places are apparently occupied by other inhabitants; in the Galápagos reptiles and in New Zealand giant wingless birds, take the place of mammals'. This implies that speciation is limited by ecological opportunity, and in particular the ability to persist in a new location that is already occupied by a superior competitor (Mayr 1947).

If filling up the environment with species represents an important impediment to range expansion, it should eventually be reflected by the failure of sister species to diverge sufficiently ecologically to enable them to spread into each other's range. Indeed, in many groups, closely related species are often geographically separated or show limited overlap (Jordan 1905; Allen 1907; Phillimore *et al.* 2008); together they form a superspecies composed of ecologically and morphologically similar species (Amadon 1966; Mayr & Diamond 2001), which may often be quite old (Weir & Schluter 2007). The best explanation for the failure of ecologically similar sister species to penetrate into each other's range seems to be competitive exclusion, coupled with adaptation to environmental factors that give the competitive edge to each species in its own range. Goldberg and Lande (2006) show that hybridization may limit expansion of sister species, because individuals at the leading edge of the range mate with members of the other species, producing low fitness offspring. However, this as well as some other nonecological explanations reviewed in Price (2008) seem less generally applicable than ecological competition.

Although dispersal may often be limited by ecological conditions in foreign locations, the rate of establishment in new locations can also vary among taxa (summarized in Price 2008, chapter 7). Phillimore *et al.* (2006) find that across families of birds a crude measure of dispersal propensity exhibits a strong positive correlation with diversification (speciation–extinction) rate. Families with many species, such as parrots and finches, include species that on average tend to be dispersive. In species-poor or monotypic families, such as the ostrich, the constituent species tend to have low dispersal capabilities. In so far as ecological conditions drive the evolution of dispersal, they have a strong influence on the probability of speciation.

Easy dispersal (stage 3) and the restriction of gene flow (stage 1) are flip sides of one another and the most favourable situation for ongoing speciation must be when dispersal and range expansion are followed by the prevention of further interchange between locations. Sometimes ecological factors may produce this combination, resulting in the generation of many species. This is seen in the evolution of flightless rails (Olson 1973; Trewick 1997; Livezey 2003). First, a flying ancestor is able to disperse into and throughout the archipelago (stage 3 in our outline), but subsequent loss of flight eliminates dispersal between the islands of the archipelago and evolution of flightlessness likely gives residents an advantage in competition with immigrants (stage 1). The result is that each island population then evolves independently (stage 2). It is possible that every reasonable-sized island in the Pacific has or had its own endemic species of flightless rail and at least five flightless species appear to have been present on Mangaia (Steadman 2006).

Finally, we note that nonecological and ecological factors may operate at different stages. For example, nonecological factors, such as some processes of sexual selection that lead to reproductive isolation (stage 2) may result in the production of ecologically similar species competitively excluded from each other's range (stage 3; West-Eberhard 1983).

## Temporal patterns of speciation

Although multiple factors clearly affect the rate at which new species are formed, in the theory of adaptive radiation, speciation rates are high early on and slow down later. This is implied in common definitions of adaptive radiation (Simpson 1953; Schluter 2000). For example, according to Schluter (2000, p. 10) adaptive radiation is 'the evolution of ecological and morphological diversity within a rapidly multiplying lineage'. Because lineages cannot multiply rapidly for long, the implication is that speciation must slow through time. On entry into a new location or after a mass extinction, rapid divergence from a single lineage to exploit a diversity of resources may accelerate ecological speciation, as well as present the opportunity for frequent successful range expansions. As the environment fills up, ecological speciation slows, and range expansions become less easy.

Studies of young adaptive radiations, such as fish in postglacial lakes and Darwin's finches, provide supporting evidence for an association of rapid speciation with ecological divergence (Schluter 2000, 2001). We ask here if the predicted slowdown develops in older species radiations.

## Tests of slowdown in molecular phylogenies

Patterns of species diversification through time can be assessed using time-calibrated molecular phylogenies. The basic idea is to tally the total number of lineages in the tree as a function of time, starting at the root (where there are two lineages) up to the present day (where the number of lineages equals the number of species). Under a null model of constant probability of branching through time, and no extinction (the 'pure-birth' model), the expected increase in the logarithm of the number of lineages against time follows a straight line (Nee *et al.* 1994a; see also Ricklefs, this volume). In the null model, variation in the so-called $\gamma$-statistic (Pybus & Harvey 2000) follows a known distribution (a normal distribution with a mean $= 0$ and a standard deviation $= 1$). A large negative value of $\gamma$ allows one to reject the pure birth null hypothesis, and is taken to imply that speciation rate has slowed towards the present. It has been suggested that the presence of a slowdown towards the present could result from accelerated extinction (Zink *et al.* 2004). However, this is only expected if extinction is highly nonrandom (e.g. if one member of a sister pair consistently goes extinct); simulations show that a slowdown is highly unlikely under a model of random (accelerating) extinction (Weir 2006, Rabosky & Lovette 2008). Thus, the $\gamma$ statistic provides a reasonably strong test for a slowdown in speciation rate, provided branch-lengths in the phylogeny adequately represent time (Phillimore & Price 2008).

A large positive value of $\gamma$ (i.e. an increasing rate of lineage accumulation towards the present in the reconstructed phylogeny) rejects a model of constant probability of speciation without extinction, but the pattern can arise in two ways. First, the speciation rate may have increased over time. Second, whenever extinction is present, reconstructed phylogenies (i.e. those that do not include fossil data) should show accelerated cladogenesis towards the present. This is because more recent extinctions remove few potential descendants in the reconstructed phylogeny, whereas older extinctions remove many more (Nee *et al.* 1994b; Ricklefs, this volume).

As described in more detail in Phillimore and Price (2008), we selected 45 bird clades at the genus to family level, for which more than 65% of all species have been sequenced for mitochondrial protein coding genes. We reconstructed each phylogenetic tree using a relaxed clock Bayesian method implemented in *Beast* (Drummond *et al.* 2006). We set the mean rate of molecular evolution to be 1% per lineage per million years and a GTR$+$I$+\Gamma$ model of nucleotide substitution (Weir & Schluter 2007). An advantage of using a relaxed clock approach is that the tree can be rooted without requiring an outgroup. We sampled trees from the posterior distribution, and for all sampled trees calculated $\gamma$ based on

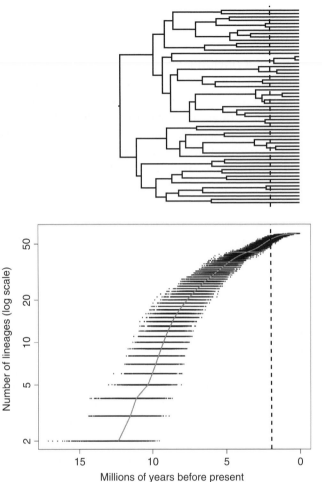

**Figure 13.1** Estimated phylogeny and a lineage-through-time plot (on a semi-log scale) for warblers in the *Phylloscopus* and *Seicercus* genera, based on mitochondrial sequences in Johansson *et al.* (2007). The phylogeny was obtained using the Bayesian relaxed clock method in the program *Beast* (Drummond *et al.* 2006; described further in Phillimore & Price 2008). The lineage-through-time plot is based on a sample of 2000 trees from the posterior distribution; the line connects the median values at each value on the *y* axis. The *γ* statistic is calculated for each of the posterior topologies from the root up until 2 million years before the present and has a median value of −3.33, indicating a highly significant slowdown (*P* < 0.01). Eleven of the 70 species in this clade are missing and, after correcting for this, *γ*=−2.96.

internode distances between the root of the tree and the last node that lies earlier than 2 million years. We set a cut-off of 2 million years because lineage splitting more recently may be under-recorded (especially under the biological species concept; Avise & Walker 1998), or over-recorded (as a result of excessive splitting of distinctive populations; Isaac *et al.* 2004). An example of a phylogeny and the corresponding lineage through time plot for the *Phylloscopus* and *Seicercus* clade is shown in Fig. 13.1. We obtained a value of *γ* for each tree sampled from the posterior distribution using the *LASER* R library (Rabosky 2006) and used the median value as an estimate of slowdown for the phylogeny. We corrected all *γ* values for the number of present-day species absent from the phylogeny using methods outlined by Pybus and Harvey (2000) and Harmon *et al.* (2003), because incomplete taxon sampling can generate apparent slowdowns (Pybus & Harvey 2000).

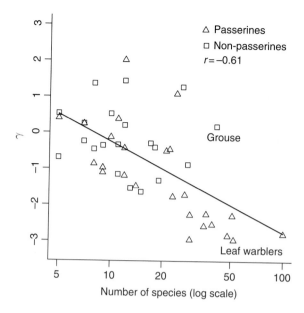

**Figure 13.2** Plot of $\gamma$ estimated from the root of the clade to the last bifurcation event prior to 2 million years versus clade size, where clade size is the total number of lineages estimated to be present 2 million years ago (and ignoring all species not sequenced) for 45 bird clades (from Phillimore & Price 2008). The least squares regression line is $b = -1.09 \pm 0.22$, $P < 0.001$. The Old World leaf warblers (*Phylloscopus* and *Seicercus*), whose phylogenetic relationships are illustrated in Figure 13.1, are indicated.

Many of the larger phylogenies have significantly negative $\gamma$ values, implying a slowdown (Fig. 13.2). The Old World leaf warbler clade in Fig. 13.1 is the most extreme example. Conversely, many of the smaller clades have $\gamma$ values that are close to zero, or even positive. In fact, if one considers that (1) many clades that have not been analysed tend to be small and (2) small clades tend to have positive $\gamma$ values, then we surmise that the average avian clade may actually show a $\gamma$ value that is close to zero or positive. Thus, across clades as a whole we find no evidence that slowdowns are the norm; very young clades show a speedup, which compensates for the slowdown in old clades.

These patterns might be expected in an adaptive radiation model; as clades get larger their available niche space becomes filled making it increasingly difficult for further speciation events. However, the same pattern is also expected from a model of uniform probability of diversification through time (Raup *et al.* 1973; Phillimore & Price 2008). Because of the exponential growth inherent in diversification, a large clade is produced when, by chance, many speciation events happen early in the clade's history, and a small clade is produced when, by chance, few speciation events happen early. Subsequently, both large and small clades regress towards a universal average rate, resulting in a slowdown in large clades and a speedup in small ones. This is a regression effect (Kelly & Price 2005) that operates over timescales of many millions of years. The patterns can be easily demonstrated by simulation (Phillimore & Price 2008).

A way to see why a correlation of slowdown with clade size should arise under the uniform probability model is through analogy with coin tossing. Here 'Heads' results in a lineage-splitting event in a given time interval and 'Tails' indicates a failure to speciate. For simplicity, extinction is not allowed. Imagine that five coin tosses result in five heads, hence five lineage-splitting events early in diversification. This lucky clade has already produced six lineages and the probability is high that the clade will quickly grow to be large. On the other hand, suppose that those five coin tosses all gave tails, hence no speciation, and the clade is on a trajectory to be small. In the next time interval, the large clade will on average produce three splitting events, whereas the small clade will on average split half the time: the rate of splitting has decreased in the large clade and increased in the small one. Regression towards the mean is particularly powerful in phylogenies that grow under a constant rate process, because as time goes on the increase in the number of lineages results in a more precise estimate of the universal diversification rate.

Different clades may well have different rates of diversification. This seems likely, if only because the number of species in a particular clade is correlated across different regions of the world (i.e. some clades consistently have more species in a region than other clades, Ricklefs, this volume). However, simulation studies show that even if net diversification rates differ between clades, a uniform probability of speciation within clades will generate a negative correlation between $\gamma$ and clade size, provided extinction rates are relatively low (Phillimore & Price 2008). Simulations also show that when the extinction rate is high the correlation between clade size and slowdown is lessened. In the extreme when extinction is set equal to speciation there is no correlation (Phillimore & Price 2008). This appears to be because, when extinction rates are high, most of the rapid early cladogenesis is eroded in the reconstructed phylogeny, and all early lineages but one or two go extinct (see Fig. 14.2 in Ricklefs, this volume). In addition to the reduction in the correlation between $\gamma$ and clade size, as extinction becomes more important, the values of $\gamma$ themselves become generally positive; i.e. there is an upturn in cladogenesis rate on the reconstructed phylogeny. Thus if extinction is high, the presence of negative $\gamma$ values in real datasets are likely to imply real slowdowns in speciation.

The pattern of slowdown in large clades is expected under constant speciation–extinction models, whenever the extinction rate is not high. Constant rate models form a simple hypothesis, so the question is, can we distinguish the observed pattern from that of constant rate? Here we suggest three approaches, based on (1) correlations with clade age, (2) correlations with geographic range attributes and (3) a comparison of the strength of slowdowns observed in large clades versus those expected by chance. All tests used the data for the bird clades shown in Fig. 13.2.

## Clade age

Among the clades analysed in Fig. 13.2, age to the root varies from an estimated 4.52 million years to 24.88 million years (mean = 11.33 ± 4.63 SD). $\gamma$ is not

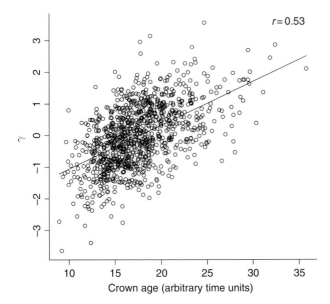

**Figure 13.3** Plot of $\gamma$ versus crown age obtained from 1000 pure birth simulations of phylogenies containing 50 species in *Phyl-o-gen* (http://evolve.zoo.ox.ac.uk/ software.html?id=phylogen; provided by A. Rambaut). The parameters used were $b=0.2$, $d=0$ and clade size $=50$ species.

correlated with clade age ($r=0.02$, $P=0.91$). However, when clade size is held constant, the partial correlation of clade age with $\gamma$ is positive ($r'=0.29$, $P<0.05$) (the correlation and partial correlation (holding age fixed) of clade size and $\gamma$ are $-0.64$ and $-0.67$, respectively). In a model of constant probability of birth or birth–death, a positive association of $\gamma$ with clade age is expected, i.e. relatively young clades should show a slowdown (see Fig. 13.3). This is because clades that reach a certain size quickly are likely to have experienced above average initial speciation rates, and hence will later show slowdowns, whereas clades that reach a certain size slowly are likely to have had below average initial speciation rates, and hence will later show a speedup. In the adaptive radiation model the opposite pattern is expected. If clades experience an initially high opportunity for speciation, which declines or ceases once a certain number of species exist, then the older the clade the stronger the expected slowdown. Thus the results of this test are in accord with the predictions of the random speciation model and do not appear to support those of the adaptive radiation model.

## Correlates with sympatry and other aspects of range size

In the adaptive radiation model an important cause of speciation slowdown is a failure of species to expand their ranges and come into sympatry, because of ecological competition from their close relatives. This model leads to three predictions. First, because it should become increasingly difficult for additional species to establish sympatric distributions, we predict a pattern of increased slowdown in association with the maximum number of co-occurring

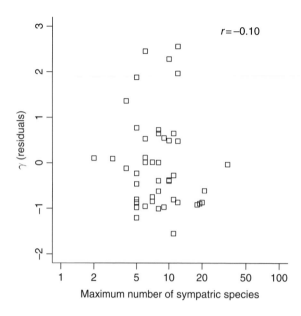

**Figure 13.4** Plot of $\gamma$ (residuals from the multiple regression on clade size [ln transformed] and clade age) versus number of sympatric species. Number of sympatric species is estimated as the maximum number of species that are present in a single grid cell (approximately 96 km by 96 km) across the world (from Orme *et al.* 2005, 2006; see Figure 13.5 legend for details).

(sympatric) species in a clade. Second, clades with a circumscribed geographic range should show strong slowdown if the total number of species that can coexist is limited (Ricklefs 2006; Ricklefs, this volume). Third, if range expansion is the rate-limiting step, then smaller average ranges for the species in a clade should correspond to a stronger slowdown (a more negative $\gamma$).

After controlling for clade size and age we found no correlation of $\gamma$ with either the maximum number of species in sympatry (Fig. 13.4; cf. Weir's (2006) finding for a sample of Amazon clades) or total geographical range size of the clade (Fig. 13.5, left). The relationship between species mean range size and $\gamma$ was significant (Fig. 13.5, right, $b = -0.09 \pm 0.041$ se, $P < 0.05$). However, the slope was of the opposite sign to that predicted under adaptive radiation reasoning; stronger slowdowns were associated with larger range sizes. We suggest that the negative relationship between average range size and $\gamma$ reflects the signature of recent allopatric speciation. Unless post-speciation range expansion is very rapid, clades that have experienced high rates of recent allopatric speciation are likely to contain species with ranges that are smaller on average (Phillimore *et al.* 2006).

The power of the tests may be low and overall evidence is weak, but more generally the tests point in the direction of a slowdown being attributable to the signature of constant probability of diversification. It would be worthwhile constructing other tests that use traits that are not *a priori* expected to correlate with speciation under the constant-probability model. For example, dispersal propensity or body size could be tested for their correlation with slowdown, after controlling for clade age and clade size.

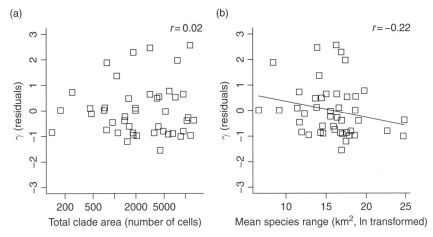

**Figure 13.5** (a) Plot of $\gamma$ (residuals from the multiple regression on clade size [ln transformed] and clade age) versus total clade area. Total clade area was calculated from a global dataset of bird species ranges projected on a Behrman equal area grid in which grid cells were approximately 96 km by 96 km in size (Orme *et al.* 2005, 2006). Area was estimated as the total number of cells in which a clade member was present. (b) Plot of $\gamma$ (residuals from the multiple regression on clade size [ln transformed] and clade age) versus the mean range size of the species in a clade. Range sizes were obtained from a global database of species distribution shape-files (Orme *et al.* 2005). Range sizes (in km$^2$) were transformed by the exponent 0.2 to normalise the data prior to obtaining the arithmetic mean (Phillimore *et al.* 2007). Note that species ranges are based on the taxonomy of Sibley and Monroe (1990, 1993), whereas in some instances our phylogenetic reconstructions are based on more recent taxonomy.

## Strength of slowdowns

A final test is to ask if the pattern of slowdown in the data (Fig. 13.2) is stronger than that predicted from the constant rate models. We asked whether the frequency of strong slowdowns (defined as $\gamma$ values $< -1.645$) in large clades (those containing $\geq 15$ species) departs from the null expectation under a constant rate model (Phillimore & Price 2008). We found that across a wide parameter space (simulation duration and the ratio of speciation to extinction were both varied), strong slowdowns were significantly more frequent than expected. Indeed, when the death rate is nonzero the probability that a constant rate model produces the observed frequency of strong slowdowns in large clades is infinitesimal. Studies on fossils indicate that extinction rates in many groups are high and may approach speciation rates (Alroy, this volume), so the slowdowns in Fig. 13.3 do seem to reflect a deterministic slowing of speciation towards the present.

## Discussion

A slowdown in cladogenesis has been regularly found in studies of single phylogenies and has often been interpreted in terms of adaptive radiation and niche filling (Nee *et al.* 1992; McKenna & Farrell 2006; Weir 2006; Price 2008). Although others have suggested slowdowns result from nonrandom extinction (Zink & Slowinski 1995; Zink *et al.* 2004), or episodic appearances of multiple barriers (Zink & Slowinski 1995; Lovette & Bermingham 1999; Kozak *et al.* 2005), niche filling limiting range expansions is also often implied in these examples. We show here that slowdowns appear when speciation follows a model of constant probability of speciation through time. Some tests yielded results that did not support the adaptive radiation model but were consistent with this model. However the presence of strong slowdowns in large clades are in accord with a model of density-dependent speciation. Ricklefs (this volume) notes that evidence for density-dependent regulation is also implied by a lack of correlation between clade age and clade numbers in birds, at least for clades considered at and around the family level. As Ricklefs also notes such regulation could be driven either by an increase in extinction or decrease in speciation as species accumulate. Our analysis of lineage-through-time plots suggests that a decrease in speciation rate is an important contributor.

There is quite strong evidence for a very high – and unsustainable over long periods – speciation rate in association with ecological opportunity, such as following mass extinctions (Foote 1997) or in young archipelagos such as the Galápagos islands (Schluter 2000, 2001). Some Pleistocene radiations on continents also seem to be exceptionally rapid, such as the production of 81 species of *Lupinus* in the Andes during the last 2 million years (Hughes & Eastwood 2006). These observations fit models of what is to be expected in the early stages of adaptive radiation (Gavrilets & Vose 2005, this volume), i.e. when a single founding species enters a new environment. Those models predict a slowdown. Our results suggest that such slowdowns are a real feature of large clades.

We draw two main conclusions from the analysis. First, several temporal patterns in diversification rates can be explained on the basis of a model of constant probability of speciation or diversification. Speciation is affected by very many factors including occasional extinctions that create new ecological opportunities, appearance of habitat that can be exploited by multiple lineages (rather than a single lineage that rapidly diversifies), the strength and persistence of barriers, chance dispersal events, and the occasional evolution of traits within lineages that affect speciation probability. If these factors operate more or less independently, then a random pattern of speciation through time is expected (Raup 1977). The second conclusion is that the presence of exceptionally rapid speciation in some young radiations, coupled with the evidence for strong slowdowns in large clades, support models of ecological controls on the rate at which new species form. It seems likely that both more or less continuous species turnover through time, plus a process of adaptive radiation followed by niche filling, contribute to patterns of speciation through history (Nee *et al.* 1992; Ricklefs, this volume).

Ecological controls on speciation may affect the speed at which reproductive isolation develops, but a main factor limiting speciation is likely to be changes in the ease of range expansions. Range expansions are an essential requirement for ongoing allopatric speciation. We suggest that those factors that affect the ability of new populations to become established and persist in local communities are likely to have an important influence on regional speciation rates.

### Acknowledgements

We thank the organizers of the Symposium for the invitation, and Robert Ricklefs, Doug Schemske and Dolph Schluter for comments on the manuscript. Research reported in this chapter was partly supported by a grant from the National Science Foundation (US) to Trevor Price.

### References

Aleixo, A. (2004) Historical diversification of a Terra-firme forest bird superspecies: a phylogeographic perspective on the role of different hypotheses of Amazonian diversification. *Evolution* **58**,1303–1317.

Allen, J. A. (1907) Mutations and the geographic distribution of nearly related species in plants and animals. *American Naturalist* **41**, 653–655.

Amadon, D. (1966) The superspecies concept. *Systematic Zoology* **15**, 245–249.

Avise, J. C. and Walker, D. (1998) Pleistocene phylogeographic effects on avian populations and the speciation process. *Proceedings of the Royal Society of London Series B–Biological Sciences* **265**, 457–463.

Barton, N. H. (1979) Gene flow past a cline. *Heredity* **43**, 333–339.

Burt, A. and Trivers, R. (2006) *Genes in Conflict: The Biology of Selfish Genetic Elements.* Harvard University Press, Cambridge, MA.

Coyne, J. A. and Orr, H. A. (2004) *Speciation.* Sinauer, Sunderland, MA.

Cracraft, J. (1986) Origin and evolution of continental biotas – speciation and historical congruence within the Australian avifauna. *Evolution* **40**, 977–996.

Darwin, C. (1859) *On the Origin of Species by Means of Natural Selection.* John Murray, London.

Drummond, A.J., Ho, S. Y. W., Phillips, M. J. and Rambaut, A. (2006) Relaxed phylogenetics and dating with confidence. *PLoS Biology* **4**, e88.

Foote, M. (1997) The evolution of morphological diversity. *Annual Review of Ecology and Systematics* **28**, 129–152.

Funk, D. J., Nosil, P. and Etges, W. J. (2006) Ecological divergence exhibits consistently positive associations with reproductive isolation across disparate taxa. *Proceedings of the National Academy of Sciences of the United States of America* **103**, 3209–3213.

Gavrilets, S. (2004) *Fitness Landscapes and the Origin of Species.* Princeton University Press, Princeton, NJ.

Gavrilets, S. and Vose, A. (2005) Dynamic patterns of adaptive radiation. *Proceedings of the National Academy of Sciences of the United States of America* **102**, 18040–18045.

Goldberg, E. E. and Lande, R. (2006) Ecological and reproductive character displacement on an environmental gradient. *Evolution* **60**, 1344–1357.

Gulick, J. T. (1890) Indiscriminate separation under the same environment, a cause of divergence. *Nature* **42**, 369–370.

Harmon, L. J., Schulte, J. A., Larson, A. and Losos, J. B. (2003) Tempo and mode of evolutionary radiation in iguanian lizards. *Science* **301**, 961–964.

Hughes, C. and Eastwood, R. (2006) Island radiation on a continental scale: exceptional rates of plant diversification after uplift of the Andes. *Proceedings of the National Academy of Sciences of the United States of America* **103**, 10334–10339.

Hurst, L. D., Atlan, A. and Bengtsson, B. O. (1996) Genetic conflicts. *Quarterly Review of Biology* **71**, 317–364.

Isaac, N. J. B., Mallet, J. and Mace, G. M. (2004) Taxonomic inflation: its influence on macroecology and conservation. *Trends in Ecology & Evolution* **19**, 464–469.

Johansson, U. S., Alström P., Olsson U., *et al.* (2007) Build-up of the Himalayan avifauna through immigration: a biogeographical analysis of the *Phylloscopus* and *Seicercus* warblers. *Evolution* **61**, 324–333.

Jordan, D. S. (1905) The origin of species through isolation. *Science* **22**, 545–562.

Keast, A. (1961) Bird speciation on the Australian continent. *Bulletin of the Museum of Comparative Zoology* **123**, 305–495.

Kelly, C. and Price, T. D. (2005) Correcting for regression to the mean in behavior and ecology. *American Naturalist* **166**, 700–707.

Kondrashov, A. S. (2003) Accumulation of Dobzhansky–Muller incompatibilities within a spatially structured population. *Evolution* **57**, 151–153.

Kozak, K. H., Weisrock, D. W. and Larson, A. (2005) Rapid lineage accumulation in a non-adaptive radiation: phylogenetic analysis of diversification rates in eastern North American woodland salamanders (Plethodontidae: *Plethodon*). *Proceedings of the Royal Society of London Series B–Biological Sciences* **273**, 539–546.

Livezey, B. C. (2003) Evolution of flightlessness in rails (Gruiformes: *Rallidae*): phylogenetic, ecomorphological, and ontogenetic perspectives. *Ornithological Monographs* **53**, 1–654.

Lovette, I. J. and Bermingham, E. (1999) Explosive speciation in the New World *Dendroica* warblers. *Proceedings of the Royal Society of London Series B–Biological Sciences* **266**, 1629–1636.

Mayr, E. (1942) *Systematics and the Origin of Species from the Viewpoint of a Zoologist*. Columbia University Press, New York.

Mayr, E. (1947) Ecological factors in speciation. *Evolution* **1**, 263–288.

Mayr, E. & Diamond, J. M. (2001) *The Birds of Northern Melanesia: Speciation, Ecology, and Biogeography*. Oxford University Press, New York.

McKenna, D. D. and Farrell, B. D. (2006) Tropical forests are both evolutionary cradles and museums of leaf beetle diversity. *Proceedings of the National Academy of Sciences of the United States of America* **103**, 10947–10951.

Moritz, C., Patton, J. L., Schneider, C. J. and Smith, T. B. (2000) Diversification of rainforest faunas: an integrated molecular approach. *Annual Review of Ecology and Systematics* **31**, 533–563.

Muller, H. (1942) Isolating mechanisms, evolution and temperature. *Biological Symposia* **6**, 71–125.

Nee, S., Mooers, A. Ø. and Harvey, P. H. (1992) Tempo and mode of evolution revealed from molecular phylogenies. *Proceedings of the National Academy of Sciences of the United States of America* **89**, 8322–8326.

Nee, S., Holmes, E. C., May, R. M. and Harvey, P. H. (1994a) Extinction rates can be estimated from molecular phylogenies. *Philosophical Transactions of the Royal Society-Series B–Biological Sciences* **344**, 77–82.

Nee, S., May, R. M. and Harvey, P. H. (1994b) The reconstructed evolutionary process. *Philosophical Transactions of the Royal Society-Series B–Biological Sciences* **344**, 305–311.

Newton, I. (2003) *The Speciation and Biogeography of Birds*. Academic Press, San Diego, CA.

Nosil, P., Vines, T. and Funk, D. J. (2005) Perspective: reproductive isolation caused by natural selection against migrants between divergent environments. *Evolution* **59**, 705–719.

Olson, S. L. (1973) *Evolution of the Rails of the South Atlantic Islands (Aves: Rallidae)*. Smithsonian Institution Press, Washington, DC.

Orme, C. D. L., Burgess, M., Eigenbrod, F., *et al.* (2005) Global hotspots of species richness are not congruent with endemism or threat. *Nature* **436**, 1016–1019.

Orme, C. D. L., Davies, R. G., Olson, V. A., *et al.* (2006) Global patterns of geographic range size in birds. *PLoS Biology* **4**, e1276.

Orr, H. A., Masly, J. P. and Presgraves, D. C. (2004) Speciation genes. *Current Opinion in Genetics and Development* **14**, 675–679.

Phillimore, A. B., Freckleton, R. P., Orme, C. D. L. and Owens, I. P. F. (2006) Ecology predicts large-scale patterns of diversification in birds. *American Naturalist* **168**, 220–229.

Phillimore, A. B., Orme, C. D. L., Davies, R. G., *et al.* (2007) Biogeographical basis of recent phenotypic divergence among birds: a global study of subspecies richness. *Evolution* **61**, 942–957.

Phillimore A. B., Orme C. D. L., Thomas G. H., *et al.* (2008) Sympatric speciation in birds is rare or absent: insights from range data and simulations. *American Naturalist* **171**, 646–657.

Phillimore, A. B. and Price, T. D. (2008) Density dependent cladogenesis in birds. *PloS Biology* **6**, e71.

Price, T. (2008) *Speciation in Birds*. Roberts and Co., Greenwood Village, CO.

Pybus, O. G. and Harvey, P. H. (2000) Testing macro-evolutionary models using incomplete molecular phylogenies. *Proceedings of the Royal Society of London Series B–Biological Sciences* **267**, 2267–2272.

Rabosky, D. L. (2006) Likelihood methods for detecting temporal shifts in diversification rates. *Evolution* **60**, 1152–1164.

Rabosky, D. L. and Lovette, I. J. (2008) Explosive evolutionary radiations: decreasing

speciation or increasing extinction through time? *Evolution* **62**, 1866–1875.

Raup, D. M. (1977) Stochastic models in evolutionary palaeontology. In: *Patterns of Evolution* (ed. A. Hallam), pp. 59–78. Elsevier, Amsterdam.

Raup, D. M., Gould, S. J., Schopf, T. J. M. and Simberloff, D. S. (1973) Stochastic models of phylogeny and the evolution of diversity. *Journal of Geology* **81**, 525–542.

Rice, W. R. (1998) Intergenomic conflict, interlocus antagonistic evolution, and the evolution of reproductive isolation. In: *Endless Forms: Species and Speciation* (ed. D. J. Howard and S. H. Berlocher), pp. 261–270. Oxford University Press, New York.

Rice, W. R. and Hostert, E. E. (1993) Laboratory experiments on speciation: what have we learned in 40 years? *Evolution* **47**, 1637–1653.

Ricklefs, R. E. (2006) Global variation in the diversification rate of passerine birds. *Ecology* **87**, 2468–2478.

Rosenzweig, M. L. (1995) *Species Diversity in Space and Time*. Cambridge University Press, Cambridge.

Rundle, H. D. and Nosil, P. (2005) Ecological speciation. *Ecology Letters* **8**, 336–352.

Schluter, D. (2000) *The Ecology of Adaptive Radiation*. Oxford University Press, Oxford.

Schluter, D. (2001) Ecology and the origin of species. *Trends in Ecology & Evolution* **16**, 372–380.

Sibley, C. G. and Monroe, B. L. (1990) *Distribution and Taxonomy of Birds of the World*. Yale University Press, New Haven.

Sibley, C. G. and Monroe, B. L. (1993) *Supplement to Distribution and Taxonomy of Birds of the World*. Yale University Press, New Haven and London.

Simpson, G. G. (1953) *The Major Features of Evolution*. Columbia University Press, New York.

Slabbekoorn, H. and Smith, T. B. (2002) Bird song, ecology and speciation. *Philosophical*

*Transactions of the Royal Society of London Series B–Biological Sciences* **357**, 493–503.

Steadman, D. W. (2006) *Extinction and Biogeography of Tropical Pacific Birds.* University of Chicago Press, Chicago.

Trewick, S. A. (1997) Sympatric flightless rails *Gallirallus dieffenbachii* and *G. modestus* on the Chatham Islands, New Zealand; morphometrics and alternative evolutionary scenarios. *Journal of the Royal Society of New Zealand* **27**, 451–464.

Weir, J. T. (2006) Divergent patterns of species accumulation in lowland and highland neotropical birds. *Evolution* **60**, 842–855.

Weir, J. T. and Schluter, D. (2007) The latitudinal gradient in recent speciation and extinction rates. *Science* **315**, 1574–1576.

West-Eberhard, M. J. (1983) Sexual selection, social competition and speciation. *Quarterly Review of Biology* **58**, 155–183.

Wiens, J. J. (2004) Speciation and ecology revisited: phylogenetic niche conservatism and the origin of species. *Evolution* **58**, 193–197.

Zink, R. M., Klicka, J. and Barber, B. R. (2004) The tempo of avian diversification during the Quaternary. *Philosophical Transactions of the Royal Society of London Series B–Biological Sciences* **359**, 215–219.

Zink, R. M. and Slowinski, J. B. (1995) Evidence from molecular systematics for decreased avian diversification in the Pleistocene Epoch. *Proceedings of the National Academy of Sciences of the United States of America* **92**, 5832–5835.

# CHAPTER FOURTEEN

# Speciation, extinction and diversity

ROBERT E. RICKLEFS

Biologists have long endeavoured to understand variation in the number of species over the surface of the earth, but have not reached a general consensus on the causes of observed patterns. Early explanations focused on history, including the effects of Ice Age climate change on diversity at northern latitudes (Wallace 1878) and the age and area of a region (Willis 1922). Paleontologists have used the fossil record to characterize vicissitudes of diversity through time, particularly the effects of catastrophic events and the replacement of older taxa by newer forms (Simpson 1944, 1953; Stanley 1979). Beginning in the 1960s, ecologists emphasized the ability of species to coexist locally in communities of interacting species, largely ignoring the effects of history (Kingsland 1985; Ricklefs 1987) and explaining variation in diversity in terms of the ability of environments to support interacting populations (MacArthur & Levins 1967; MacArthur 1970; Vandermeer 1972; May 1975). This approach proved to be compelling, and paleontologists soon integrated population thinking in their work, constructing models of diversification to explain patterns in taxon richness through time (Raup & Gould 1974). MacArthur and Wilson's (1967) equilibrium theory of island biogeography was influential in this regard, particularly the application of models of species formation and extinction within regions to understand both long-term stasis in diversity and variation in diversity at global scales (Rosenzweig 1995).

More recently, ecologists have begun to appreciate (again) the importance of regional/historical processes in shaping large-scale patterns of diversity (Ricklefs & Schluter 1993; Currie *et al.* 2004; Wiens & Donoghue 2004; Fine & Ree 2006; Wiens *et al.* 2006). In particular, the development of phylogenetic reconstruction from molecular data has provided new tools for analysing the history of diversification and estimating rates of speciation and extinction. However, the phylogenetic framework also has created a mindset among many biologists in which diversity patterns largely reflect the relative rate of increase within monophyletic groups (Slowinski & Guyer 1989, 1993; Hey 1992; Haesler & Seehausen 2005) and the period over which they have accumulated species (Nee *et al.* 1992; Harvey *et al.* 1994; Barraclough *et al.* 1999; Magallón &

*Speciation and Patterns of Diversity*, ed. Roger K. Butlin, Jon R. Bridle and Dolph Schluter. Published by Cambridge University Press. © British Ecological Society 2009.

Sanderson 2001; Barraclough & Nee 2001; Fine & Ree 2006; McPeek & Brown 2007; Mittelbach *et al.* 2007). In this chapter, I explore some of the implications of this diversification focus and suggest that it cannot provide a complete explanation for variation in diversity without taking into account the influence of regional characteristics, and of diversity itself, on speciation and extinction. The fossil record suggests that, barring mass extinction and recovery from such events, established clades of species reach relatively stable levels within regions (regional 'carrying capacities') at which they remain for long periods (Simpson 1953; Sepkoski 1979; Stanley 1979). Thus, global patterns of diversity appear to be relatively insensitive to rates of diversification, but rather reflect the capacity of regions to support species realized through diversity-dependence of speciation and extinction.

Biologists and paleobiologists differ in their perspectives on diversity in ways that are instructive for our present purpose (Nee *et al.* 1992; Nee 2006). Paleontologists have reconciled their views with the fossil record, which often reveals rapid diversification as a new clade fills ecological space, followed by a period of relatively stable taxonomic diversity and eventual decline to extinction, sometimes hastened by calamitous events (Stanley 1979; Foote 1993, 2000). Biologists focus on the contemporary diversity of modern groups, within each of which the species have descended from a single common ancestor. Thus, their focus has been on variation in the rate of increase in species numbers.

### Rates of speciation and extinction in diversifying clades

One approach to understanding variation in diversity makes use of the size and age of clades to estimate relative rates of diversification, which can then be related to potential influences, including traits of species and environments. The simplest application of this approach is to assume a constant rate of speciation ($\lambda$) and no extinction, in which case the number of species ($N$) increases with time, or clade age ($t$), according to $N = e^{\lambda t}$, and the rate of speciation can be estimated by $\lambda = \ln N / t$ (Stanley 1979; Wilson 1983) or by $(N - 1)/S$, where $S$ is the total branch length in a phylogeny, including the root edge (Purvis *et al.* 1995; Nee 2001; McPeek & Brown 2007). Clade age can be determined from the fossil record or from time-calibrated molecular phylogenies (Renner 2005; Won & Renner 2006; Benton & Donoghue 2007; Donoghue & Benton 2007).

Magallón and Sanderson (2001) incorporated rate of extinction ($\mu$) into analyses of the relationship between clade size and age based on a random-walk model of diversification (Bailey 1964), in which species number increases as

$$N(t) = \frac{\lambda E(n) - \mu}{\lambda - \mu},$$
(14.1)

where $E(n) = e^{(\lambda - \mu)t}$. When extinction is expressed as a multiple of the rate of speciation, i.e. $\mu = \kappa\lambda$, then this equation may be written as

$$N(t) = \frac{E(n) - \kappa}{1 - \kappa}.$$   (14.2)

Magallón and Sanderson estimated the net diversification rate, $(\lambda - \mu)$ or $\lambda(1 - \kappa)$, in orders and higher taxa of flowering plants for $\kappa = 0$ and $\kappa = 0.9$. They emphasized that although the estimated rate of speciation differed considerably when using $\kappa = 0$ and $\kappa = 0.9$, this was paralleled by variation in $\mu$ so that $(\lambda - \mu)$ varied little. In their analysis, the rate of diversification across many clades of angiosperms averaged 0.089 per million years $(My^{-1})$ for $\kappa = 0$ and 0.077 $My^{-1}$ for $\kappa = 0.9$.

Clearly, each clade of flowering plants has increased from a single common ancestor. However, Magallón and Sanderson's treatment implied that the underlying process embodies a positive net diversification rate. Moreover, they regarded $\kappa = 0.9$ as arbitrary but an upper bound for the relative extinction rate, for two reasons (p. 1765; their epsilon ($\varepsilon$) is the same as $\kappa$):

First, for large clades, the probability of survival to the present is very closely approximated by $1 - \varepsilon$ ..., which means that values of $\varepsilon$ greater than 0.9 correspond to clades having less than a 10% chance of surviving to the present. For angiosperms as a whole, with a standing diversity of more than 260,000 species, retrospectively, it seems improbable that $\varepsilon$ could have been much higher than 0.9 on average. Second, as $\varepsilon \to 1$, the magnitude of speciation and extinction rates, $\lambda$ and $\mu$, increases rapidly to maintain the same net diversification rate, $r$. At values above $\varepsilon = 0.9$, estimated values of $\lambda$ and $\mu$ begin to exceed 1.0 events per million years, which is approximately the upper limit estimated from real data from a variety of taxa (Stanley 1979; Hulbert 1993). When relative extinction has higher values than this, diversification quickly becomes a highly chaotic process dominated by stochastic extinction and extremely rapid turnover.

The rate of speciation ($\lambda$) and the relative rate of extinction ($\kappa$) can be estimated directly for a sample of known-age clades by maximum likelihood (Bokma 2003). The likelihood of clade size $n$, given rates $\lambda$ and $\kappa$ over time $t$, is

$$P(n|t) = (1 - \kappa) \frac{[E(n) - 1]^{n-1}}{[E(n) - \kappa]^n},$$   (14.3)

where $E(n) = e^{\lambda(1-\kappa)t}$ $(= e^{(\lambda - \mu)t})$. The likelihood $l_i$ for clade $i$ is $P(n_i|t_i)$, and the maximum likelihood for a particular sample is the combination of $\lambda$ and $\kappa$ that maximize $\ln L = \sum \ln(l_i)$. Maximum likelihood estimates fail to converge for $\lambda$ and $\kappa$ in Magallón and Sanderson's (2001) data. However, in an analysis of New World clades of passerine birds, maximum likelihood estimates of $\lambda = 3.16$ and $\kappa = 0.995$ pertained to a sample of 18 North American (primarily temperate) clades, and $\lambda = 5.32$ and $\kappa = 0.954$ to 14 South American (primarily tropical) clades (Ricklefs 2007). Although confidence limits on these estimates are broad, and the estimate of $\kappa$ for North America is at the upper bound of 1 for persisting clades, the values suggest more rapid speciation and net diversification in the tropics, and a high relative rate of extinction in both regions.

## Reconstructed phylogenies and lineage-through-time plots

Phylogenetic reconstructions, which are becoming available for many taxa, depict the increase in number of ancestral lineages through time. These provide detailed information about diversification in individual clades by describing the splitting times of ancestral lineages. However, phylogenies are retrospective, working from contemporary species – the tips of a phylogeny – back through time to their common ancestor. Because of this, extinctions are not apparent and we are presented with a picture of continuous lineage splitting.

Harvey *et al.* (1991) and Nee *et al.* (1992) were the first to use phylogenetic data formally to analyze patterns of diversification, suggesting that such data could complement the fossil record, which is incompletely sampled and often lacks species-level distinctions. Phylogenies provide data on lineage splitting in two complementary forms. The first is the lineage-through-time (or LTT) plot, which portrays the logarithm of the number of ancestral lineages as a function of time – a monotonically increasing function. The second is the sequence of internode intervals from the common ancestor to the present. An internode interval is the time between successive branching events anywhere in the tree, which decreases on average with time as the number of branches in the tree increases. Although the LTT plot provides a more intuitive representation of diversification within a clade, successive points are cumulative and lack independence. Therefore, statistical characterization of diversification generally must be based on internode intervals, which are independent when they are measured without error. More recently, branch lengths themselves have been used to estimate diversification rates over the history of a clade (Bininda-Emonds *et al.* 2007).

When speciation and extinction are time-homogeneous – i.e. constant over time – the number of lineages increases at an average exponential rate of $\lambda - \mu$. However, because the LTT plot is bounded at both the beginning and the end, the result is not completely time-homogeneous. The number of lineages in a clade increases more rapidly at the beginning when lineages are young and have yet to suffer extinction; some time is needed to establish an equilibrium lineage age structure. The number of lineages in a clade at any one time is described by equation (1). As the expected value of $N$ increases, the term $\lambda E(n)$ becomes large relative to $\mu$ and the equation approaches $N(t) = \lambda E(n)/(\lambda - \mu)$, or $\ln N(t) = \ln(\lambda/(\lambda - \mu)) + (\lambda - \mu)t$, which describes a linear increase with time.

Harvey *et al.* (1994) showed that the number of ancestral lineages (of modern taxa) at any given time in the past ($\hat{N}$) follows a curve that is complementary to $N$, rising in slope towards the present as fewer lineages have had time to suffer extinction. The early, approximately asymptotic portion of the $\hat{N}(t)$ curve has slope $(\lambda - \mu)$. When extrapolated to the present, this line is amount $a = \ln(\lambda/(\lambda - \mu))$ less than the logarithm of the contemporary number of species. Thus, $e^a = \lambda/(\lambda - \mu)$ and, knowing $(\lambda - \mu)$ from the slope of the early portion of the curve, one can calculate $\lambda$ and then $\mu$. Methods for calculating $\lambda$ and $\mu$ from statistically independent internode distances are described by Nee *et al.* (1994) and in more recent references, including Bokma (2003), Rabosky (2006) and Bininda-Emonds *et al.* (2007). As an example of this approach, Purvis *et al.* (1995) estimated speciation and extinction rates for four major clades of primates, assuming time-homogeneity, using maximum likelihood (Nee *et al.* 1994). In each case, speciation considerably exceeded extinction ($\kappa = 0$, 0, 0.14, 0.28), as one might expect of large clades.

A difficulty with the analysis of whole phylogenies is that alternative underlying models of diversification sometimes cannot be distinguished because extinction in a time-homogeneous model produces a pattern similar to a model with increasing speciation rate. Pybus and Harvey (2000) devised the gamma ($\gamma$) statistic based on the distribution of internodal intervals to test for departures from a time-homogeneous pure birth process, for which the expectation is $\gamma = 0$ with a standard deviation of 1. If $\gamma$ significantly exceeds 0, nodes are concentrated closer to the present than expected, which could result from the introduction of time-homogeneous extinction or an increase in the speciation rate towards the present. If, however, nodes were more concentrated towards the base of the clade than expected ($\gamma < 0$), the most plausible explanation would be a slowing in the rate of diversification (see Price and Phillimore, this volume). Pybus and Harvey (2000) found significant slowing of diversification in North American *Dendroica* warblers (Lovette & Bermingham 1999) and Strepsirhini primates of Madagascar, but not Platyrhini primates of South America (Purvis *et al.* 1995). They suggested that diversification in the Strepsirhini might have been constrained by the relatively small area of Madagascar, without specifying a mechanism. Rabosky (2006) used a maximum likelihood approach applied to branching times to show a significant slowing in the diversification rate of Australian agamid lizards about 13 My ago, which might have been related to significant late-Tertiary increase in aridity (see also Ricklefs 2005a, for Australian corvid birds). Bininda-Emonds *et al.* (2007) have recently shown how one can use branch lengths to examine fine-scale changes in diversification rates (see also Purvis and Orme, this volume).

## Diversification bias

Two aspects of diversification in large clades are striking. First, rate of diversification frequently shows evidence of slowing towards the present (Pybus & Harvey 2000; Kozak *et al.* 2006; Weir 2006). This suggests that diversification might be diversity-dependent, in the sense that geographic or ecological regions can be filled by species, which then constrains the further building of species richness. This pattern is a common feature of diversity perceived in the fossil records of large groups (Simpson 1953; Sepkoski 1979, 1998; Stanley, 1979). Second, in many analyses of samples of known-age clades, clade size and age are unrelated (McPeek & Brown 2007, figs. 3 and 4; Ricklefs 2006a; Ricklefs *et al.*

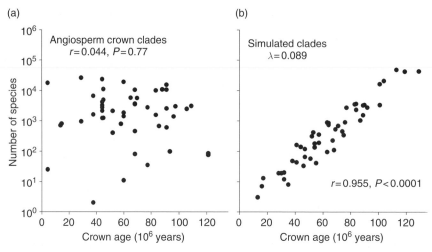

(a)                                              (b)

**Figure 14.1** (a) Relationship of number of species to crown age of 49 clades of flowering plants for which ages could be estimated from earliest fossils (Magallón & Sanderson 2001). Eudicots and core eudicots were not included. Size and age are not significantly correlated. (b) Relationship of number of species to crown age in 49 randomly generated clades using a net speciation rate of 0.089 ($\lambda=0.089$, $\mu=0$) and clade ages distributed with a mean (63 My) and standard deviation (28.5 My) equal to that of the observed clades. Adding variation to the rate of diversification does not change the strength of the species–age correlation appreciably. For example, a variable speciation rate among clades ($\lambda=0.089\pm0.020$) gave $r=0.79$ for ln(N) versus time and $r=0.48$ for ln(N) versus $\lambda$. Adding extinction also does not change this pattern. In one simulation with $\lambda=0.77$ and $\kappa=0.90$, the correlation of the logarithm of species and clade age was $r=0.87$ for 51 clades surviving out of an initial 90.

2007). Under a model of constant diversification, clades grow, and this should be apparent in samples of clades diversifying under approximately the same process but differing in age. As we saw above, Magallón and Sanderson (2001) fitted a speciation–extinction model to clades of flowering plants, obtaining an average diversification rate of 0.077 ($\kappa=0.9$) or 0.089 ($\kappa=0$) My$^{-1}$. However, these clades show no relationship between clade size and age, in contrast to clades simulated under the same parameters (Fig. 14.1).

Even though a sample of clades shows no indication of continued diversification, phylogenetic analysis of each of these clades would suggest consistent growth in the number of lineages descended from the ancestral node (Linder *et al.* 2005). The discrepancy between retrospective phylogenetic analysis and the absence of a diversity–age relationship in samples of clades might be resolved if we could examine the paleontological record of diversity. When diversity increases steadily, positive diversification within clades is a reasonable conclusion. However, if diversity remains steady over long periods, one must conclude that increase in some clades is balanced by decrease in others and eventual extinction, which we cannot observe in phylogenetic reconstructions of contemporary species (see Alroy, this volume).

Ricklefs (2006b) estimated time-homogeneous speciation and extinction rates from a LTT plot of the suboscine passerines of South America constructed from the Sibley and Ahlquist (1990) DNA-hybridization phylogeny. Depending on the time calibration used, the speciation rate was $\lambda = 0.43$–$0.86\,\mathrm{Myr}^{-1}$, and extinction rate was $\mu = 0.35$–$0.71\,\mathrm{Myr}^{-1}$ ($\kappa = 0.82$). A strongly positive rate of diversification is not surprising for a clade that has increased to almost 1000 species, probably within the Tertiary. A question that arises, however, is whether the common ancestor of the suboscines was alone in South America or might have shared the continent with 1000 other species, none of which has left living descendants. Unfortunately, small land birds leave a very poor fossil record. One must also recognize that the upturn in the LTT plot towards the present, which provides an estimate of the extinction rate in a time-homogeneous process, could also represent an increase in the speciation rate, as Fjeldså (1994) and Price (2008, pp. 115ff.) have argued for South American birds.

Species-level records of the paleontological history of clades are available for few groups. Clearly, new groups with key innovations can diversify rapidly (Lupia et al. 1999). However, two examples show that species richness in established clades can be relatively constant over long periods. The terrestrial mammals of North America have an excellent fossil record. Number of genera (Stuckey 1990) and species (Alroy 1998, 2000) of all mammals, and number of species of carnivores and herbivores (Van Valkenburgh & Janis 1993) fluctuated through much of the Tertiary, but remained approximately constant for tens of millions of years. According to the record of first and last appearances, turnover rates of genera averaged approximately 0.2 per million years (a 5 My average life span), suggesting that species turnover rate likely was much higher (Stuckey 1990). Both records show a dramatic increase in taxa associated with the drying out of the continental interior and the spread of extensive grasslands in the late Tertiary, only to be reversed by the cooling trend towards the present. Alroy's (2000) refined analysis of species of fossilized mammals in Western North America shows a nearly flat trend for the past 40 Mya.

Jaramillo et al. (2006) traced the history of plant diversity based on pollen grain morphotypes in northwestern South America from the beginning of the Tertiary until about 20 million years ago. The number of morphospecies varied from ca. 200, both at the beginning of the record in the early Paleocene and at the end of the record in the early Miocene, to a mid-Eocene peak of ca. 350 morphospecies. Pollen types turned over completely during this interval, particularly rapidly in the late Paleocene and at the beginning of the Oligocene. The record is not one of time homogeneity (Wing & Harrington 2001). However, it also does not support continuously positive increase in species richness through the

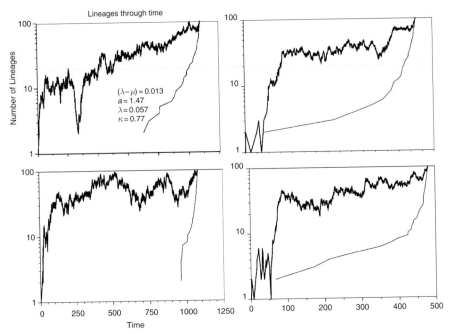

**Figure 14.2** Lineage-through-time plots produced by simulation using PhyloGen (http://evolve.zoo.ox.ac.uk/software/PhyloGen/main.html) in which speciation and extinction rates were equal to 0.1 and simulations were continued until 10 clades of 100 species had been produced. Of 812 simulations required for this, 802 went extinct before reaching 100 species. Thus 812 original lineages resulted in 1000 descendant species from 10 ancestors. The plots show the total number of species as a function of clade age and the number of lineages ancestral to contemporary species for four of the 10 completed simulations. Fitted values for the LTT plot are presented for the upper left-hand panel.

period – quite the opposite. On the whole, through the Tertiary up the Miocene, net diversification rate was close to zero, while turnover was high.

Each of these cases, smoothed and simplified, represents an extreme scenario of approximately constant overall species richness with speciation and extinction balanced, whereby some clades increase and others dwindle to extinction just by chance. Nonetheless, estimates of net diversification rate in the larger remaining clades would exceed zero even though the underlying process was $\lambda = \mu$, the so-called critical case (Fig. 14.2).

### Balanced diversification

Faced with paleontological evidence pointing to relative constancy of species numbers over long periods in many groups, and the apparently biased view of diversification revealed by the phylogenies of living clades, it seems necessary to consider processes that predict steady state, or at least slowly changing, species richness. Three kinds of process have been proposed: (1) a balanced random walk, in which speciation and extinction are equal but stochastic, and the total number of lineages can increase or decrease at random, often called the critical speciation – extinction process; (2) a Moran process, such as genetic drift or Hubbell's (2001) community drift, with fixed numbers of individuals or species and random replacement following death or extinction; and (3) diversity-dependent speciation and/or extinction (MacArthur & Wilson 1967; Rosenzweig 1975, 1995; Sepkoski 1979; Stanley 1979; Harvey et al. 1994).

Under a random-walk speciation–extinction process with $\lambda = \mu$, the probability that a single ancestral lineage is extinct after time $t$ is $P(n=0|t)=\lambda t/(\lambda t+1)$, and the time required to reach probability ($P$) of a lineage being extinct is $t=P/\lambda(1-P)$. Using a survival approach, the probability that all but one of $N$ lineages are extinct is $P=(N-1)/N$. Substituting $P=(N-1)/N$ into the expression for time, we get $t=(N-1)/\lambda$ or $\lambda t=N-1$. Thus, the time required for replacement of all but one lineage is approximately equal to the number of lineages times the individual survival time $(1/\mu=1/\lambda)$.

Another approach to estimating turnover time in a steady-state clade is to calculate the binomial probability that $N-1$ lineages initiated with a single individual are extinct at time $t$. This is $P(1|N,t)=N\cdot P(n=0|t)^{N-1}(1-P(n=1|t))$, which can be expressed as

$$P(1|N,t) = \frac{N}{\lambda t}\left(\frac{\lambda t}{\lambda t+1}\right)^N.$$

(14.4)

The maximum value of $P(1|N,t)$, hence the most likely estimate, occurs at a value of approximately $\lambda t=N$. A simple simulation shows this relationship: in 1000 replicates of $N=10$ stem clades with $\lambda=\mu=0.1$, nine of the ancestral lineages were extinct in an average replacement time of $106.5\pm268.7$ SD (range 7–4600) time units ($\lambda t=10.65$), and the size of the remaining clade averaged $10.8\pm24.0$ SD (0–346) terminal lineages.

Under a Moran process, the expected coalescence time, after which a single ancestral clade remains, is $N$ 'generations' (Hudson 1990). For a diversifying clade, each generation is the expected life span of a single lineage, or $1/\mu$. Regardless of the model, the replacement of $N$ lineages under a random-walk speciation–extinction process requires approximately $N$ multiples of the waiting time to extinction. Thus, for 1000 species of South American suboscine passerines with an average time to extinction of 1–2 My (Ricklefs 2006b), the replacement time would be about 1–2 billion years. Similar calculations for other groups lead to the inescapable conclusion that random-walk processes cannot explain the replacement of lineages through time.

An alternative view is that a clade initially experiences rapid diversification but eventually reaches a limit owing to diversity-dependent changes in speciation and/or extinction rates, much as a population is limited by density-dependent changes in birth and death rates (Raup et al. 1973; Sepkoski 1979; Head & Rodgers

1997). Suppose that the per-species speciation rate is $b - \beta N$ and the per-species extinction rate is $d + \delta N$. In this case, the equilibrium number of species would be $N = (b - d)/(\beta + \delta)$ $[b > d]$ and the rate of extinction at equilibrium would be $(\beta d + \delta b)/(\beta + \delta)$. If diversity were controlled by a decrease in speciation rate without changing the extinction rate (i.e. $\delta = 0$), then the extinction rate at equilibrium would equal $d$, the extinction rate at low diversity (i.e. low turnover). If diversity were controlled only by an increase in the extinction rate (i.e. $\beta = 0$), then extinction rate at equilibrium would equal $b$, the speciation rate at low diversity (i.e. high turnover).

In general, random walks with balanced speciation and extinction are too slow to account for observed lineage replacement over time. Diversity-dependent speciation and/or extinction provide a reasonable alternative, implying that clades are constrained to a particular number of species, which appears to be higher in larger regions and at lower latitudes (Ricklefs 2006a). The absence of an age–size relationship among clades of passerine birds, squamate reptiles, flowering plants and other groups suggests that most clades are at or close to equilibrium sizes. If clade size were regulated in a diversity-dependent manner, however, the approximately geometric distribution of number of species (Ricklefs 2003) could not result from a homogeneous random-walk speciation–extinction process, but must reflect the way in which ecological and geographic space is apportioned among diversifying clades.

If random-walk speciation and extinction rates of individual clades varied over time as a result of environmental change or evolutionary adaptation, then some clades would assume positive rates of diversification while other, disfavoured clades would decline. In addition, as the net diversification rate decreases, the time for replacement of ancestral lineages by the descendants of a single ancestor increases. Simulations, such as that portrayed in Fig. 14.2, in which a single clade diversifies until it either goes extinct or reaches 100 species, demonstrate this pattern. With $\lambda = 0.10$ and $\mu$ varying from 0 to 0.10, the time for clade size to reach 100 species varied from 53.7 ($\pm 12.1$ SD) time units for $\kappa = 0$, to 64.4 (14.5, $\kappa = 0.2$), 90.9 (22.2, 0.50), 291.6 (127.3, 0.90), 425 (223, 0.95), 531 (288, 0.99), 588 (355, 0.999) and 554 (315, 1.000). In this particular example, the time for replacement compared to that for balanced diversification is reduced only by about one-half for $\kappa = 0.90$, one-quarter for $\kappa = 0.70$ and to about one-tenth in the absence of extinction. Accordingly, rapid turnover of lineages would require substantial departures from $\lambda = \mu$.

As clade size must result from the interplay between intrinsic clade characteristics and properties of the environment, we should expect a relationship between clade size and attributes of the species that make up the clade, as has been found in several comparative studies (Barraclough et al. 1998; Phillimore et al. 2006; Ricklefs & Renner 1994). Kozak et al. (2005) emphasized the correspondence between rapid ecomorphological divergence and rapid diversification

**Table 14.1** *Product–moment correlations between the numbers of genera (upper right) and numbers of species (lower left) of trees in orders of flowering plants in five regions*

| | Tropical | | | Temperate | |
| --- | --- | --- | --- | --- | --- |
| | Ecuador | Madagascar | Malaysia | Eastern Asia | North America |
| Ecuador | 20 | **0.798** | **0.794** | 0.546 | 0.509 |
| Madagascar | **0.645** | 22 | **0.791** | 0.065 | 0.052 |
| Malaysia | **0.805** | **0.704** | 26 | 0.535 | 0.419 |
| Eastern Asia | 0.516 | −0.023 | 0.527 | 22 | **0.770** |
| North America | 0.547 | −0.150 | 0.337 | **0.730** | 17 |

*Note:* The number of orders in each region is indicated on the diagonal, although correlations might involve fewer comparisons because absent orders in any pair of regions were not included. Boldface numbers indicate $P < 0.01$.

*Source:* Ecuador (Renner *et al.* 1990); Madagascar, compiled by E.M. Friis from Leroy (1978); Malaysia, compiled by E.M. Friis from Whitmore (1972–1973); temperate floras (Latham & Ricklefs 1993).

in the early evolution of plethodontid salamanders in eastern North America (also see Schluter 2000). The existence of clade-specific attributes that influence number of species could be tested further in comparisons of the sizes of individual clades independently realized in different regions. To illustrate this approach, the number of species and genera of forest trees in orders of flowering plants (Angiosperm Phylogeny Group 1998) is compared among two temperate and three tropical floras in Table 14.1. Correlation coefficients are strong and highly significant among the tropical floras and among the temperate floras, but not between them, as one might expect. A similar analysis of the number of species in clades of small land birds produced a similar pattern (Table 14.2), with 9 of 10 correlation coefficients >0, and 4 of 10 significant at $P < 0.05$.

The absence of a correlation between clade age and clade size leads to the conclusion that species richness has been more or less constant over long periods. The correlation of clade size between regions suggests that the size of each clade is regulated independently, possibly by the size and character of the region, including other clades occupying the same region, but certainly by the adaptations shared by clade members. Nee *et al.* (1992, p. 8323) envisioned such a unification of paleontological and evolutionary perspectives on species richness:

One may conjecture that instantaneous cladogenesis reflects two distinct processes: evolutionary processes such as sexual selection producing, from a coarse view, ecological equivalents, and ecological processes producing diversification into an adaptive landscape of distinct ecological niches or ways of life. A lineage that invades an empty niche,

**Table 14.2** *Product–moment correlation coefficients (upper triangle; P-values, lower triangle) for the number of species of birds in 25 clades of small land birds compared between biogeographic regions*

|      | Biogeographic Region | | | | |
|------|------|------|------|------|------|
|      | NA | EU | OR | NT | AF |
| NA   | 20    | **0.738** | **0.656** | **0.542** | 0.465 |
| EU   | 0.002 | 22        | 0.409     | 0.423     | **0.570** |
| OR   | 0.004 | 0.082     | 22        | 0.424     | 0.319 |
| NT   | 0.037 | 0.091     | 0.080     | 20        | −0.039 |
| AF   | 0.081 | 0.017     | 0.171     | 0.878     | 22 |

*Note:* NA = North America; EU = Eurasia; OR = Oriental, or South and Southeast Asia; NT = Neotropics; AF = Africa. The number of clades in each region is indicated on the diagonal, although not all were represented in both members of each pair of regions.
*Source:* Sibley and Monroe (1990).

coarsely defined, may be assured of leaving descendants in the distant future, barring calamities such as comet impacts or niche usurpation by a quite different sort of beast. However, the subsequent history of cladogenesis *within* the coarsely defined niche may be well described by random speciation and extinction models in which the overall number of lineages is roughly constant (24). [Reference (24) is to Raup *et al.* (1973).]

Two issues raised by Nee *et al.* (1992) invite comment here: first, that the size of a clade within a region is determined by adaptations to fill niche space, i.e. ecological adaptations primarily concerned with resource use; and second, that regulated diversity results in a random speciation–extinction process with $\lambda = \mu$. In regard to ecological determination of clade size, while local resource supply undoubtedly limits the total number of individuals in a clade, the way in which those individuals are apportioned among species might be independent of resources to some degree. For example, the niche space occupied by a population appears to be compressible or expandable depending on the number of competing populations – the phenomena of ecological compression and ecological release. Presumably, ecological breadth or niche space is driven in part by the pressure of species formation within a region, just as habitat distribution and abundance of populations on islands is influenced by the number of coexisting species determined by the colonization–extinction balance (Cox & Ricklefs 1977).

The relationship between geographic or ecological distribution of species and the size of the clade to which they belong has not been addressed. A first assessment of this relationship for passerine birds in South America can be made from data in Stotz *et al.* (1996) on habitat distribution and number of zoogeographic

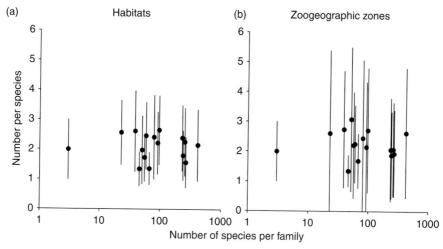

**Figure 14.3** Relationship of the average number of habitats (a) and zoogeographic regions (b) per species in 16 families of passerine birds, portrayed as a function of clade size. Four monotypic families are not included. Vertical lines are ±1 SD. Regressions weighted by the square root of the number of species were not significant for either habitats ($F_{1,14}=0.06$, $P=0.81$) or regions ($F_{1,14}=0.09$, $P=0.76$).

zones occupied (Fig. 14.3). For 16 family-level clades ranging from 3 (Phytotomidae) to 417 (Tyrannidae) species, neither the mean number of habitats nor the mean number of zoological regions per species was related to the logarithm of the number of species. Thus, species in the larger families either overlap more in ecological and geographic space, or the larger families occupy larger spaces.

Stotz *et al.* (1996) recognized 14 types of forest habitats and 14 types of open habitats within South America. I summed the incidences of the passerine species in each family in each of the habitat types and calculated a Simpson index of habitat diversity for each family: $S=1/\sum p_i^2$, where $p_i$ is the proportion of occurrences in habitat $i$. Number of forested habitats increased with the common logarithm of number of species per family ($S_F=1.80$ (±0.22 SD) + 0.38 (±0.12) ln (species); $F_{1,17}=9.5$, $P=0.0067$, $R^2=0.36$) as did number of open habitats ($S_O=0.97$ (±0.42 SD) + 0.80 (±0.21) ln(species); $F_{1,12}=13.9$, $P=0.0029$, $R^2=0.54$). However, because the number of habitats did not increase as rapidly as number of species (0.4 or 0.8 habitats per 10-fold increase in species richness), more diverse families might have to accommodate more overlap and denser packing of species into local ecological niche space.

Denser packing or increased ecological overlap should result in lower population density locally. However, in three censuses of birds in lowland Neotropical habitats (Terborgh *et al.* 1990; Thiollay 1994; Robinson *et al.* 2000), family averages for the densities of species of passerine birds was unrelated to the number of species in each family locally (all $P>0.20$). The sums of population densities for each family increased in direct relation to the number of species (all $P < 0.01$) and were correlated between the census plots (Pearson $r = 0.90–0.97$, $P < 0.001$; Spearman $r = 0.62–0.90$, $P = 0.0002–0.025$).

These census results suggest that more species-rich clades use larger proportions of the available ecological space within a locality. Accordingly, larger clades should exhibit greater ecological diversity. Assuming that ecological diversity parallels morphological diversity (Miles & Ricklefs 1984; Ricklefs & Miles 1994), I related species richness in 81 tribe-to-family level clades of passerine birds to the area of the primary regions occupied by the clade, representation in tropical regions, and the standard deviations of species scores on eight morphological principal component (PC) axes (see Ricklefs 2004, 2005b). Species richness increased with area and was higher in tropical regions, as found previously (Ricklefs 2006a), but species richness also increased with variation within clades in PC3 (7% of the total morphological variance, roughly tarsus × mid toe/ beak width; $F_{1,76} = 9.4$, $P = 0.0030$) and PC7 (1%, tarsus/toe; $F_{1,76} = 8.6$, $P = 0.0044$) (total $R^2 = 0.45$). These morphological axes are strongly related to foraging method and substrate (Ricklefs 2005b), and provide an indication of the occupation of some part of ecological space.

If clade size were determined by diversity-dependence of speciation and/or extinction, how would diversity influence these processes? Much has been written on this topic, summarized by Rosenzweig (1995). The example of South American passerine birds suggests that the average geographic and ecological extent of populations does not vary between clades of different size. Thus, the 'space' occupied by all the species in a clade as a whole appears to vary in direct proportion to the number of species in the clade. Either population size and ecological extent is regulated independently of clade identity, or species in each clade diversify to fill space to approximately the same density, with the total available space depending on the clade. In either case, nothing about the ecology or geography of individual species would indicate why rates of speciation or extinction might vary between between clades.

Within a particular clade, increasing number of species would presumably lead to smaller average population size, which might increase extinction rate. Larger populations characteristic of small clades might split or produce small isolates – incipient species – more readily (Rosenzweig 1975, 1995). If variation in the number of species in a clade depended primarily on the rate at which extinction increased with diversity, then the turnover rate of species within a clade should be higher in smaller clades. Conversely, if clade size variation resulted primarily from the rate at which speciation rate decreased with diversity, then smaller clades should experience a lower turnover rate. Presumably the direction of this relationship could be evaluated by the distribution of branch lengths subtending the terminal taxa in a sample of phylogenetic trees (Weir & Schluter 2007).

   The influence of region area on diversity could reflect either the lower extinction rates of larger populations or higher speciation rates resulting from greater distances and opportunities for genetic isolation. Again, whether one or the other of these trends predominated might appear in the relationship of recent branch lengths (Weir & Schluter 2007) to region area. Many hypotheses have linked tropical environments to higher rates of species formation and lower rates of extinction (Mittelbach et al. 2007). Tropical species are thought to be more sedentary and specialized with respect to the physical environment than those in more seasonal environments, which could promote genetic isolation of populations. Benign environmental conditions buffered against large-scale change could also reduce the probability of extinction. In any case, this type of explanation links diversity to region area and conditions through regional, rather than ecological, influences on speciation and extinction. Accordingly, the ecology of a region, as well as adaptations of populations to ecological conditions, would influence these processes primarily through their effects on the distribution and mobility of individuals within populations. Thus, while we may adopt the concept of a carrying capacity for clade size, which parallels the concept of density-dependent determination of the carrying capacity for populations, we must distinguish the regional basis of the first from the ecological basis of the second.

   Finally, we must resolve the turnover of species within clades and the replacement of clades within regions. If we assume that the overall number of species is relatively constant, random-walk speciation–extinction processes are too slow to account for the turnover of species within regions. Accordingly, although speciation and extinction are approximately balanced, environmental change and adaptive evolution would give some individual clades positive diversification rates that lead to increase in clade size more rapidly than would occur by chance. The adaptive advantage held by such clades could cause, through ecological competition, negative diversification rates in less favoured clades. Thus, either adaptive or environmental change favouring any one clade and disfavouring another would influence both the size and ecological extent of individual populations, but also their propensity to produce additional evolutionary lineages or to go extinct, thereby linking ecology and diversification.

## Conclusions

The points made here can be summarized as follows:

(1)   In several analyses, clade size is unrelated to clade age implying that, on average, rates of extinction approximately equal rates of speciation and, barring dramatic environmental changes on catastrophes, species richness remains relatively constant. This is seen in the fossil record for several groups following the initial rise in diversity that typically accompanies their origins.

(2)   A common perspective that diversification has resulted in an increase in number of species has been promoted by the retrospective nature of phylogeny reconstruction based on molecular or other characters of contemporary species. Moreover, a focus, out of necessity, on extant clades in general, and on large clades in particular, produces a positive diversification bias.

(3)   Turnover of clades through time is too rapid to have resulted from a balanced ($\lambda = \mu$) random-walk speciation–extinction process. To account for observed replacement of species through time, diversification rates of clades must vary over relatively long periods owing to changes in environment or evolutionary adaptation. Random-walk speciation–extinction models are inappropriate for explaining variation in species richness.

(4)   We can surmise that clades are continuously expanding and contracting within regions competitively. Accordingly, equilibrium clade size, which reflects diversity-dependent variation in speciation and extinction rates, changes in response to environmental conditions and the origin of adaptive innovation. This reflects both lineage and regional properties and probably works through large-scale, rather than local-ecological, processes.

(5)   The sizes of clades also are sensitive to the area of the region within which they have diversified, and to general environmental conditions, as revealed by comparisons between temperate and tropical groups.

(6)   Number of species is partly intrinsic to each clade, as illustrated by correlations between the sizes of clades represented independently in different regions of the earth. Thus, clade size is to some extent conservative over time. It would be interesting to know whether these properties are maintained in the case of larger and smaller clades, i.e. at higher or lower taxonomic levels than examined in this analysis. Thus, at what level does a clade exhibit heritable properties that influence speciation and extinction? We might even ask, What makes a particular clade interesting to us?

(7)   In Neotropical passerine birds, the geographic and ecological extent of individual species/populations appears to be independent of clade size. This is also true of local population density. This implies that large clades occupy more geographic and ecological space overall, and possibly that diversification is constrained by the minimum viable population size and extent of individual species. It would appear, then, that diversity-dependent speciation and extinction equalize population parameters of species that make up a clade, regardless of the size of the clade.

(8)   How population characteristics interact with regional properties to determine rates of speciation and extinction remain an unsolved issue. In particular, we would like to know what factors determine the carrying capacity of a region for a particular clade? With respect to processes working on large scales of time and space, a regional speciation–extinction balance might

be influenced by the way in which continent area and climatic conditions affect both species formation and extinction. However, the ability of a clade to occupy ecological space through adaptive radiation also is important.

(9)   Patterns of species richness among clades integrate evolutionary and eco-logical processes and interactions within the physiographic backdrop of a region and thus provide a focus for synthetic analysis in biology.

## Acknowledgements

This paper was completed while the author was a Visiting Fellow at the Centre for Population Biology, Imperial College London. The author is also grateful for discussion with members of the NCEAS (US National Science Foundation) Working Group in Latitudinal Gradients of Diversification. Paul Fine, Trevor Price, Andy Purvis, Dan Rabosky, Dolph Schluter and Kathy Willis commented on the manuscript. Else Marie Friis compiled some of the data used to produce Table 14.1.

## References

Alroy, J. (1998) Equilibrial diversity dynamics in North American mammals. In: *Biodiversity Dynamics. Turnover of Populations, Taxa, and Communities* (ed. M. L. McKinney), pp. 232–287. Columbia University Press, New York.

Alroy, J. (2000) Successive approximations of diversity curves: ten more years in the library. *Geology* **28**, 1023–1026.

Angiosperm Phylogeny Group (1998) An ordinal classification for the families of flowering plants. *Annals of the Missouri Botanical Garden* **85**, 531–553.

Bailey, N. T. J. (1964) *The Elements of Stochastic Processes with Applications to the Natural Sciences*. John Wiley & Sons, New York.

Barraclough, T. G. and Nee, S. (2001) Phylogenetics and speciation. *Trends in Ecology & Evolution* **16**, 391–399.

Barraclough, T. G., Vogler, A. P. and Harvey, P. H. (1998) Revealing the factors that promote speciation. *Philosophical Transactions of the Royal Society of London Series B: Biological Sciences* **353**, 241–249.

Barraclough, T. G., Vogler, A. P. and Harvey, P. H. (1999) Revealing the factors that promote speciation. In: *Evolution of Biological*

*Diversity* (ed. A. E. Magurran and R. M. May), pp. 202–219. Oxford University Press, Oxford.

Benton, M. J. and Donoghue, P. C. J. (2007) Paleontological evidence to date the tree of life. *Molecular Biology & Evolution* **24**, 26–53.

Bininda-Emonds, O. R. P., Cardillo, M., Jones, K. E., *et al.* (2007) The delayed rise of present-day mammals. *Nature* **446**, 507–512.

Bokma, F. (2003) Testing for equal rates of cladogenesis in diverse taxa. *Evolution* **57**, 2469–2474.

Cox, G. W. and Ricklefs, R. E. (1977) Species diversity, ecological release, and community structuring in Caribbean land bird faunas. *Oikos* **29**, 60–66.

Currie, D. J., Mittelbach, G. G., Cornell, H. V., *et al.* (2004) Predictions and tests of climate-based hypotheses of broad-scale variation in taxonomic richness. *Ecology Letters* **7**, 1121–1134.

Donoghue, P. C. J. and Benton, M. J. (2007) Rocks and clocks: calibrating the Tree of Life using fossils and molecules. *Trends in Ecology and Evolution* **22**, 424–431.

Fine, P. V. A. and Ree, R. H. (2006) Evidence for a time-integrated species-area effect on the latitudinal gradient in tree diversity. *American Naturalist* **168**, 796–804.

Fjeldså, J. (1994) Geographical patterns for relict and young species of birds in Africa and South-America and implications for conservation priorities. *Biodiversity & Conservation* **3**, 207–226.

Foote, M. (1993) Discordance and concordance between morphological and taxonomic diversity. *Paleobiology* **19**, 185–204.

Foote, M. (2000) Origination and extinction components of taxonomic diversity: Paleozoic and post-Paleozoic dynamics. *Paleobiology* **26**, 578–605.

Haesler, M. P. and Seehausen, O. (2005) Inheritance of female mating preference in a sympatric sibling species pair of Lake Victoria cichlids: implications for speciation. *Proceedings of the Royal Society of London Series B: Biological Sciences* **272**, 237–245.

Harvey, P. H., May, R. M., and Nee, S. (1994) Phylogenies without fossils. *Evolution* **48**, 523–529.

Harvey, P. H., Nee, S., Mooers, A. Ø. and Partridge, L. (1991). These hierarchical views of life: phylogenies and metapopulations. In: *Genes in Ecology* (ed. R. J. Berry, T. J. Crawford and G. M. Hewitt), pp. 127–137. Blackwell Scientific, Oxford.

Head, D. A. and Rodgers, G. J. (1997) Speciation and extinction in a simple model of evolution. *Physical Review E* **55**, 3312–3319.

Hey, J. (1992) Using phylogenetic trees to study speciation and extinction. *Evolution* **46**, 627–640.

Hubbell, S. P. (2001) *The Unified Neutral Theory of Biodiversity and Biogeography*. Princeton University Press, Princeton, NJ.

Hudson, R. R. (1990) Gene genealogies and the coalescent process. *Oxford Surveys in Evolutionary Biology* **7**, 1–44.

Jaramillo, C., Rueda, M. J. and Mora, G. (2006) Cenozoic plant diversity in the Neotropics. *Science* **311**, 1893–1896.

Kingsland, S. E. (1985) *Modeling Nature. Episodes in the History of Population Ecology*. University of Chicago Press, Chicago.

Kozak, K. H., Larson, A. A., Bonett, R. M. and Harmon, L. J. (2005) Phylogenetic analysis of ecomorphological divergence, community structure, and diversification rates in dusky salamanders (Plethodontidae: *Desmognathus*). *Evolution* **59**, 2000–2016.

Kozak, K. H., Weisrock, D. W. and Larson, A. (2006) Rapid lineage accumulation in a non-adaptive radiation: phylogenetic analysis of diversification rates in eastern North American woodland salamanders (Plethodontidae: *Plethodon*). *Proceedings of the Royal Society of London Series B: Biological Sciences* **273**, 539–546.

Latham, R. E. and Ricklefs, R. E. (1993). Continental comparisons of temperate-zone tree species diversity. In: *Species Diversity: Historical and Geographical Perspectives* (ed. R. E. Ricklefs and D. Schluter), pp. 294–314. University of Chicago Press, Chicago.

Leroy, J.-F. (1978) Composition, origin and affinities of the Madagascan vascular flora. *Annals of the Missouri Botanical Garden* **65**, 535–589.

Linder, H. P., Hardy, C. R. and Rutschmann, F. (2005) Taxon sampling effects in molecular clock dating: an example from the African Restionaceae. *Molecular Phylogenetics and Evolution* **35**, 569–582.

Lovette, I. J. and Bermingham, E. (1999) Explosive speciation in the New World *Dendroica* warblers. *Proceedings of the Royal Society of London Series B: Biological Sciences* **266**, 1629–1636.

Lupia, R., Lidgard, S. and Crane, P. R. (1999) Comparing palynological abundance and diversity: implications for biotic replacement during the Cretaceous angiosperm radiation. *Paleobiology* **25**, 305–340.

MacArthur, R. H. (1970) Species packing and competitive equilibrium for many species. *Theoretical Population Biology* **1**, 1–11.

MacArthur, R. H. and Levins, R. (1967) The limiting similarity, convergence, and divergence of coexisting species. *American Naturalist* **101**, 377–385.

MacArthur, R. H. and Wilson, E. O. (1967) *The Theory of Island Biogeography*. Princeton University Press, Princeton, NJ.

Magallón, S. and Sanderson, M. J. (2001) Absolute diversification rates in angiosperm clades. *Evolution* **55**, 1762–1780.

May, R. M. (1975) Patterns of species abundance and diversity. In: *Ecology and Evolution of Communities* (ed. M. L. Cody and J. M. Diamond), pp. 81–120. Belnap Press, Harvard University, Cambridge, MA.

McPeek, M. A. and Brown, J. M. (2007) Clade age and not diversification rate explains species richness among animal taxa. *American Naturalist* **169**, E97–E106.

Miles, D. B. and Ricklefs, R. E. (1984) The correlation between ecology and morphology in deciduous forest passerine birds. *Ecology* **65**, 1629–1640.

Mittelbach, G. G., Schemske, D. W., Cornell, H. V., *et al.* (2007) Evolution and the latitudinal diversity gradient: speciation, extinction and biogeography. *Ecology Letters* **10**, 315–331.

Nee, S. (2001) Inferring speciation rates from phylogenies. *Evolution* **55**, 661–668.

Nee, S. (2006) Birth-death models in macroevolution. *Annual Review of Ecology Evolution & Systematics* **37**, 1–17.

Nee, S., May, R. M. and Harvey, P. H. (1994) The reconstructed evolutionary process. *Philosophical Transactions of the Royal Society of London Series B: Biological Sciences* **344**, 305–311.

Nee, S., Mooers, A. O. and Harvey, P. H. (1992) Tempo and mode of evolution revealed from molecular phylogenies. *Proceedings of the National Academy of Sciences of the United States of America* **89**, 8322–8326.

Phillimore, A. B., Freckleton, R. P., Orme, C. D. L. and Owens, I. P. F. (2006) Ecology predicts large-scale patterns of phylogenetic diversification in birds. *American Naturalist* **168**, 220–229.

Price, T. (2008) *Speciation in Birds*. Roberts and Company, Greenwood Village, CO.

Purvis, A., Nee, S. and Harvey, P. H. (1995) Macroevolutionary inferences from primate phylogeny. *Proceedings of the Royal Society of London Series B: Biological Sciences* **260**, 329–333.

Pybus, O. G. and Harvey, P. H. (2000) Testing macro-evolutionary models using incomplete molecular phylogenies. *Proceedings of the Royal Society of London Series B: Biological Sciences* **267**, 2267–2272.

Rabosky, D. L. (2006) Likelihood methods for detecting temporal shifts in diversification rates. *Evolution* **60**, 1152–1164.

Raup, D. M. and Gould, S. J. (1974) Stochastic simulation and evolution of morphology – towards a nomothetic paleontology. *Systematic Zoology* **23**, 305–322.

Raup, D. M., Gould, S. J., Schopf, T. J. M. and Simberloff, D. S. (1973) Stochastic models of phylogeny and the evolution of diversity. *Journal of Geology* **81**, 525–542.

Renner, S. S. (2005) Relaxed molecular clocks for dating historical plant dispersal events. *Trends in Plant Science* **10**, 550–558.

Renner, S. S., Balslev, H. and Holm-Nielsen, L. B. (1990) Flowering plants of Amazonian Ecuador – a checklist. *AAU Reports* **24**, 1–241.

Ricklefs, R. E. (1987) Community diversity: relative roles of local and regional processes. *Science* **235**, 167–171.

Ricklefs, R. E. (2003) Global diversification rates of passerine birds. *Proceedings of the Royal Society of London Series B: Biological Sciences* **270**, 2285–2291.

Ricklefs, R. E. (2004) Cladogenesis and morphological diversification in passerine birds. *Nature* **430**, 338–341.

Ricklefs, R. E. (2005a) Phylogenetic perspectives on patterns of regional and local species richness. In: *Tropical Rainforests. Past, Present, and Future* (ed. E. Bermingham, C. W. Dick and C. Moritz), pp. 16–40. University of Chicago Press, Chicago and London.

Ricklefs, R. E. (2005b) Small clades at the periphery of passerine morphological space. *American Naturalist* **165**, 43–659.

Ricklefs, R. E. (2006a) Global variation in the diversification rate of passerine birds. *Ecology* **87**, 2468–2478.

Ricklefs, R. E. (2006b) The unified neutral theory of biodiversity: do the numbers add up? *Ecology* **87**, 1424–1431.

Ricklefs, R. E. (2007) Estimating diversification rates from phylogenetic information. *Trends in Ecology and Evolution* **22**, 601–610.

Ricklefs, R. E., Losos, J. B. and Townsend, T. M. (2007) Evolutionary diversification of clades of squamate reptiles. *Journal of Evolutionary Biology* **20**, 1751–1762.

Ricklefs, R. E. and Miles, D. B. (1994) Ecological and evolutionary inferences from morphology: an ecological perspective. In: *Ecological Morphology: Integrative Organismal Biology* (ed. P. C. Wainwright and S. M. Reilly), pp. 13–41. University of Chicago Press, Chicago.

Ricklefs, R. E. and Renner, S. S. (1994) Species richness within families of flowering plants. *Evolution* **48**, 1619–1636.

Ricklefs, R. E. and Schluter, D. (eds) (1993) *Species Diversity in Ecological Communities*. University of Chicago Press, Chicago.

Robinson, W. D., Brawn, J. D. and Robinson, S. K. (2000) Forest bird community structure in central Panama: influence of spatial scale and biogeography. *Ecological Monographs* **70**, 209–235.

Rosenzweig, M. L. (1975) On continental steady states of species diversity. In: *Ecology and Evolution of Communities* (ed. M. L. Cody and J. M. Diamond), pp. 121–140. Harvard University Press, Cambridge, MA.

Rosenzweig, M. L. (1995) *Species Diversity in Space and Time*. Cambridge University Press, Cambridge.

Schluter, D. (2000) *The Ecology of Adaptive Radiation*. Oxford University Press, Oxford.

Sepkoski, J. J., Jr. (1979) A kinematic model of Phanerozoic taxonomic diversity II. Early Phanerozoic families and multiple equilibria. *Paleobiology* **5**, 222–251.

Sepkoski, J. J., Jr. (1998) Rates of speciation in the fossil record. *Philosophical Transactions of the Royal Society B: Biological Sciences* **353**, 315–326.

Sibley, C. G. and Ahlquist, J. E. (1990) *Phylogeny and Classification of the Birds of the World*. Yale University Press, New Haven, CT.

Sibley, C. G. and Monroe, B. L., Jr. (1990) *Distribution and Taxonomy of Birds of the World*. Yale University Press, New Haven, CT.

Simpson, G. G. (1944) *Tempo and Mode in Evolution*. Columbia University Press, New York.

Simpson, G. G. (1953) *The Major Features of Evolution*. Columbia University Press, New York.

Slowinski, J. B. and Guyer, C. (1989) Testing the stochasticity of patterns of organismal diversity: an improved null model. *American Naturalist* **134**, 907–921.

Slowinski, J. B. and Guyer, C. (1993) Testing whether certain traits have caused amplified diversification: an improved method based on a model of random speciation and extinction. *American Naturalist* **142**, 1019–1024.

Stanley, S. M. (1979) *Macroevolution: Pattern and Process*. W. H. Freeman, San Francisco.

Stotz, D. F., Fitzpatrick, J. W., Parker, T. A., III. and Moskovits, D. K. (1996) *Neotropical Birds. Ecology and Conservation. With Ecological and Distributional Databases by Theodore A. Parker III, Douglas F. Stotz, and John W. Fitzpatrick*. University of Chicago Press, Chicago and London.

Stuckey, R. K. (1990) Evolution of land mammal diversity in North America during the Cenozoic. *Current Mammalogy* **2**, 375–432.

Terborgh, J., Robinson, S. K., Parker, T. A., III, Munn, C. A. and Pierpont, N. (1990) Structure and organization of an Amazonian forest bird community. *Ecological Monographs* **60**, 213–238.

Thiollay, J.-M. (1994) Structure, density and rarity in an Amazonian rainforest bird community. *Journal of Tropical Ecology* **10**, 449–481.

Vandermeer, J. H. (1972) Niche theory. *Annual Review of Ecology and Systematics* **3**, 107–132.

Van Valkenburgh, B. and Janis, C. M. (1993) Historical diversity patterns in North American large herbivores and carnivores. In: *Species Diversity in Ecological Communities. Historical and Geographical Perspectives* (ed. R. E. Ricklefs and D. Schluter), pp. 330–340. University of Chicago Press, Chicago.

Wallace, A. R. (1878) *Tropical Nature and Other Essays*. Macmillan, New York and London.

Weir, J. T. (2006) Divergent timing and patterns of species accumulation in lowland and highland neotropical birds. *Evolution* **60**, 842–855.

Weir, J. T. and Schluter, D. (2007) The latitudinal gradient in recent speciation and extinction rates of birds and mammals. *Science* **315**, 1574–1576.

Whitmore, T. C. (ed.) (1972–1973) *Tree Flora of Malaya. A Manual for Foresters. Volumes 1 and 2.* Longman Malaysia, Kuala Lumpur.

Wiens, J. J. and Donoghue, M. J. (2004) Historical biogeography, ecology and species richness. *Trends in Ecology & Evolution* **19**, 639–644.

Wiens, J. J., Graham, C. H., Moen, D. S., Smith, S. A. and Reeder, T. W. (2006) Evolutionary and ecological causes of the latitudinal diversity gradient in hylid frogs: treefrog trees unearth the roots of high tropical diversity. *American Naturalist* **168**, 579–596.

Willis, J. C. (1922) *Age and Area. A Study in Geographical Distribution and Origin of Species.* Cambridge University Press, Cambridge.

Wilson, M. V. H. (1983) Is there a characteristic rate of radiation for insects? *Paleobiology* **9**, 79–85.

Wing, S. L. and Harrington, G. J. (2001) Floral response to rapid warming in the earliest Eocene and implications for concurrent faunal change. *Paleobiology* **27**, 539–563.

Won, H. and Renner, S. S. (2006) Dating dispersal and radiation in the gymnosperm *Gnetum* (Gnetales) – clock calibration when outgroup relationships are uncertain. *Systematic Biology* **55**, 610–622.

CHAPTER FIFTEEN

# Temporal patterns in diversification rates

ANDY PURVIS, C. DAVID L. ORME, NICOLA H.
TOOMEY AND PAUL N. PEARSON

## Introduction

The study of rates of speciation and extinction, and how these have changed over time, has traditionally mainly been the preserve of paleontology (Simpson 1953; Stanley 1979; Raup 1985). More recently, phylogenies of extant species have been shown to contain information on these rates and how they may have changed, under the assumption that the same rules have applied in all contemporaneous lineages (Harvey *et al.* 1994; Kubo & Iwasa 1995; see Nee 2006 for a recent review). The first section of this chapter contrasts the strengths and weaknesses of these two approaches – paleontological and phylogenetic – to the study of macroevolution in general.

Moving to a specific macroevolutionary hypothesis, we then outline some tests of the hypothesis that diversification rates have declined in the recent past, either in response to changed abiotic conditions or as a result of density-dependence or diversity-dependence. It has long been appreciated that incomplete species-level sampling can cause a bias in favour of this hypothesis at the expense of the null hypothesis of no change (Pybus & Harvey 2000), but we highlight a further sort of incompleteness that is likely to be very widespread and which is not widely appreciated – products of recent lineage splits are unlikely to be considered as distinct species. We reanalyze the data from a key early paper (Zink & Slowinski 1995) to show how this incompleteness, which is inevitable when taxonomy and phylogeny meet, is sufficiently strong to account for much (though not all) of the apparent tendency for rates to have declined through time.

Attempts to infer the deeper history of diversification rates are relatively free of this problem but instead encounter others: in particular, the assumption that the underlying probabilities of speciation and extinction per unit time have changed in the same way in all lineages becomes increasingly unlikely as a wider and wider clade is considered. We explain how analyses of a near-complete species-level phylogeny of extant mammals show a complex pattern of temporal variation in the rate of effective cladogenesis (Bininda-Emonds *et al.* 2007), and compare the mammalian picture with that obtained some years previously for birds (Nee *et al.* 1992).

*Speciation and Patterns of Diversity*, ed. Roger K. Butlin, Jon R. Bridle and Dolph Schluter. Published by Cambridge University Press. © British Ecological Society 2009.

Phylogenies of extant species do not contain information to decide whether such temporal variations reflect changes in speciation rates, extinction rates or both. This information can only ever be obtained from the relatively few fossil records that are sufficiently well sampled to permit sophisticated analysis at the species level. The final section of the chapter shows how a model system approach to macroevolution would inform and develop the field as a whole. We propose Tertiary planktonic foraminifera as the most promising model system, and show how such systems have the potential to provide insights into how clades wax and wane that is simply not available from data on extant species alone.

## Two approaches to macroevolution

Figure 15.1a is a cartoon of the ideal data set for macroevolutionary research. The fossil record of this obliging clade is comprehensive and continuous, permitting lineages to be traced accurately through the rocks: lineage splits are speciations, and the sampling is so complete that the disappearance of a lineage marks its extinction. This level of detail greatly facilitates identification of non-random patterns in the clade's history. For example, four of the lineages in Figure 15.1a went extinct within a very narrow time window; and the clade comprising the two extant species on the left has had both lower speciation rate and a lower extinction rate than the clade comprising the remaining extant species. Furthermore, the ideal data set also contains the history of many species attributes: morphometric features such as body size and shape can be read directly at any point in time, as can each species' geographical range, paleo-temperature and so on. Provided that such a data set is large enough to give reasonable statistical power, hypothesis-testing is easy: for instance, rates (whether of speciation, extinction or anagenetic change) can be correlated with attributes of both species and their environments (Foote 1996; Jablonski & Roy 2003); character evolution can be traced simply along lineages to test hypotheses about trends or stable attractors (Alroy 2000).

Unfortunately, this ideal is never reached. Paleontologists typically have a data set that is more like the caricature in Fig. 15.1b, so testing hypotheses reliably is more difficult. More – usually much more – of the fossil record is missing than present, and what is present is not fully representative. Some lineages (e.g. the two right-hand species) have no fossil record at all, and the same is true of many time periods (some short, some long); sample completeness is likely to vary over time in a complicated fashion that can be hard to correct for (Alroy *et al.* 2001; Smith *et al.* 2001; Peters & Foote 2002). Occasionally a lineage can be traced for a long time, but the record for any lineage is usually very fragmentary. Consequently, speciation events and extinction events are much harder to infer correctly: many will be missed, because the species is entirely absent from the record, and others may be incorrectly inferred to have

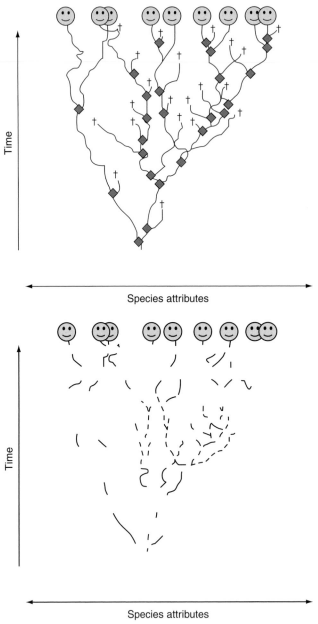

Species attributes

Species attributes

**Figure 15.1** Caricatures of data sets for macroevolution. (a) The ideal data set; speciations (diamonds), extinctions (crosses), relationships and attributes are all directly available throughout the group's history. (b) A paleontological data set. The record is too fragmentary to confidently infer dates of speciation or extinction, and has gaps that are non-random with respect to both lineage and time. (c) A data set restricted to present-day diversity. Relationships can be inferred (straight lines), and the history of attributes can be inferred (position along attribute axis), but neither extinctions nor attribute history are recoverable directly.

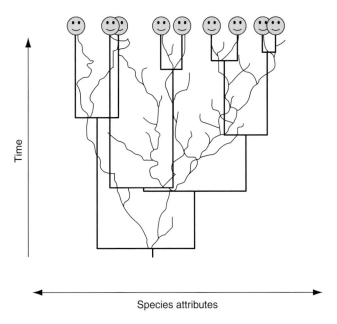

Figure 15.1 (cont)

taken place. Such pseudoextinction and pseudospeciation arise when a single evolving lineage of ancestor–descendant populations is split more or less arbitrarily into multiple morphospecies, the type specimens of which may be very distinct (Stanley 1979; Pearson 1998a). Analyses of evolutionary dynamics that are based on morphospecies therefore risk conflating rates of speciation with rates of anagenetic change (Pearson 1998b). It can often be difficult to cross-reference timescales among geographic regions, adding uncertainty to the timing of inferred events. Additionally, the attribute data are likely to be even more incomplete than the figure shows: most groups have geographically patchy coverage, for instance, complicating inferences relating biogeography to clade dynamics.

The phylogenetic approach has become extremely popular in recent years, with the explosion in molecular systematics making it very much easier to produce more or less complete species-level phylogenies than more or less complete species-level fossil records. The sophistication of phylogeny estimation procedures has increased greatly in recent years, though there is still ongoing development of 'relaxed clock' methods for obtaining relative dates of branching points, and of procedures for using the fossil record to calibrate these to absolute time (Benton & Donoghue 2007). Attributes of present-day species – including those that do not fossilize – can also be measured relatively easily and directly. However, even if the data and phylogeny were absolutely perfect, the neontologist's view of a clade's history is still woefully incomplete

(Fig. 15.1c). One obvious problem, though not our main focus in this chapter, is that the attributes of ancestors have not been observed, and so must be inferred from the attributes of their extant descendants, using the phylogeny and some model of how attributes change through time. It is hard to know what model to use, and estimates of ancestral characteristics will often be unreliable (see, e.g. Cunningham *et al*. 1998; Oakley & Cunningham 2000; Webster & Purvis 2002).

Another serious problem is that extinctions are completely missing from the picture in Figure 15.1c. Nodes in the phylogeny represent speciations, but only those speciations from which more than one daughter branch has an extant descendant: other speciations are missed. This obviously gives a limited view of what has happened, and has also led to awkward terminology: the rate at which nodes occur on the tree is not the rate of speciation, nor of cladogenesis, so is typically termed either the net rate of diversification (Zink & Slowinski 1995) or the rate of effective cladogenesis (Nee *et al*. 1992).

Faced with an absence of direct information about the past state of the system under study, researchers typically must resort to simple models of what might have happened. When the issue is diversification rates through time, the simple models are likely to be pure birth or birth–death models, in which all species, or all contemporaneous species, have the same chances of speciation and, in a birth–death process, extinction. When the issue is character evolution, the model is likely to be a continuous-time Markov model (Felsenstein 1985; Pagel 1994; Schluter *et al*. 1997). These models are either fitted, to produce parameter estimates, or used as null models to test whether some marginally more complex alternative provides a better fit to the data. The next two sections consider how such approaches have been used to test for temporal patterns in the diversification rate of clades.

### Tests for simple changes in the net rate of diversification

Extinctions are missing from the data in Fig. 15.1c, and are also missing from one of the models most commonly used in macroevolutionary analyses of molecular phylogenies. In the pure birth process, or Yule process (after Yule 1924, who first developed the model mathematically for the study of macroevolution), each species has the same constant instantaneous speciation rate, $\lambda$, and there is no extinction. Cladogenesis is modelled as a stochastic process: the waiting time before a given species speciates has a mean of $1/\lambda$, but is exponentially distributed. This simple model leads to many straightforward but powerful results (see Nee 2001, 2006). Two are of particular interest here. One is that a clade should grow exponentially, such that a plot of ln(lineage number) against time should produce a straight line with slope $\lambda$. Such lineages-through-time plots (Nee *et al*. 1992) have been widely used to give a first impression of clade dynamics. A closely related result concerns the timing between successive speciation events as a clade diversifies under this process. The expected wait before any of $n_t$ species in a clade speciates is $1/n_t \lambda$. So successive speciation events are separated by ever less time. However, with $n_t$ species, the expected amount by which the phylogeny's total branch length increases during the wait for the next speciation is $n_t/n_t \lambda = 1/\lambda$ (Purvis *et al.* 1995). Under a Yule process, then, the phylogeny of a clade grows by (on average) the same amount between each pair of speciation events, and this amount is simply the reciprocal of the per-lineage speciation rate.

These two results provide baselines against which to judge the pattern of node ages in molecular phylogenies. If $\lambda$ has not been constant, but has been increasing through time, then the lineages-through-time plot will steepen towards the present, and successive speciations will have been accompanied by less and less growth in the total branch length of the phylogeny. Conversely, if $\lambda$ has been decreasing with time, the lineages-through-time plot will tend to flatten, and the phylogeny will on average grow by more and more between successive speciation events.

Extinction changes the picture. At least, it *should*. However, a common way that researchers have tried to generalize the Yule process to include extinction makes no difference. If speciation rate is $\lambda$ and extinction rate is $\mu$, then it is tempting to model the resulting birth–death process as a pure birth process with speciation rate $\lambda - \mu$, and statistical tests have been developed that use this model to test whether diversification rates have changed through time (Paradis 1997, 1998; Price *et al.* 1998). However, the birth–death process cannot be modelled in this way (Nee 1994). If $\mu > 0$ then, even if $\lambda$ and $\mu$ are constant through time, the lineages-through-time plot steepens towards the present: the expected slope for much of a clade's history is indeed $\lambda - \mu$ but, starting around $1/(\lambda - \mu)$ time units ago, it starts to steepen until, at the present day, the slope estimates $\lambda$ (Nee *et al.* 1994). Equally, the phylogeny is expected to grow by ever smaller amounts between successive speciation events. Consequently, most tests are unable to discriminate between a true increase in the net rate of diversification and a constant-rates birth–death process (Pybus & Harvey 2000).

Decreases in the net rate of diversification are another matter, however. They are not expected from a complete species-level phylogeny that has grown under a birth–death process. They are also fascinating theoretically, because they are consistent with density-dependent diversification, which some (though not all) large-scale paleontological analyses support (see, e.g. Alroy 1996, 1998; Sepkoski 1996; Foote 2000; Kirchner & Weil 2000).

In a key paper, Zink and Slowinski (1995) found that small ($N = 3$ to 11 species) but mostly complete species-level molecular phylogenies of 11 North American bird genera mostly showed diversification rate to have been decreasing rather than increasing. Ten genera were large enough to assess the direction of curvature in the lineages-through-time plot, and nine of these (significantly more than half) showed a decrease rather than an upturn. Zink and Slowinski also developed a test statistic that uses the branch length increases between successive speciations, and which was intended to follow a standard Normal distribution under a pure birth process; 10 of the 11 genera (again, significantly more than half) had a negative value of this statistic. Some individual genera showed significant slowdown (though one-tailed tests were used incorrectly, so the $p$-values should all be doubled), but the startling feature of the analysis was that such a high proportion of the data sets showed a tendency towards slowdown.

Pybus and Harvey (2000) corrected an error in Zink and Slowinski's test statistic, calling the corrected version $\gamma$. They explicitly drew attention to how incomplete sampling biases $\gamma$ downwards (i.e. towards a result that supports slowdown) and proposed using Monte Carlo simulations to test whether an apparent slowdown remained significant when the (known) level of sample incompleteness was accounted for, under an assumption that species are missing from the phylogeny at random. They noted that overdispersed and underdispersed samples would bias $\gamma$ downwards and upwards, respectively. This method has been widely used since, with results often showing significant slowdown of diversification (see Price & Phillimore, this volume, for a review).

There is one important way in which sampling is often likely to be non-randomly incomplete. Because speciation is usually a gradual process, there are unlikely to be many pairs of sister species that split in the very recent past. Such recent lineage splits are unlikely to be deemed speciation events: a node is likely to be designated as a speciation event only if both lineages persist long enough to evolve differences that attract taxonomic attention.

In order to illustrate how species designation depends on node age, we have used published data on the levels of *cyt b* sequence divergence for splits that were within species (Avise & Walker 1998; Avise *et al.* 1998) or between putatively sister species (Johns & Avise 1998). Due to sample incompleteness, some of these latter nodes probably delimit clades of more than two species, and the estimated 'sister'-species sequence divergence does indeed decrease linearly as the proportion of recognized species sampled with each genus increases. We have used residuals around this relationship, converted to units of time using the 2% per million years calibration, as corrected estimates of sister-species divergence. As the intraspecific divergence times presented in Avise and Walker (1998), and Avise *et al.* (1998) represent the upper limit of possible between-population divergence, we have used the reported numbers of populations for each study as additional estimates of intraspecific divergence. The actual divergence estimates for these between-population splits are not reported, so we have made the assumption that these splits have zero sequence divergence; this assumption is conservative with respect to the point we wish to make here. Divergences were converted to units of time as in the papers from which they came. The full data set was then used in a binomial general linear model (logistic regression) to predict whether divergence time and vertebrate group (mammal, bird, herpetofauna or

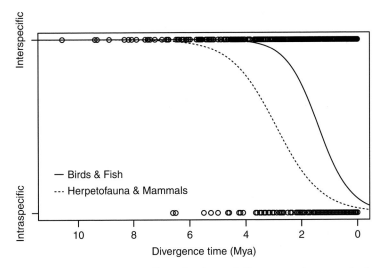

**Figure 15.2** Plot showing effect of estimated divergence time on whether a branching event is deemed a speciation event. Curves show predictions from a binary general linear model including vertebrate group (fish, herpetofauna, mammal or bird) as an additional predictor; model simplification indicates birds and fish share a model, as do mammals and herpetofauna.

fish) predict whether a given divergence represents a speciation event, with model simplification used as appropriate (Crawley 2002). The resulting model is shown in Fig. 15.2. In each vertebrate group, the data suggest that recent splits are very unlikely to be designated as speciation events, especially in birds and fish. Other data sets would doubtless give quantitatively different results, but the qualitative pattern is an inevitable consequence of grafting species-level taxonomy and lineage phylogeny together.

Figure 15.3 shows the implication of the pattern for studies of slowdown in species-level phylogenies. If the truth has been a Yule process, but recent splits are unlikely to be included in the phylogeny (which is nonetheless complete, inasmuch as it contains all the recognized species in a group), then the lineages-through-time plot or $\gamma$ are biased downwards (i.e. towards slowdown).

To what extent could this bias be responsible for the prevalence of negative $\gamma$ values in published studies? We combined simulations of a Yule process with the above species designation model to produce a preliminary assessment. The value of $\lambda$ came from pooling an illustrative set of 14 avian species-level phylogenies (incorporating those of Zink and Slowinski that we could replicate) to find the average $\lambda$ from early in a clade's history. We used the 14 series of internode distances (as percentage sequence divergence) and calculated the mean of each successive internode distance from the root across the groups (i.e. averaging all the first internodes, then all the second internodes, and so on) to obtain a

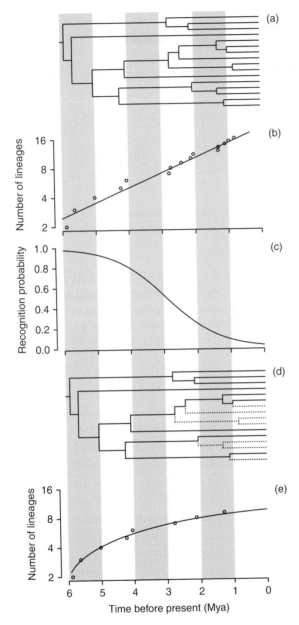

**Figure 15.3** Cartoon illustrating how apparent slowdown in diversification can arise simply through age-dependency in whether nodes are deemed to be speciation events. A phylogeny that has grown under a constant rate, pure birth model (a) yields a lineages-through-time plot with a constant slope (b). If the probability that cladogenetic events within the phylogeny are recognized as speciation events varies as a function of the age of the event (c), then the resulting species phylogeny (d; dotted branches are not viewed as separate species) will tend to yield lineages-through-time plots that exhibit apparent decrease in net diversification (e).

lineages-through-time plot averaged across the 14 groups. Our estimate of early $\lambda$ was then the slope of the first segment in a breakpoint regression of ln(N) against time (constrained to have an initial intercept of ln(2), and with the breakpoint chosen by least-squares). One thousand Yule phylogenies were then simulated with this value of $\lambda$, with crown groups the same age as the average among our 14 clades. We computed $\gamma$ for each of these 1000 simulated trees.

**Table 15.1** *Slowdown in 14 avian phylogenies, showing $\gamma$ and p for each group when analysed at face value, along with the number of species (n) and the sources*

| Taxon | $\gamma$ | p | n | Source |
|---|---|---|---|---|
| *Ammodramus*[†] | −1.449 | 0.074 | 8 | Zink and Avise (1990) |
| *Anas*[a] | −0.296 | 0.383 | 27 | Johnson and Sorenson (1998) |
| *Dendroica*[a] | −4.173 | <0.001 | 26 | Lovette and Bermingham (1999) |
| Diomedeidae | −0.509 | 0.305 | 14 | Nunn *et al.* (1996) |
| Gruidae | −2.413 | 0.008 | 15 | Krajewski and Fetzner (1994) |
| *Melospiza*[†] | −1.410 | 0.079 | 3 | Kessler and Avise (1985) |
| *Parus* (Chickadees)[†] | −2.645 | 0.004 | 7 | Gill *et al.* (1993) |
| *Parus* (Titmice)[†] | 0.288 | 0.613 | 4 | Gill and Slikas (1992) |
| *Passerella*[†] | −1.633 | 0.051 | 4 | Zink (1994) |
| *Pipilo*[†] | −1.255 | 0.105 | 4 | Zink and Dittman (1991) |
| *Quiscalus*[†] | −1.182 | 0.119 | 4 | Zink *et al.* (1991b) |
| *Spizella*[†b] | −0.924 | 0.178 | 6 | Zink and Dittman (1993) |
| *Toxostoma*[†] | −2.490 | 0.006 | 11 | Zink *et al.* (1999) |
| *Zonotrichia*[†] | 0.486 | 0.687 | 6 | Zink *et al.* (1991a) |

Note: Phylogenies used by Zink and Slowinski (1995) are indicated (†). The phylogenies contain all described species except where marked: (a) >90% described species; (b) <90% described species.

To simulate the effect of age-biased sampling, we used the species designation model to give the probability that nodes of a given age would be sampled from the simulated trees: older nodes within a tree are more likely to be sampled than younger nodes, as in Fig. 15.2. Unsampled nodes were pruned from the trees, and each tree's $\gamma$ was computed. By comparing the values of $\gamma$ from before and after pruning, we were able to estimate the likely effect of age-dependent sample incompleteness on the $\gamma$ values in the original analyses of these phylogenies. Before pruning, the mean $\gamma$ was −0.19 (significantly negative, as pointed out by Price & Phillimore, this volume); pruning reduces the mean to −0.75.

Table 15.1 shows the $\gamma$ for each of the 14 phylogenies; the centre of the distribution is significantly less than zero (Wilcoxon signed ranks test: $p=0.0004$), but only marginally significantly less than −0.75, the mean for the pruned simulations (Wilcoxon signed-ranks test: $p=0.045$). This result suggests that the age-dependent probability of designating a node as a speciation event could be responsible for much of the apparent slowdown in diversification in these avian phylogenies, and particularly for much of the general tendency for slowdown.

One of the phylogenies does show strong evidence for slowdown: *Dendroica*, with 24 species, is the largest of the set, and is large enough to attempt to fit the same sort of model to it on its own. However, the initial phase of diversification in this genus is so rapid that Yule processes with the estimated $\lambda$ grow too large for our software before they reach the present day, implying that the slowdown is much more severe than age-dependent incompleteness could produce (Orme 2002).

If age-dependent sample incompleteness biases tests of slowdown, what can be done about it? Increasingly, sequence data are available for many populations within species, which may permit a solution. Pons *et al.* (2006) show how the difference in expected branching pattern between within- and between-species phylogenies can be used to identify the depth in the tree that best represents the species level. For groups with sufficiently rich data sets, it is therefore possible to prune the phylogeny back to this depth, and used the pruned phylogeny as the basis for testing.

## Tests for more complex temporal patterns

The $\gamma$ test reduces the information available in a molecular phylogeny to a single number, which provides insight into whether the net rate of diversification decreased or increased over time. In reality, clade dynamics may have shown a more complex temporal pattern of diversification, so there is a need for an approach that can be used to test more complex hypotheses. Two sorts of approaches have been developed recently. The first views a clade's history as having had multiple (usually two) different birth–death regimes, with a sudden transition between them. The timing of the transition(s), and the different speciation and extinction rates, are estimated from the timings of nodes in the phylogeny, and a range of approaches are available to test whether the additional complexity is merited by the data (Barraclough & Vogler 2002; Turgeon *et al.* 2005; Rabosky 2006).

Bininda-Emonds *et al.* (2007) developed a second approach, using generalized additive models (GAMs; Wood 2006) to model the net rate of diversification as a smooth function of time – a curve rather than a set of steps. Unlike standard regression approaches, GAMs do not require a detailed specification of the nature of the relationship being modelled. This feature is particularly attractive here, because there is generally no *a priori* reason for preferring a particular form. The GAM provides a model with minimum complexity necessary to capture the relationship, and also provides the facility for testing whether the relationship is consistent among clades (as an ANCOVA would do in a linear model framework). Their analysis took its data from a near-complete and dated species-level composite 'supertree' phylogeny of extant mammals (Bininda-Emonds *et al.* 2007). Although the phylogeny is almost complete, it is only around 46% as resolved as a fully bifurcating tree would be, with most of the lack of resolution being near the tips (especially within genera; the phylogeny prior to 32 MYA is 75% resolved) and within a few poorly studied groups (notably Muridae). Also, the dates of around one third of the nodes in the tree are not estimated directly from sequence data or fossils, but interpolated based on the dates of surrounding nodes and on the diversities of the clades involved.

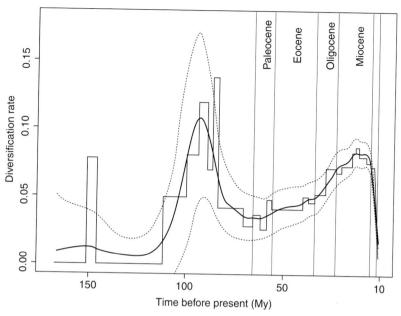

**Figure 15.4** Mammalian net diversification rate through time. The set of grey steps represent the average rates within subepochs. The solid curve is the fitted rate from a generalized additive model (GAM); the dashed lines are 95% confidence intervals. The thick upright grey line at 65.5 Mya represents the Cretaceous-Tertiary transition; other vertical lines demarcate Tertiary epochs. After Bininda-Emonds *et al.* (2007).

Consequently, it would be unwise to use all the branches in the phylogeny at face value. The analysis was therefore restricted to only those branches whose start and end dates were known most reliably, i.e. branches that did not start at a polytomy, and whose start and end dates were estimated from data rather than interpolated. Over 3100 branches, many of them ending in a present-day species, met these criteria. These are not a random sample of the branches in the tree – resolution is poorest within genera – which leads to predictable biases in the results; we describe and discuss these below.

The net rate of diversification was first estimated in each geological subepoch using a survival analysis (where a node, counter-intuitively, marks the 'failure' of the parent lineage), as shown in Fig. 15.4. This analysis made clear that the temporal pattern of net diversification rate was indeed likely to be complex (stepped line in Fig. 15.4). The analysis then moved to a 0.1 My timescale, because this was the level of resolution in the dates in the phylogeny, and all the branches in the analysis are at least this long. Within each 0.1 My interval, the lineages present that did and did not result in a node at the end of the interval were counted, and turned into a binomial response variable. The advantage of this approach over simply computing a rate is that the binomial response variable contains sample size information, so the subsequent model is appropriately weighted. The response variable was then modelled as a smooth function of time. Because the relationship could in principle be very complex, with peaks or troughs lasting only one or a few million years (very short compared to the full 166 My time scale), the basis dimension ($k$, a parameter that sets the maximum complexity of the curve) for the fitted curves was set high enough to permit a kink in the curve every 1 My; using the default value of $k$ permitted only ten kinks, which made the curve noticeably less responsive to the data. Further details of the analysis can be found in Bininda-Emonds *et al.* (2007). The curved lines in Fig. 15.4 show the fitted rates and confidence intervals plotted against time.

The right-hand side of the curve is hard to interpret for three reasons. First, the fact that recent lineages are unlikely to be designated as species is at least partly responsible for the very low rates of diversification in the very recent past. Second, this low rate is likely to also be partly due to a tendency for only species-poor genera to have well-resolved phylogenies: branches that are in reality long are more likely to meet the criteria for inclusion in the analysis. This bias will be much less marked deeper in the tree, but is likely to suppress estimates of the net diversification rate for the last few million years. Third, some of the preceding increase in rate will be the upturn expected under a birth–death process. This artifact has the longest reach into the past, but is expected to subside within about $1/(\lambda - \mu)$ time units ago (Harvey *et al.* 1994); this is basically because, if a clade survives for $1/(\lambda - \mu)$ and $\lambda > \mu$, it will very probably have attained a high enough diversity to make future stochastic extinction unlikely. Bininda-Emonds *et al.* (2007) therefore restricted their interpretation to the period earlier than 25 My ago (the reciprocal of their lowest net rate of diversification), focusing on the peak in rates from 100–85 My ago, the decline in rates that persisted through the end-Cretaceous mass extinction, and then a subsequent upturn that did not start until around the early Eocene, about 55 My ago. Their analysis showed no indication of any significant pulse or other increase in the rate of origin of extant mammalian lineages after the end-Cretaceous event, except for a possible increase in marsupials: there, lineage number doubled around the Cretaceous–Tertiary boundary, but the numbers of lineages involved are too small (rising from 3 to 6) for any firm conclusion to be reached. Intriguingly, the lack of an upturn in the earliest Tertiary agrees completely with a much earlier analysis of Sibley and Ahlquist's (1990) DNA–DNA hybridization phylogeny of birds. Nee *et al.* (1992) analysed the portion of the tree for which lineage sampling was probably complete; according to the time calibration used by Sibley and Ahlquist, this was prior to 45 MY ago. They used a range of tests to demonstrate that the per-lineage net rate of diversification had decreased throughout the preceding history of bird diversification, with the lineages-through-time plot showing a similar shape to the one reported by Bininda-Emonds *et al.* (2007). The phylogeny underpinning this analysis is controversial, but it will be interesting to see whether the similarity in temporal pattern persists when more robust phylogenies are available for birds. It is also important to note that both sets of dates, along with many other timescales derived from molecular phylogenies, are often considerably older than corresponding estimates taken directly from the fossil record, which would have modern orders originating after the Cretaceous–Tertiary boundary (Wible *et al.* 2007). In dating the tree, Bininda-Emonds *et al.* (2007) took the view that a fossil provides only a minimum estimate for the age of the crown-group clade within which it is nested. This is because, with an incomplete fossil record, the early days of the crown group are likely to have left no fossils that are both known and correctly identified; the more incomplete the record, the longer such gaps will tend to be (Tavaré *et al.* 2002). An alternative view is that the fossil record is complete enough that at least some of the fossils used in calibration provide more than a minimum estimate, indicating that a clade must have originated only soon before the time those species lived. Reconciling these views is a challenge for future research (Cifelli & Gordon 2007; Penny & Phillips 2007).

The GAM approach permits great flexibility in modelling. It is possible to test for significant differences among clades, analogous to the use on ANCOVAs as an extension of standard regressions: the two major superorders, Euarchontoglires and Laurasiatheria, show broadly similar patterns, while the other two placental superorders combined (Xenarthra and Afrotheria) show the early high rate but do not show the rate increase at 55 MY ago that these two groups show (Bininda-Emonds *et al.* 2007). It is also possible to use GAMs to test whether the response variable shows a sharp disjunction at a time thought in advance to be important (Wood 2006), such as ends of epochs. Further, it has the advantage that it can be used even if the start and end dates are not known for all the branches in the phylogeny, provided that it is reasonable to assume that the branches for which these dates are available are a random or otherwise representative sample of the whole phylogeny. Also, the focus on deeper branches in the phylogeny make it easier in principle to meet the requirement for completeness, and make it possible to test for increases as well as decreases in the net rate of diversification.

Although these methods have some advantages over earlier ones developed to answer similar questions, they are still inherently limited by their reliance on data from present-day diversity. For example, there is excellent fossil evidence from North America indicating that mammalian speciation rates spiked in the immediate aftermath of the end-Cretaceous event (Alroy 1999, 2000), but the phylogeny of present-day species contains no significant signal of this event, because the great majority of the new species were in groups that have subsequently declined or gone extinct (Alroy 2000; Bininda-Emonds *et al.* 2007). In this instance, the paleontological and present-day approaches give complementary insights. The North American fossil record reveals the speciation pulse, but leaves open the possibility that still-extant lineages were diversifying in other, less well-studied, parts of the world at the same time. The phylogeny contains no evidence of the early Tertiary radiations that died out completely, and is mute about whether speciation rate or extinction rate fluctuations have mattered more, but shows that the extant lineages were not diversifying in some as-yet-unworked part of the world. One obvious priority for future macroevolutionary research is to develop statistical methods that more formally integrate data from the fossil record and present-day diversity: both should reflect the same underlying reality, and so frameworks combining these both should lead to a more accurate and precise picture of what actually happened (Jablonski *et al.* 2003).

### Problems with the phylogenetic approach

Mammals have a good fossil record, compared with that of most groups, so it is easy to see the ways in which data from present-day diversity alone give an incomplete or even misleading picture of what happened. What of other groups? The reliability of any inferences will depend upon the adequacy of the models used to bridge the gap left by the lack of direct information about the history of the system. Unfortunately, the data from present-day diversity do not provide a good basis for testing whether the models are adequate. They are not informative about whether, if net diversification rates have varied, variation is driven by speciation or extinction rates (Barraclough & Nee 2001). They do not directly record past diversity, weakening tests for density-dependence. They are mute about whether individual characters have shown evolutionary trends (Oakley & Cunningham 2000; Webster & Purvis 2002). These problems are exacerbated when testing hypotheses in which diversification and character evolution are linked: data from extant diversity alone may be unable to distinguish between very different scenarios (Purvis 2004; Maddison 2006).

### A model system approach for macroevolution?

The combined weaknesses of the paleontological and phylogenetic approaches greatly hamper macroevolutionary research, because there is currently no synthetic global overview of macroevolution in any large clade. However, a few clades do at least begin to approach the ideal depicted in Fig. 15.1, and so might usefully be developed as model systems for macroevolutionary research. The planktonic foraminifera have the best species-level fossil record of any group over the past 65 MY. These sexual unicellular marine zooplankton secrete ornate calcareous shells (Fig. 15.5), and have morphologies that are specific to biological species or closely related groups of cryptic species, of which around 50 are extant today (Hemleben *et al.* 1989; Norris 2000; Kucera & Darling 2002). There is ongoing controversy about the meaning of species level categories both in this group and in general, much of it semantic or arising from the desire to apply hard and fast concepts to what is a loose natural category (Hey *et al.* 2003).

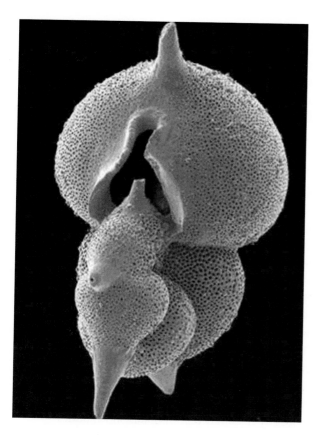

**Figure 15.5** Fossil shell of a planktonic foraminifera from 33.7 Mya (*Hantkenina nanggulanensis*). Such shells occur in vast numbers in stratified deep sea sediments and provide an unrivalled opportunity for examining the macroevolution of a major taxonomic group by sampling at the species level. Scale: about 1 mm long.

The 'species level' in this context refers to clusters of related genotypes that secrete shells of a recognizable unimodal type, sometimes changing through anagenetic evolution but not by diversifying into more than one type, that can often be traced for millions of years from their origin to either a branching event (speciation) or their (often sudden) final extinction. The shells are deposited in vast numbers in ocean sediments, where in favourable circumstances they can accumulate continuously over many millions of years, thereby providing a continuous record of the species' existence from their origin to their extinction (Parker *et al.* 1999). Many sites have now been cored in each of the world's major ocean basins and samples are also available from a wide range of reference sections now exposed on land. The taxonomy of the group is well-developed and mature (Olsson *et al.* 1999; Pearson *et al.* 2006), as is knowledge of their biostratigraphic distribution (Stewart & Pearson 2002). Furthermore, although the group does not match the high levels of diversity of some other groups commonly used in paleontological research, the fossil record is sufficiently complete that it is even possible to sample large populations at will to focus on lineages or times of particular interest, rather than rely on more or less haphazard finds of a few specimens as is typical of most paleontological research.

The fossil record of planktonic foraminifera has been widely used for evolutionary research, including studies of morphometrics of individual lineages (Malmgren & Kennett 1981; Malmgren *et al.* 1983), species survivorship (Arnold 1982; Stanley *et al.* 1988; Pearson 1996; Parker & Arnold 1997), overall size trends (Schmidt *et al.* 2004), diversity trends (Parker *et al.* 1999), tree shape (Pearson 1998b; MacLeod 2001), shell chirality (Norris & Nishi 2001) and evolutionary rates (Prokoph *et al.* 2000; Allen *et al.* 2006). The group is ripe for a thorough reinvestigation using the full range of phylogeny-based analytical tools that have been developed recently.

For such a set of analyses to be successful, they must be based on a phylogeny of evolutionary species-lineages rather than morphospecies. Morphospecies (which are the standard units of taxonomy and biostratigraphy) are recognized on the basis of similarity to the type specimen, whereas a species-lineage (or, more simply, species) is a branch on a valid phylogenetic tree at the species level (Fordham 1986; Pearson 1998b). The distinction is needed because a species may evolve sufficiently that its members become recognized under different taxonomic names; morphospecies boundaries and even generic distinctions may in principle be crossed without any new lineage being formed. Some recent macroevolutionary studies of the group (Prokoph *et al.* 2000; Allen *et al.* 2006; Doran *et al.* 2006) have not made this distinction, so may have conflated taxonomic turnover with anagenetic change.

Pearson (1993) constructed such a lineage phylogeny for Paleogene planktonic foraminifera. Although this phylogeny is somewhat out-of-date now, both taxonomically and in terms of the timescale, it serves to illustrate the potential that an updated and extended lineage phylogeny would have to shed light on macroevolutionary dynamics (see also Pearson 1996, 1998b). Here, we use it for a preliminary test a key assumption of the models at the heart of most neontological macroevolution, namely that a species' age has no bearing on its chances of speciation or extinction (Kendall 1949; Hey 1992). Previous paleontological tests have used superspecific taxa, used morphospecies, been geographically restricted and/or been equivocal (Van Valen 1973; Parker & Arnold 1997; Alroy 1998; Pearson 1998b). We focused on only those internodes in the phylogeny that terminated, either in speciation or extinction, before the end of the period covered by the phylogeny. A logistic regression of fate (speciation or extinction) against the length of the branch suggests that the balance between speciation and extinction changed significantly with species age, with young species being more likely to speciate and old species more likely to go extinct; the balance point is at around 7 MY (Fig. 15.6). Further analyses could show whether this change reflects age-dependency in speciation rates, extinction rates or both, and could also test whether it is driven by differences in the balance between speciation and extinction through geological time, or by clade differences in rates. The system can also provide answers to a range of

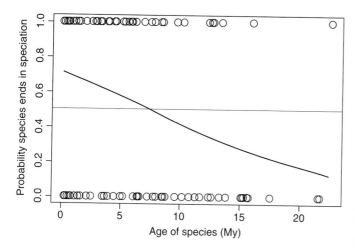

Figure 15.6 The relative chances of speciation or extinction change as species' age, for Paleogene planktonic foraminifera. Circles: branches in the phylogeny that end in either speciation ($y = 1$) or extinction ($y = 0$). Thick line: fit from logistic regression model, which is highly significant (increase in residual deviance on term deletion = 10.502, 1 d.f, $p = 0.001$). Thin grey line: speciation and extinction are equally likely, for any age of species.

other questions which, although often addressed in the paleontological literature, cannot be tackled as unambiguously in many fossil groups (or in any groups based solely on data from extant organisms). Does density-dependence act by reducing speciation rates or raising extinction rates at high density? Does the trend towards increasing size through time (Cope's Rule: Arnold *et al.* 1995; Webster & Purvis 2002) result from within-lineage changes alone, or do size-correlated speciation and extinction play a significant role (Alroy 2000)? And have changes in environmental features such as climate or ocean chemistry driven macroevolution? This hypothesis is long-standing (Stenseth & Maynard Smith 1984) but remains controversial (Alroy *et al.* 2000; Gingerich 2006; Jackson & Erwin 2006).

The model system approach has been very successful in many other areas of biology, revealing many generalities that might never have emerged from less intensive studies of a wider range of systems. We argue that the field of macro-evolution would similarly benefit greatly from gaining detailed knowledge and understanding of the most tractable systems, because of the light they may be able to shed on some of the key processes that underpin diversity patterns in all other groups. Much of the research based solely on extant diversity is reaching a crucial stage, in which the very simple models of clade growth and character change that gave the field its initial impetus are no longer providing useful new directions for research. Demonstrating that clades reject these simple null models is of some interest, but the problem is that there are many ways in which the simple models might be made more complex, and data from extant species alone provide little basis for choosing among them. Clades, such as planktonic foraminifera, that have exceptional fossil records can provide the impetus for the next generation of models, by highlighting which aspects of complexity seem to be most needed.

## Acknowledgements

We are very grateful to the editors for the invitation to the Symposium, and to J. Alroy, O.R.P. Bininda-Emonds and R.E. Ricklefs for very helpful comments on the manuscript. This work was funded by the Natural Environment Research Council (UK) and the Leverhulme Trust.

## References

Allen, A. P., Gillooly, J. F., Savage, V. M. and Brown, J. H. (2006) Kinetic effects of temperature on rates of genetic divergence and speciation. *Proceedings of the National Academy of Sciences of the United States of America* **103**, 9130–9135.

Alroy, J. (1996) Constant extinction, constrained diversification, and uncoordinated stasis in North American mammals. *Palaeogeography, Palaeoclimatology, Palaeoecology* **127**, 285–311.

Alroy, J. (1998) Equilibrial diversity dynamics in North American mammals. In: *Biodiversity Dynamics: Turnover of Populations, Taxa and Communities* (ed. M. L. McKinney and J. A. Drake), pp. 232–287. Columbia University Press, New York.

Alroy, J. (1999) The fossil record of North American mammals: evidence for a Paleocene evolutionary radiation. *Systematic Biology* **48**, 107–118.

Alroy, J. (2000) Understanding the dynamics of trends within evolving lineages. *Paleobiology* **26**, 319–329.

Alroy, J., Koch, P. L. & Zachos, J. C. (2000). Global climate change and North American mammalian evolution. *Paleobiology* **26S**, 259–288.

Alroy, J., Marshall, C. R., Bambach, R. K., *et al.* (2001) Effects of sampling standardization on estimates of Phanerozoic marine diversification. *Proceedings of the National Academy of Sciences of the United States of America* **98**, 6261–6266.

Arnold, A. J. (1982) Species survivorship in the Cenozoic Globigerinida. *Proceedings of the 3rd North American Paleontology Convention* **1**, 9–12.

Arnold, A. J., Kelly, D. C. and Parker, W. C. (1995) Causality and Cope's Rule: evidence from the planktonic Foraminifera. *Journal of Paleontology* **69**, 203–210.

Avise, J. C. and Walker, D. (1998) Pleistocene phylogeographic effects on avian populations and the speciation process. *Proceedings of the Royal Society of London Series B-Biological Sciences* **265**, 457–463.

Avise, J. C., Walker, D. and Johns, G. C. (1998) Speciation durations and Pleistocene effects on vertebrate phylogeography. *Proceedings of the Royal Society B* **265**, 1707–1712.

Barraclough, T. G. and Nee, S. (2001) Phylogenetics and speciation. *Trends in Ecology & Evolution* **16**, 391–399.

Barraclough, T. G. and Vogler, A. P. (2002) Recent diversification rates in North American tiger beetles estimated from a dated mtDNA phylogenetic tree. *Molecular Biology and Evolution* **19**, 1706–1716.

Benton, M. J. and Donoghue, P. C. J. (2007) Paleontological evidence to date the tree of life. *Molecular Biology and Evolution* **24**, 26–53.

Bininda-Emonds, O. R. P., Cardillo, M., Jones, K. E., *et al.* (2007) The delayed rise of present-day mammals. *Nature* **446**, 507–512.

Cifelli, R. L. and Gordon, C. L. (2007) Evolutionary biology – re-crowning mammals. *Nature* **447**, 918–920.

Crawley, M. J. (2002) *Statistical Computing: An Introduction to Data Analysis Using S-Plus*. John Wiley & Sons, New York and Chichester.

Cunningham, C.W., Omland, K.W. and Oakley, T. H. (1998) Reconstructing ancestral character states: a critical reappraisal. *Trends in Ecology & Evolution* 13, 361–366.

Doran, N. A., Arnold, A. J., Parker, W. C. and Huffer, F. W. (2006) Is extinction age-independent? *PALAIOS* 21, 571–579.

Felsenstein, J. (1985) Phylogenies and the comparative method. *American Naturalist* 125, 1–15.

Foote, M. (1996) Perspective: evolutionary patterns in the fossil record. *Evolution* 50, 1–11.

Foote, M. (2000) Origination and extinction components of taxonomic diversity: Paleozoic and post-Paleozoic dynamics. *Paleobiology* 26, 578–605.

Fordham, B. G. (1986) Miocene-Pleistocene planktic Foraminifers from DSDP sites 208 and 77, and phylogeny and classification of Cenozoic species. *Evolutionary Monographs* 6, 1–200.

Gill, F. B., Mostrom, A. M. and Mack, A. L. (1993) Speciation in North American Chickadees: I. Patterns of mtDNA divergence. *Evolution* 47, 195–212.

Gill, F. B. and Slikas, B. (1992) Patterns of mitochondrial DNA divergence in North American Crested Titmice. *The Condor* 94, 20–28.

Gingerich, P. D. (2006) Environment and evolution through the Paleocene-Eocene thermal maximum. *Trends in Ecology & Evolution* 21, 246–253.

Harvey, P. H., May, R. M. and Nee, S. (1994) Phylogenies without fossils. *Evolution* 48, 523–529.

Hemleben, C., Spindler, M. and Anderson, O. R. (1989) *Modern Planktonic Foraminifera*. Springer-Verlag, New York.

Hey, J. (1992) Using phylogenetic trees to study speciation and extinction. *Evolution* 46, 627–640.

Hey, J., Waples, R. S., Arnold, M. L., Butlin, R. K. and Harrison, R. G. (2003) Understanding and confronting species uncertainty in biology and conservation. *Trends in Ecology & Evolution* 18, 597–603.

Jablonski, D. and Roy, K. (2003) Geographical range and speciation in fossil and living molluscs. *Proceedings of the Royal Society of London Series B-Biological Sciences* 270, 401–406.

Jablonski, D., Roy, K. and Valentine, J. W. (2003) Evolutionary macroecology and the fossil record. In: *Macroecology: Concepts and Consequences* (ed. T. M. Blackburn and K. J. Gaston), pp. 368–390. Blackwell Science, Oxford.

Jackson, J. B. C. and Erwin, D. H. (2006) What can we learn about ecology and evolution from the fossil record? *Trends in Ecology & Evolution* 21, 322–328.

Johns, G. C. and Avise, J. C. (1998) A comparative summary of genetic distances in te vertebrates from the mitochondrial cytochrome *b* gene. *Molecular Biology and Evolution* 15, 1481–1490.

Johnson, K. P. and Sorenson, M. D. (1998) Comparing molecular evolution in two mitochondrial protein coding genes (cytochrome *b* and ND2) in the dabbling ducks (Tribe: Anatini). *Molecular Phylogenetics and Evolution* 10, 82–94.

Kendall, D. G. (1949) Stochastic processes and population growth. *Journal of the Royal Statistical Society, Series B* 11, 230–264.

Kessler, L. G. and Avise, J. C. (1985) A comparative description of mitochondrial DNA differentiation in selected avian and other vertebrate genera. *Molecular Biology and Evolution* 2, 109–125.

Kirchner, J. W. and Weil, A. (2000) Delayed biological recovery from extinctions throughout the fossil record. *Nature* 404, 177–180.

Krajewski, C. and Fetzner, J. W., Jr. (1994) Phylogeny of Ganes (Gruiformes: Gruidae) based on cytochrome *b* DNA sequences. *The Auk* **111**, 351–365.

Kubo, T. and Iwasa, Y. (1995) Inferring the rates of branching and extinction from molecular phylogenies. *Evolution* **49**, 694–704.

Kucera, M. and Darling, K. F. (2002) Cryptic species of planktonic foraminifera: their effect on palaeoceanographic reconstructions. *Philosophical Transactions of the Royal Society of London A* **360**, 695–718.

Lovette, I. J. and Bermingham, E. (1999) Explosive speciation in the New World Dendroica warblers. *Proceedings of the Royal Society of London Series B-Biological Sciences* **266**, 1629–1636.

MacLeod, N. (2001) The role of phylogeny in quantitative paleobiological data analysis. *Paleobiology* **27**, 226–240.

Maddison, W. P. (2006) Confounding asymmetries in evolutionary diversification and character change. *Evolution* **60**, 1743–1746.

Malmgren, M. A., Berggren, W. A. and Lohmann, G. P. (1983) Evidence for punctuated gradualism in the Late Neogene *Globorotalia tumida* lineage of planktonic foraminifera. *Paleobiology* **9**, 377–389.

Malmgren, M. A. and Kennett, J. P. (1981) Phyletic gradualism in a late Cenozoic planktonic foraminiferal lineage, DSDP Site 284, Southwest Pacific. *Paleobiology* **7**, 230–240.

Nee, S. (1994) How populations persist. *Nature* **367**, 123–124.

Nee, S. (2001) Inferring speciation rates from phylogenies. *Evolution* **55**, 661–668.

Nee, S. (2006) Birth-death models in macroevolution. *Annual Review in Ecology, Evolution and Systematics* **37**, 1–17.

Nee, S., May, R. M. and Harvey, P. H. (1994) The reconstructed evolutionary process. *Philosophical Transactions of the Royal Society of London Series B* **344**, 305–311.

Nee, S., Mooers, A. Ø. and Harvey, P. H. (1992) The tempo and mode of evolution revealed from molecular phylogenies. *Proceedings of the National Academy of Sciences of the United States of America* **89**, 8322–8326.

Norris, R. D. (2000) Pelagic species diversity, biogeography, and evolution. *Paleobiology* **26**, 236–258.

Norris, R. D. and Nishi, H. (2001) Evolutionary trends in coiling of tropical planktic foraminifera. *Paleobiology* **27**, 327–347.

Nunn, G. B., Cooper, J., Jouventin, P., Robertson, C. R. J. and Robertson, G. G. (1996) Evolutionary relationships among extant albatrossses (Procellaroiformes: Diomedeidae) established from complete cytochrome *b* gene sequences. *The Auk* **113**, 784–801.

Oakley, T. H. and Cunningham, C. W. (2000) Independent contrasts succeed where ancestor reconstruction fails in a known bacteriophage phylogeny. *Evolution* **54**, 397–405.

Olsson, R. K., Hemleben, C., Berggren, W. A. and Huber, B. T. (eds) (1999) *Atlas of Paleocene Planktonic Foraminifera*. Smithsonian Contributions to Paleontology. Smithsonian, New York.

Orme, C. D. L. (2002) Body size and macroevolutionary patterns of species-richness. In *Department of Biological Science, Imperial College of Science, Technology and Medicine*, vol. PhD. University of London, London.

Pagel, M. (1994) Detecting correlated evolution on phylogenies: a general model for the comparative analysis of discrete characters. *Proceedings of the Royal Society of London Series B* **255**, 37–45.

Paradis, E. (1997) Assessing temporal variations in diversification rates from phylogenies: estimates and hypothesis testing. *Proceedings of the Royal Society of London Series B-Biological Sciences* **264**, 1141–1147.

Paradis, E. (1998) Detecting shifts in diversification rates without fossils. *American Naturalist* **152**, 176–187.

Parker, W. C. and Arnold, A. J. (1997) Species survivorship in the Cenozoic planktonic foraminifera: a test of exponential and Weibull models. *PALAIOS 12*, 3–11.

Parker, W. C., Feldman, A. and Arnold, A. J. (1999) Paleobiogeographic patterns in the morphologic diversification of the Neogene planktonic foraminifera. *Palaeogeography, Palaeoclimatology, Palaeoecology* **152**, 1–14.

Pearson, P. N. (1993) A lineage phylogeny for the Paleogene planktonic foraminifera. *Micropaleontology* **39**, 193–232.

Pearson, P. N. (1996) Cladogenetic, extinction and survivorship patterns from a lineage phylogeny: the Paleogene planktonic foraminifera. *Micropaleontology* **42**, 179–188.

Pearson, P. N. (1998a) Evolutionary concepts in biostratigraphy. In: *Unlocking the Stratigraphical Record* (ed. P. Doyle and M. R. Bennett), pp. 123–144. John Wiley, Chichester.

Pearson, P. N. (1998b) Speciation and extinction asymmetries in paleontological phylogenies: evidence for evolutionary progress? *Paleobiology* **24**, 305–335.

Pearson, P. N., Olsson, R. K., Huber, B. T., Berggren, W. A. and Hemleben, C. (eds) (2006) *Atlas of Eocene Planktonic Foraminifera*. Cushman Foundation, Cushman Foundation Special Publications.

Penny, D. and Phillips, M. J. (2007) Evolutionary biology – Mass survivals. *Nature* **446**, 501–502.

Peters, S. E. and Foote, M. (2002) Determinants of extinction in the fossil record. *Nature* **416**, 420–424.

Pons, J., Barraclough, T. G., Gomez-Zurita, J., *et al.* (2006) Sequence-based species delimitation for the DNA taxonomy of undescribed insects. *Systematic Biology* **55**, 595–609.

Price, T., Gibbs, H. L., de Sousa, L. and Richman, A. D. (1998) Different timing of the adaptive radiations of North American and Asian warblers. *Proceedings of the Royal Society of London Series B-Biological Sciences* **265**, 1969–1975.

Prokoph, A., Fowler, A. D. and Patterson, R. T. (2000) Evidence for nonlinearity in a high-resolution fossil record of long-term evolution. *Geology* **28**, 867–870.

Purvis, A. (2004) How do characters evolve? *Nature* **432**, doi:10.1038/nature03092.

Purvis, A., Nee, S. and Harvey, P. H. (1995) Macroevolutionary inferences from primate phylogeny. *Proceedings of the Royal Society of London Series B* **260**, 329–333.

Pybus, O. G. and Harvey, P. H. (2000) Testing macro-evolutionary models using incomplete molecular phylogenies. *Proceedings of the Royal Society of London Series B* **267**, 2267–2272.

Rabosky, D. L. (2006) Likelihood methods for detecting temporal shifts in diversification rates. *Evolution* **60**, 1152–1164.

Raup, D. M. (1985) Mathematical models of cladogenesis. *Paleobiology* **11**, 42–52.

Schluter, D., Price, T., Mooers, A. Ø. and Ludwig, D. (1997). Likelihood of ancestor states in adaptive radiation. *Evolution* **51**, 1699–1711.

Schmidt, D. N., Thierstein, H. R., Bollman, J. and Schiebel, R. (2004) Abiotic forcing of plankton evolution in the Cenozoic. *Science* **303**, 207–210.

Sepkoski, J. J. (1996) Competition in macroevolution: the double-wedge revisited. In: *Evolutionary Paleobiology* (ed. D. Jablonski, D. H. Erwin and J. H. Lipps), pp. 213–255. University of Chicago Press, Chicago.

Sibley, C. G. and Ahlquist, J. E. (1990) *Phylogeny and Classification of Birds: A Case Study of Molecular Evolution*. Yale University Press, New Haven, CT.

Simpson, G. G. (1953) *The Major Features of Evolution*. Columbia University Press, New York.

Smith, A. B., Gale, A. S. and Monks, N. E. A. (2001) Sea-level change and rock-record bias in the Cretaceous: a problem for extinction and biodiversity studies. *Paleobiology* **27**, 241–253.

Stanley, S. M. (1979) *Macroevolution*. W H Freeman, San Francisco.

Stanley, S. M., Wetmore, K. L. and Kennett, J. P. (1988) Macroevolutionary differences between the two major clades of Neogene planktonic foraminifera. *Paleobiology* **14**, 235–249.

Stenseth, N. C. and Maynard Smith, J. (1984) Coevolution in ecosystems: red Queen evolution or stasis? *Evolution* **38**, 870–880.

Stewart, D. R. M. and Pearson, P. N. (2002) Plankrange: a database of planktonic foraminiferal ranges. http://palaeo.gly.bris.ac.uk/Data/plankrange.html.

Tavaré, S., Marshall, C. R., Will, O., Soligo, C. and Martin, R. D. (2002) Using the fossil record to estimate the age of the last common ancestor of extant primates. *Nature* **416**, 726–729.

Turgeon, J., Stoks, R., Thum, R. A., Brown, J. M. and McPeek, M. A. (2005) Simultaneous Quaternary radiations of three damselfly clades across the Holarctic. *American Naturalist* **165**, E78–E107.

Van Valen, L. (1973) A new evolutionary law. *Evolutionary Theory* **1**, 1–30.

Webster, A. J. and Purvis, A. (2002) Testing the accuracy of methods for reconstructing ancestral states of continuous characters. *Proceedings of the Royal Society of London Series B* **214**, 143–149.

Wible, J. R., Rougier, G. W., Novacek, M. J. and Asher, R. J. (2007) Cretaceous eutherians and Laurasian origin for placental mammals near the K/T boundary. *Nature* **447**, 1003–1006.

Wood, S. N. (2006) *Generalized Additive Models: An Introduction with R.* Boca Chapman & Hall/CRC, Raton, FL.

Yule, G. U. (1924) A mathematical theory of evolution, based on the conclusions of Dr. J. C. Willis F.R.S. *Philosophical Transactions of the Royal Society of London A* **213**, 21–87.

Zink, R. M. (1994) The geography of mitochondrial DNA variation, popoulation structure, hybridization, and species limits in the Fox Sparrow (*Passerella iliaca*). *Evolution* **48**, 96–111.

Zink, R. M. and Avise, J. C. (1990) Patterns of mitochondrial DNA and allozyme evolution in the avian genus *Ammodramus*. *Systematic Zoology* **39**, 148–161.

Zink, R. M. and Dittman, D. L. (1991) Evolution of Brown Towhees: mitochondrial DNA evidence. *The Condor* **93**, 98–105.

Zink, R. M. and Dittman, D. L. (1993) Population structure and gene flow in the Chipping Sparrow and a hypothesis for evolution in the genus *Spizella*. *The Wilson Bulletin* **105**, 399–413.

Zink, R. M., Dittman, D. L., Klicka, J. and Blackwell-Rago, R. C. (1999) Evolutionary patterns of morphometrics, allozymes and mitochondrial DNA in Thrashers (Genus *Toxostoma*). *The Auk* **116**, 1021–1038.

Zink, R. M., Dittman, D. L. and Rootes, W. L. (1991a) Mitochondrial DNA variation and the phylogeny of *Zonotrichia*. *The Auk* **108**, 578–584.

Zink, R. M., Rootes, W. L. and Dittman, D. L. (1991b) Mitochondrial DNA variation, population structure, and evolution of the Common Grackle (*Quiscalus quiscula*). *The Condor* **93**, 318–329.

Zink, R. M. and Slowinski, J. B. (1995) Evidence from molecular systematics for decreased avian diversification in the Pleistocene epoch. *Proceedings of the National Academy of Sciences of the United States of America* **92**, 5832–5835.

# CHAPTER SIXTEEN

# Speciation and extinction in the fossil record of North American mammals

JOHN ALROY

## Introduction

Paleontological data have been used for decades to address a series of very general and intrinsically interesting questions concerning speciation. Many of them are essentially microevolutionary, morphological or both. What is the relative prevalence of anagenesis and cladogensis (Wagner & Erwin 1995)? Do constraints on morphology cause occupation of morphospace to slow down as diversity increases (Foote 1993)? Is morphological change gradual or punctuated across speciation events (Simpson 1944)?

A survey of the analytical paleobiology literature would reveal, however, that interest in all of these questions has waned over the last decade or two. The one topic relating to speciation that remains very popular is the quantification and modelling of turnover rates (Foote 1994b, 2000, 2003; Sepkoski 1998; Newman & Eble 1999; Kirchner and Weil 2000; Allen *et al.* 2006; Alroy 2008). Coincidentally and fortuitously, the explosion of molecular data sets and great improvements in phylogenetic methods have led to quantifying speciation rates by tracking the accumulation of lineages through time (Nee *et al.* 1992, 1994; Purvis *et al.* 1995; Magallon & Sanderson 2001; Roelants *et al.* 2007).

Nonetheless, paleontological research has focused far more strongly on taxonomic diversity than on speciation in recent years (Alroy 1996, 1998b, 2000; Miller & Foote 1996; Sepkoski 1997; Alroy *et al.* 2001, 2008; Connolly & Miller 2001; Peters & Foote 2001; Smith 2001; Jablonski *et al.* 2003; Bush *et al.* 2004; Krug & Patzkowsky 2004; Allen *et al.* 2006; Crampton *et al.* 2006). In addition to research on diversity at the global scale, there has been an explosion of studies on diversity at the community scale (Powell & Kowalewski 2002; Bush & Bambach 2004; Olszewski 2004; Peters 2004; Kosnik 2005; Kowalewski *et al.* 2006; Wagner *et al.* 2006; and many others). Because the latter kind of work concerns the richness and evenness of individual fossil collections from precise locations and stratigraphic horizons, it has no direct bearing on turnover.

When turnover rates do get attention, the paleontological literature has focused on two overriding issues: whether they can be explained using intrinsic dynamic mechanisms, such as either density dependence (Sepkoski 1978, 1979,

*Speciation and Patterns of Diversity*, ed. Roger K. Butlin, Jon R. Bridle and Dolph Schluter. Published by Cambridge University Press. © British Ecological Society 2009.

1984) or random sorting or competition between taxa that leads to long-term declines in average rates (Raup & Sepkoski 1982; Gilinsky 1994), or by extrinsic environmental controls, including catastrophic perturbations such as asteroid impacts and more gradual changes in sea level, climate and so forth (Vrba 1985; Brett & Baird 1995).

Much of this literature, and arguably a large majority of it, has depended on Jack Sepkoski's two monumental, global compendia of all marine animal age ranges, first of families (Sepkoski 1982) and later of genera (Sepkoski 2002). However, it has become apparent that these data are compromised by a variety of sampling biases (Raup 1976; Alroy *et al.* 2001, 2008; Peters & Foote 2001; Smith 2001). A steady progression of papers has emphasized this point (Foote 2003; Krug & Patzkowsky 2004; Crampton *et al.* 2006). Defenses of these compendia have focused on side-arguments, such as an attack (Bush *et al.* 2004) on sampling-standardization methods that fail to show a large exponential radiation after the Jurassic (Alroy *et al.* 2001); the logically insufficient claim that because an increase in local-scale richness matches the difference between the ends of Sepkoski's curves, those curves are likely to be correct (Bush & Bambach 2004); and the entirely predictable high quality of the Plio-Pleistocene bivalve record (Jablonski *et al.* 2003), a group whose exceptional preservation is unrepresentative of marine invertebrates in general (Foote & Sepkoski 1999).

The Paleobiology Database (http://paleodb.org) has more than tripled since an initial, preliminary analysis of two large Phanerozoic intervals (Alroy *et al.* 2001) suggested that corrections for sampling intensity bias might alter key patterns in Sepkoski's genus-level data. Without any corrections, the full height of Sepkoski's curve can now be replicated with reasonable corrections, a dramatically different and much flatter curve is indeed recovered (Alroy *et al.* 2008). This paper asks similar questions of a smaller model system: the North American record of Cenozoic mammals (Alroy 1996, 1998a,b, 1999, 2000, 2002; Alroy *et al.* 2000).

Three things make this system fruitful, and even necessary, to reanalyse. First, another decade of information has accumulated, including large amounts of data on reptiles that are used below to generate the time scale undergirding the analysis. Second, many of the statistical methods have improved greatly, as discussed below. Third, this system is much more tractable than Phanerozoic marine invertebrate data sets, because:

1.  The high turnover rates of mammals compared to other groups (Stanley 1979) mean that the mammalian time scale (Alroy 1996, 2000) is about far more precise than traditional marine stages and epochs. The scale can be broken down into uniform and objectively defined intervals.
2.  Identifications of fossil taxa within collections (i.e. occurrences) are taxonomically standardized, although unrecognized synonyms will always

remain (Alroy 2002). Such a thing is almost unimaginable of the overall marine record, which involves at least 30 000 genera and far more species.

3. The data largely derive from the Western Interior of the United States and neighbouring parts of Mexico and Canada (Alroy 1998b), so very large-scale shifts in the geography of sampling do not exist.

4. Mammals have reasonably constant taphonomic properties across time and taxonomic groups: a tooth is a tooth, but body parts such as trilobite cranidia, crinoid stem ossicles, and bivalve tests that are all thrown together in marine invertebrate studies have little in common. Taphonomy remains of great interest, especially because large and small mammals preserve and are collected in quite different ways. However, this issue is put aside for now, because large changes in the distribution of body masses (Alroy 1998a, 2000) make it difficult to address without going into much detail.

5. Finally, the timing of global climate changes is well understood in the Cenozoic (Zachos *et al.* 2001), if not much farther back than the late Cretaceous, so climate proxy data can be compared directly to turnover patterns (Alroy *et al.* 2000).

The following sections present new and powerful methods for preparing the data, and then test a series of straightforward and general hypotheses. How many intervals witness markedly low or high turnover rates? Do rates fall through the Cenozoic? Are origination and extinction rates correlated? Are origination rates depressed or extinction rates inflated when diversity reaches high levels? Finally, does diversity vary within a narrow enough range to suggest an equilibrium?

## Time-scale analysis

There are two key steps in preparing the data. The first is to define a numerical time scale, as opposed to traditional categories such as land mammal ages, and the second is to produce a diversity curve.

The time-scale analysis uses information on all North American Cretaceous and Cenozoic turtles, crocodylians, dinosaurs, and mammals, not just mammals as in previous studies (Alroy 1996, 1998b, 2000). These taxa were selected because of their high preservation potential. The resulting data set includes 7642 fossil collections that document 39 743 occurrences of 1493 genera and 3694 species. The last published analysis (Alroy 2000) captured 4978 collections, 30 951 occurrences, 1241 genera, and 3243 species.

The quantitative biochronological method used here is appearance event ordination (Alroy 1992, 2000), which generates absolute estimates of the ages of fossil collections in millions of years ago (Mya). It shuffles first and last appearances until it maximizes a likelihood function involving hypothesized but as yet undemonstrated overlaps of age ranges (conjunctions). Conjunctions are demonstrated by co-occurrence within collections or by mutual stratigraphic superposition. In contrast to earlier analyses that used graph theory to resolve geographic biases, the current one uses the simpler and apparently more accurate method of ignoring cases where disjunct pairs of taxa also do not overlap geographically for the purpose of computing likelihoods.

The resulting sequence of first and last appearance events is numbered from oldest to youngest by counting consecutive runs of first or last appearances. The counts are compared to geochronological age estimates to translate them into an absolute, numerical time scale. The position of each geochronological date in the event sequence is equated with that of the particular collection it dates. A collection's position in the event sequence is just the span between the oldest first appearance and youngest last appearance of any taxa in it, i.e. its assemblage zone. Line segments in the calibration plot are created by selecting 41 tie points from a larger set of 127 high precision $^{40}$Ar/$^{39}$Ar, uranium series, and paleomagnetic stratigraphy dates. After calibration, the event sequence is subdivided into uniform, as in previous studies (Alroy 2000). Collections only are used in the diversity analysis if their zones fall entirely within one bin.

A previous nonparametric method for selecting tie points called shrinkwrapping (Alroy 2000) is replaced here with a simpler set of winnowing criteria that is less sensitive to outliers: (1) remove 22 dates with assemblage zones entirely spanning those of others; (2) create averages from 31 dates that form 12 sets of identical assemblage zones; (3) use a simple greedy algorithm to remove 11 dates that unambiguously conflict with the position of other dates in the event sequence; (4) remove 10 dates immediately following younger geochronological estimates that the event sequences implies are older, but only if (a) the preceding date is a maximum, or (b) the removed date is a minimum; and (5) remove 23 points that create abrupt and temporary changes in the slope. The latter are identifed by dividing the current slope by the preceding slope and the current slope by the following slope; multiplying the two ratios, which should each be 1.0 if there is no change; and seeing if this product is more than the current slope, or less than 1 divided by the current slope. An exception is made for the short segment spanning the rapid pulse of turnover at the Cretaceous–Tertiary boundary, which is well documented from independent evidence.

An alternative analysis employing all non-volant, terrestrial tetrapods resulted in a higher residual standard deviation (square root of the residual standard error) at the midpoint of each assemblage zone (1.34 Myr versus 1.06 Myr with the subset of four groups). Note that the error term computation forgives cases where the best interpolated age estimate is older than a minimum geochronological date or younger than a maximum, but not those where the date is a maximum or minimum but falls within an assemblage zone's range of age estimates, in which case the appropriate endpoint of the range is used to compute the residual.

The error figure is higher than before (Alroy 2000) because far fewer tie points are selected, so more of the dates fall off the calibration line. The difference might suggest that short sampling bins are no better than traditional land mammal ages. However, a simple nonparametric bootstrap simulation shows that short bins are robust. It draws from the observed ranges of age estimates for the calibration points and the observed offsets between these estimates and the geochronological dates. At a bin size of 0.1 Myr, just 37% of the few collections that have been assigned randomly to an exact bin have been assigned to the right one. This fraction rises swiftly to 64% at a bin size of 1.0 Myr and then slowly asymptotes, reaching 72, 77, and 84% at bin sizes of 1.5, 2.0, and 3.0 Myr. An entirely different, also nonparametric analysis with an earlier version of the data set showed much the same tradeoff of precision and accuracy (Alroy 1996).

Earlier studies (Alroy 1996, 1998b, 2000) used a bin size of 1.0 Myr, but doing so with this data set would create a noisy diversity curve with low autocorrelation, large changes between adjacent points, and low potential sampling quotas. These problems are essentially resolved by using a 1.635 Myr bin length that divides the Cenozoic into exactly 40 sampling intervals.

## Diversity estimation

The next task is to produce a diversity curve (Fig. 16.1) that is a proper statistical estimate based on comparably sized, randomly drawn subsets of the data, instead of a simple summation (Alroy 1996, 2000). Subsets only capture ecological dominant species that are abundant in samples, but no part of the fossil record is complete, so any count of actual species will reflect both richness and dominance. Methods of extrapolating instead of interpolating diversity, such as mark–recapture estimation (Connolly & Miller 2001), do seek to estimate overall species richness. However, they are not applied here because they

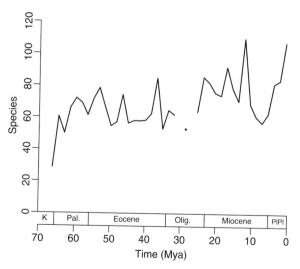

Figure 16.1 Cenozoic trend in the standing diversity of mammalian species from western North America. Data are sampling standarized by randomly drawing enough entire fossil collections to reach a quota of 890 estimated individual specimens, with values averaged over 100 trials. The estimate for each collection is made by squaring the number of species it contains (Alroy 2000).

introduce noise while failing to remove the sampling overprint from this particular data set. Common species are ecologically important and biologically interesting, and it is better to have an accurate and precise measure of diversity in the general sense than an inaccurate and imprecise measure of species richness in the strict sense.

All of the steps involved in subsampling are improved upon here. They are drawing a uniform set of collections in each temporal interval (called by-list subsampling by Alroy 1996, 2000, and sample-based rarefaction by Gotelli & Colwell 2001); counting taxa from the subsamples; and estimating turnover rates, as discussed in the following section. Although these methods may seem difficult, the logic is sound, dramatically different results would be obtained by omitting any one step, and the estimates appear to be highly robust.

The argument has been made (Alroy 2000) that collections should be counted not one by one (Gotelli & Colwell 2001; Alroy *et al.* 2001, 2008; Allen *et al.* 2006), but by estimating the number of specimens in each, summing these estimates as collections are drawn, and stopping at a quota set in terms of specimens instead of collections. A good rule of thumb for fossil mammals is that the number of specimens is about equal to the square of the number of species occurrences within a collection (Alroy 2000). The simpler method of counting the occurrences with no transformation (Alroy 1996) is just a special case in which the relationship is assumed to be linear instead of exponential.

The problem with methods that use approximated specimen counts is underestimation of overall diversity when large collections predominate. In such cases, only a few collections representing a fraction of the landscape need be drawn to fill the quota. Thus, considerable beta diversity may go unsampled. The solution is to inversely weight the probability of drawing a collection by its estimated specimen count: if two collections have 5 and 10 taxa respectively, and therefore estimates of 25 and 100 specimens, it should be four times more probable to draw the smaller one. By doing this, sampling is dispersed spatially and environmentally, and on average each collection contributes exactly the same estimated number of specimens to each randomized draw.

Whether to estimate the specimen count for a collection by raising the number of taxa to a power 2 or, say, 1.4 has been debated (Bush *et al.* 2004). This problem turns out to be unimportant after dispersing the sampling by inverse weighting, because nearly identical curves are produced. The real issue is thus the spatial concentration of data, not the accuracy of the collection size estimates.

Until recently, most data sets such as Sepkoski's compendia (Sepkoski 1982, 2002) were simply lists of families or genera with geological first and last appearance dates. Such data only could be turned into diversity curves by assuming the presence of each taxon within every interval spanned by its age range. Thus, curves were based on range-through counts, which represent the minimum

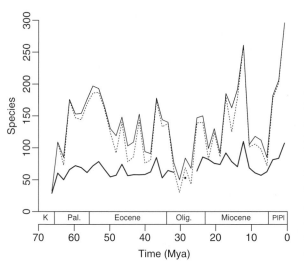

**Figure 16.2** Alternative curves showing the diversity of North American mammals. Upper, thin line: range-through diversity, the sum of species actually sampled and species not sampled but inferred to exist because they are seen before and after a bin. Data are not standardized by subsampling. Dotted line: directly sampled diversity with no standardization. Thick line: sampled diversity after standardization (data used to construct Fig. 16.1).

numbers of taxa that must actually have existed in the bins. The effects of using these counts and not standardizing the data are great (Fig. 16.2; see also Alroy 1998b). For example, a very large apparent spike in the Pleistocene simply reflects the exceptional fossil record in that interval.

As sampling degrades, taxa found only in one interval (singletons) come to be more common in relative terms (Foote & Raup 1996). This problem was addressed previously by counting not the taxa within each interval, but those spanning the boundaries between intervals (Alroy 1996, 1998b, 2000). Recent simulation analyses to be presented elsewhere show that, in fact, these boundary-crosser counts are just as biased as range-through counts (Fig. 16.2): they fall at the edges of time series and both well before and after mass extinction events, and they increase as turnover rates decrease. They also rise sharply as the data approach a very thoroughly sampled interval, such as the Recent in many pale-ontological data sets (the Pull of the Recent: Raup 1979) combined with the Pleistocene in this one (Fig. 16.2).

Instead, simple counts of the number of taxa actually found within an inte-rval's random subsample (sampled in bin or SIB diversity: Miller & Foote 1996; Alroy *et al.* 2001) turn out to be quite robust to all such biases, as long as sampling really is rendered uniform by drawing constant amounts of data, and the geographic, environmental, and taxonomic scope of sampling does not change much.

That said, SIB counts can be misleading because they scale to the sum of all taxa to have existed during each interval, so they may rise with rising turnover rates. Therefore, they may present a composite signal of diversity and turnover. For purposes of comparing diversity to the turnover rates, the interaction can be removed by obtaining an estimate of the average number of taxa existing at any one time within a bin (standing diversity). Raup (1985, eqn. A29) showed that total diversity (the number tracked by SIB) can be approximated by multiplying a simple function of the origination and extinction rates ($\lambda$ and $\mu$) either by diversity at the base of a bin or diversity at the top of a bin. Therefore, standing diversity at either the bin bottom or bin top can be extracted by rearranging the expression.

Because separate standing diversity estimates are obtained from each turnover rate, their geometric mean is used to arrive at a final curve. These values are very congruent, with the median difference on a log scale being 0.020. Values from one calculation or the other are used in cases where only one estimate is available because of gaps in the subsampled data. The $\lambda$ and $\mu$ calculations respectively contribute two and one of these values. The raw SIB and standing diversity curves are nearly mirror images, with the median standing diversity estimate being 28% lower.

## Turnover rates

Turnover rates are notoriously difficult to estimate (Foote 1999, 2003), and numerous equations with different biases have been proposed (Foote 1994b). One robust method is to count the taxa at the base of a bin (i.e. the lower boundary crossers), see how many are present at the upper boundary, and compute an instantaneous extinction rate by taking the ratio of the logged values (Foote 1999; Alroy 2000). The origination rate is just the opposite: the log of the upper boundary crosser count divided by the lower boundary count of the same cohort. These exponential rates are directly equivalent to the ones computed in molecular lineage through time studies (Nee et al. 1992, 1994). The term 'origination' is used here because a relatively small number of appearances are of immigrants from Eurasia or South America.

For standing diversity counts, the rate can be computed analogously: the extinction rate is the absolute value of the logged standing diversity estimate for a bin $i$ divided by the count of those same taxa sampled once again in the following bin $i+1$ (the two timers), and the origination rate is the mirror image (Alroy 2008). Taxa in the cohort must only be counted if they are actually sampled in bin $i+1$ because their later appearances in bins are subject to all the above counting method biases.

Such rates are biased in three ways that must be accounted for. First, the rates have a de facto upper bound imposed by the finite number of specimens drawn during subsampling. For example, if the quota is 100 specimens and two consecutive bins share 50 taxa, the upper bound on the origination rate is the absolute value of $\log(100/50) = 0.693$, and the realistic limit is much lower, because many taxa will be sampled repeatedly. The bound, and therefore the bias, can be removed easily by dividing the counts of sampled and shared sampled taxa by the sampling quota, logit-transforming these proportions, and taking their difference. The difference of logits is just the log of the odds ratio, a common measure of effect size. With the current data the correction has little effect because the number of sampled species never closely approaches the number of sampled specimens.

Second, the rates are biased upwards by failure to resample the cohort. A reasonable correction involves a variant of gap analysis (Paul 1982), in which all taxa ranging across any given bin $i$ are examined, and the proportion that is not sampled is computed. Such counts are inaccurate due to the above-mentioned problems with simple range data, such as edge effects. The solution is to only examine the group of taxa found in both bins $i-1$ and $i+1$. The count of those not found (the part timers) is divided by that count plus the count of taxa found in all three intervals (the three timers). The probability of sampling a taxon $s$ is therefore one minus this proportion (Alroy 2008). The corrected extinction rate can be shown with simple algebra to be approximately, although not exactly, $\mu + \log p$.

Small sample sizes may create high noise in separate estimates of $s$ for each bin. Thus, the part and three timer counts are summed across all the data and put through this simple equation to produce an average $s$ value (here 0.561) for use in correcting each bin's rates. Simulation analyses again not detailed here show that the estimates are unbiased regardless of edge effects and the like, as is desired.

Third, standing diversity figures in the turnover rate equations, but standing diversity itself is estimated from the rates, as mentioned in the previous section. Both things therefore must be generated recursively. First, standing diversity is estimated using the simplified expressions $SIB/(1+\lambda)$ or $SIB/(1+\mu)$, as appropriate (Raup 1985, eqn. A28). These equations assume that $\lambda$ and $\mu$ are about equal. The above-mentioned ratio of the logit-tranformed standing and two timer diversity values is then used to estimate $\lambda$ or $\mu$, and the procedure is iterated repeatedly. In a second round, the full equation for SIB allowing $\lambda$ and $\mu$ to differ (Raup 1985, eqn. A29) is used instead. Here the estimate of $\lambda$ at each iteration uses the median value of $\mu$ determined in the first round, and vice versa.

## Data

The data employed in diversity analyses include all mammals except the volant order Chiroptera and four fully marine groups (Cetacea, Desmostylia, Pinnipedimorpha, and Sirenia). The geographic spread of sampling extends to the eastern USA during the late Neogene, most importantly with the addition of many fossil localities in southeastern Texas and Florida (Alroy 1998b). To remove the resulting biogeographic signal, the data are restricted to a rectangle spanning 20–60°N and 95–130°W, i.e. the entire western region of the Canada, the USA, and Mexico that yields substantial numbers of Paleogene in addition to Neogene vertebrates. This restriction leaves 24 690 out of 27 014 occurrences (91.4%), which include 3088 out of 3315 Cenozoic mammals

(93.2%). Several analyses also included a bin representing the terminal Cretaceous.

Two points within the Oligocene were removed because their estimated specimen counts would not meet any realistic sampling quota. The other bins all meet a quota of 890, which it is generally high enough to sample 60–80 species, comparable to earlier studies (Alroy 1996, 1998b, 2000). Similar patterns are seen in a curve based on a doubled specimen quota. The geometric mean of raw subsampled diversity across bins in this larger data set is 90.6 instead of 68.3. However, the curves are visually almost identical, the rank-order correlation between them (Spearman's $\rho$) is 0.966, and after logging both data sets the slope of a least-squares regression of the lower curve on the higher one is 0.997+/−0.038. Thus, there is no significant change in shape as the quota is raised because the increase in diversity is essentially uniform.

### General patterns

The new sampling standardized diversity curve (Fig. 16.1) and turnover rates (Fig. 16.3) differ in some ways from the previous ones (Alroy 2000), most of which are not biologically important. These are: (1) the curve does not show the crash at the Cretaceous–Tertiary boundary because it represents diversity within bins instead of at bin boundaries, and the recovery is fast; (2) likewise, the end-Cretaceous extinction rate is high relative to the confidences intervals, but not very impressive, because almost all turnover actually was concentrated at the boundary but the equations assume continuous turnover through 1.635 Myr-long bins; (3) there is a much less pronounced diversity drop in the late Paleocene, an earlier pattern considered to be possibly artifactual (Alroy et al. 2000); (4) although this drop is followed by a small rebound, there is no large earliest Eocene origination spike and the Eocene is not consistently higher than the Paleocene; (5) there are consecutive extinction and origination peaks around 39 Ma that relate to poor sampling within the Duchesnean land mammal age; (6) there are brief diversity peaks in the late middle Miocene and Plio-Pleistocene that relate to intense sampling of small mammals, the former matched with concurrent origination and extinction rate excursions; and (7) there is trough throughout the late Miocene, between about 11 and 5 Ma, instead of a short downwards excursion near the Mio–Pliocene boundary.

Despite these differences, the previous curve's most robust and biologically important key features (Alroy 2000) are still evident: (1) a weak overall trend; (2) a high earliest Paleocene origination rate; (3) a substantial offset between the Paleogene and Miocene; and (4) a full rebound from the late Miocene drop.

As noted before (Alroy et al. 2000), the most severe climate changes throughout the Cenozoic (Zachos et al. 2001) seem to have inconsistent effects, or perhaps none at all. A rapid global warming event at the Paleocene–Eocene boundary does trigger the invasion of North America by key groups such as crown-group

(a)

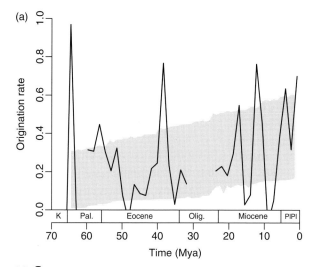

(b)

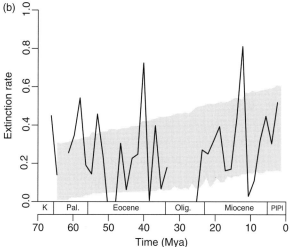

**Figure 16.3** Cenozoic origination (speciation plus immigration) and extinction rates for mammals from western North America, based on the same sampling standardized data shown in Figs. 16.1 and 16.2. Shaded areas show 95% confidence limits based a parametric bootstrap analysis employing 10 000 trials. (a) Instantaneous origination rates, based on the log ratio of estimated standing diversity and the number of these species also sampled in the previous bin (the two timers). (b) Instantaneous extinction rates computed using the log ratio of standing diversity and the following two timer count.

primates, artiodactyls, and perissodactyls (Clyde & Gingerich 1998; Smith *et al.* 2006), but the new turnover and diversity data show nothing dramatic at this scale of resolution. Likewise, there is no evidence for a response to the abrupt growth of ice sheets in the earliest Oligocene. Two missing data points in the data set (Fig. 16.1) make it hard to say whether the global late Oligocene warming event corresponds with a jump in the curve, although a match is certainly plausible. Finally, the late Miocene low comes during by a relatively minor global cooling trend (Zachos *et al.* 2001), and starts well before a major shift from $C_3$ to $C_4$ vegetation in the Western Interior that signals high seasonality (Fox & Koch 2004). Moreover, the possibility that this drop reflects poor sampling of small mammals cannot be dismissed, because it does not correspond clearly to changes in turnover rates (Fig. 16.3). As before (Alroy 2000), the certainly anthropogenic end-Pleistocene megafaunal mass extinction (Alroy 1999) is not addressed, because an extinction rate for the last bin cannot be computed with the current methodology.

## Hypothesis tests

All reported correlations are nonparametric (Spearman's $\rho$). The basic results are:

1. Cenozoic diversity has a trend (Fig. 1): there is a moderate correlation of subsampled diversity with time ($\rho = -0.391$; $p = 0.015$). The $p$-value shows nothing about possible mechanisms because the data are autocorrelated. It is useful simply to illustrate that there is indeed an upwards drift, justifying the use of an exponential trend as a null model in the following tests. A model assuming constant diversity would be easier to reject and therefore not as conservative.
2. The turnover rates (Fig. 16.3) show a weak upwards trend if any instead of declining, as found with less accurate methods (Alroy 2000). Regressions of origination and extinction rates against time are insignificant ($\rho = -0.096$, $-0183$; $p = 0.584$, $0.298$). The apparent decline was generated by high Paleocene rates, but here it makes no difference if the Paleocene points (the first six) are removed ($\rho = -0.259$, $-0.329$; $p = 0.166$, $0.082$). Note that all tests reported here exclude large, paired outliers in the rates that are associated with an unusual burst of pseudoextinction between the second and third bins, which straddle the Puercan–Torrejonian boundary.
3. The rates are not predictable from one interval to the next. Each rate shows no serial correlation, i.e. a correlation with itself lagged by one interval ($\rho = 0.136$, $-0.028$; $p = 0.455$, $0.880$). These tests are liberal because the data do trend very weakly.

From here on, a simple parametric bootstrap simulation iterated 10 000 times is used to compute significance levels. The null model is that the unseen, overall counts underlying the diversity and turnover rate estimates follow a long-term, exponential trend with no noise. The slopes and intercepts are estimated by linear regressions against time of SIB and two timer counts. The counts are divided by the average sampling probability $s$ to obtain estimated figures for the entire species pool. In each interval, two timer and SIB subsamples are drawn at random with the same probability $s$ to mimic the empirical data, and these values are used to generate standing diversity estimates and turnover rates. The results are:

1. The turnover rates show no great variability. Almost all the rates other than the few discussed above fall within or very near the confidence intervals generated by bootstrapping (Fig. 16.3). The standard deviations of the real

origination and extinction rates on a log scale are respectively 0.913 and 0.979, nearly equal to each other. The median origination and extinction rates are also subequal, being 0.228 and 0.249 species per species (spp/sp) per Myr. The extinction figure implies that the median species duration is 2.78 Myr. In combination with the preceding tests, these results indicate that the turnover rates are mostly white noise.

2. The rates seem not to correlate with each other. The bootstrapped 95% confidence limit of $\rho$ is −0.241 to 0.493, and the observed value is 0.199. Shifting the origination rates forward by one bin produces a seemingly great but insignificant cross-correlation of 0.754 (95% limits 0.433 to 0.885). The reason for the high correlations after lagging in both the real and simulated data is that in this case the rate estimate in each interval involves the same two timer count, which is small and therefore quite variable. High shared error is the price paid for removing such biases as edge effects.

3. Density dependence of either origination or extinction would produce logistic growth (Sepkoski 1978; Alroy 1996, 1998b; Foote 2000), and a diversity-origination relationship was found previously (Alroy 2000) with less robust data preparation methods. Here, however, there is no suggestion that origination rates fall with diversity or that extinction rates rise with diversity. Even after an outlier in the Paleocene, there are weak correlations (0.201 and −0.008) of standing diversity with origination and extinction during the next interval, and these correlations are insignificant based either on bootstrapping or on conventional tests. Here all of the time series are largely independent because either turnover rate in the next bin is computed from the diversity estimate in that bin, and in any case the simulation would build in any spurious correlation that might exist. Additionally, no detrending of the rates (Foote 2000) is necessary because the simulation includes the weak trend as an assumed parameter.

## Diversity curve simulation

In contrast to the turnover rates, the shape of the diversity curve itself provides strong evidence for logistic dynamics. The trend is much more flat than would be expected of a random walk with no density dependence, as shown by a simple, realistically parametrized simulation (Fig. 16.4) in the tradition of Raup (1977).

In this analysis, the starting count of species is the value predicted by the regression of log subsampled diversity against time. The count is multiplied at each step by a random number drawn from all the ratios of neighbouring points in the real diversity curve. The real curve's standard deviation is compared to the standard deviations of the simulated curves to test the hypothesis that density dependence flattens the trajectory. In a second test, the variance around a regression of log diversity against time is compared to the variance around similar

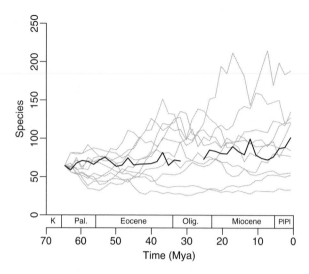

**Figure 16.4** Simulated diversity trends. Thick black line shows empirical data (Fig. 16.1); thin gray lines show ten representative random walks generated by drawing from the observed distribution of ratios between neighbouring diversity values.

regression lines fit to the random walks. All of the curves are logged before computing these figures.

Random counting error increases the real curve's variance, which makes the simulated curves more variable and therefore less likely to match the real one. The raw data curve is therefore dampened by (1) taking residuals to remove the temporal trend; (2) squaring the residuals to compute a variance measure; (3) separately estimating the variance of each data point that is created by sampling error as sqrt(SIB)/SIB (Foote 1994a); (4) taking one minus the ratio of the sampling-related and observed variances; (5) rescaling the residuals from step 1 with this factor; (6) and adding back the trend to reconstitute the diversity curve. To make the test even more conservative, the ceiling on diversity imposed by the sampling quota of 890 estimated specimens is lifted by multiplying each diversity point by the ratio 890 SIB/(890 − SIB), which is analogous to the logit tranformation of turnover rates discussed above. This correction is applied to the raw data immediately before the step that rescales the residuals.

Based on 10 000 trials, there is a 0.030 probability of generating a standard deviation less than the 0.126 log units seen in the rescaled data, with a median simulated standard deviation of 0.269. Thus, 53% of the expected variance is suppressed. Using the test that accounts for temporal trends, the probability of seeing such a smooth progression is 0.010.

## Discussion
### Turnover rates

As mentioned, the median extinction rate of 0.249 spp/sp/Myr implies a median expected duration of 2.78 Myr, which is substantially but not greatly more than the estimate of 2.14 produced by the last study of these data using very different equations (Alroy 2000). It also is not very different from the rate of 1.7 Myr derived using yet another, even more divergent method, and an even older version of the same data set (Foote & Raup 1996).

However, the almost identical origination rate of 0.228 spp/sp/Myr is much higher than rates found in most molecular studies. Nee *et al.* (1992) estimated net speciation rates for all birds equivalent to 0.090 spp/sp/Myr. Magallon & Sanderson (2001) estimated a speciation rate of no more than 0.089 spp/sp/Myr for angiosperms over a wide range of possible extinction rates. Roelants *et al.* (2007) performed similar calculations for amphibians that produced a remarkably low maximal speciation rate of 0.0217 spp/sp/Myr. Of more direct relevance, Purvis *et al.* (1995) found negligible extinction rates and net diversification rates ranging from 0.070 – 0.342 spp/sp/Myr, but mostly on the low end of the range, in an analysis of all four major primate groups.

To some extent these differences are unsurprising, because mammals have long been recognized by paleontologists as evolving far more quickly than angiosperms and amphibians (Stanley 1979), although paleontological evidence is lacking for birds. However, the differences are large enough to suggest a real conflict.

One could argue that the molecular analyses are less biased because they work directly with dated branching events. However, the current analysis uses an almost unbiased method for computing origination rates, because edge effects are eliminated by working with counts of species that survive into and are sampled within an immediately following interval. The method also employs sampling-standardized data and takes failure to resample cohort members into account.

Nonetheless, there is still a legimate reason to doubt that the rates derived in this study are comparable to true rates of cladogenesis and extinction. Specifically, perhaps much of the apparent turnover is due to pseudoextinction, as appears to be the case at the Puercan–Torrejonian boundary (see above). Because each pseudoextinction must lead to a pseudospeciation, this problem would tend not just to lower but to equalize calculated speciation and extinction rates. However, phylogenetic analyses of mammalian fossil clades suggest that pseudoextinction rates are normally around 10–25%. For example, Hulbert (1993) tallied 15 pseudoextinctions and 62 real extinctions in Neogene North American horses (19%), whereas Wang *et al.* (1999) inferred nine pseudoextinctions and 57 true extinctions (14%). Pseudoextinction rates that low would not create a large gap between paleontological and molecular results, or remove a large gap between speciation and extinction rates if any existed.

Although the current analyses are strictly at the species level, previously an ad hoc attempt was made to account for pseudoextinction. The correction involved identifying possible anagenetic lineages by finding cases in which two species of the same genus did not overlap in time and jointly spanned an interval that did not otherwise include the genus (Alroy 1996, 2000). In other words, it estimated the absolute minimum possible diversity level. This correction is not used here, because it has a small effect; it does not employ real phylogenetic data; it works within the flawed range-through paradigm; and it assumes that genera are circumscribed in a similar way in all parts of the time scale.

A related objection is that the nominal turnover rates may largely reflect changes in preservability that cause lineages to enter and leave the pool of potentially sampled species. However, these evolutionary shifts are unlikely to be common, because all of the factors governing preservability are strongly heritable between species: local abundance, geographic range size, geographic range location relative to the area being sampled, taphonomic robustness of parts, ease of collection of parts, and taxonomic specificity of preserved parts.

## Macroevolutionary theories

The data are relevant to several important macroevolutionary hypotheses. First, Raup and Sepkoski (1982) showed that the average extinction rate across all marine animal families declined substantially through the Phanerozoic, and the same was later shown for genus-level origination rates (Sepkoski 1998; Foote 2000; Alroy 2008). There has been much interest in explaining these patterns with such mechanisms as simple natural selection (Raup & Sepkoski 1982) or sorting out of higher taxa (Sepkoski 1984; Gilinsky 1994), as well as interest in exploring the trends' implications for theories of diversity dynamics (Newman & Eble 1999). A similar pattern was seen earlier in the North American mammal data, and attributed to high volatility during the Paleocene (Alroy 1998b). The new data show either no real trend, or perhaps the opposite pattern.

Meanwhile, two established theories predict that turnover rates should be cross-correlated, one developed based on African Plio-Pleistocene mammals and the other on eastern North American early Paleozoic invertebrates. First, the 'turnover pulse' hypothesis (Vrba 1985, 1992) suggests that global climate cycles on the Milankovitch scale restrict and fragment the ranges of species that are poorly adapted to certain climate phases, driving bouts of speciation and extinction. Second, the 'coordinated stasis' hypothesis (Brett & Baird 1995) posits that entire regional biotas show little change not just in terms of taxonomic composition, but morphology and local-scale community composition. These patterns are maintained for millions of years, not within Milankovitch cycles, by strong species interactions and stabilizing selection. They are disrupted only by major environmental shifts. The observed lack of a cross-correlation in the current data, despite a long time series and presumably robust methods, casts real doubt on both theories.

It could be argued that the 1.635 Myr time scale used here is not relevant to the turnover pulse theory because multiple Milankovitch cycles will fall in each bin. However, there is no reason that the cycles should have exactly the same cumulative effect in each bin, and the data set also spans several important changes in the relative intensity of the cycles (Zachos *et al.* 2001).

Indeed, there is nothing in the first place to motivate an argument that climate change has an important effect on a bin to bin basis, because most of the time the rates merely show random variation (Fig. 16.3; Alroy *et al.* 2000). The only clear exceptions are the extinction and origination pulses at either side of the Cretaceous–Tertiary boundary, because the few other excursions are all plausibly explained as sampling artifacts. Certain climate changes may still have caused changes in the biota's composition, if not in diversity per se. For example, there was an immigration pulse in the earliest Eocene that was coincident with a global warming event and does not register in the current data set. However, this dispersal episode was driven directly by biogeographic shifts and therefore only indirectly by climate (Clyde & Gingerich 1998; Smith *et al.* 2006). Together, the data suggest that routine climate cycles and changes in these cycles might be less important causes of extinction than biotic interactions such as competition, predation, and epidemic disease that are altered by adaptation and immigration.

Finally, the data do seem to confirm the Red Queen hypothesis (Van Valen 1973), which assumes that the constant evolution of predators, prey, and competitors prevents any species from establishing strong incumbency. Thus, extinction rates are predicted to show no trend through time, as in the current data set (Fig. 16.3), instead of declining, as in the marine data (Raup & Sepkoski 1982; Alroy 2008). Van Valen's theory does not directly address adaptive radiations and instead assumes a crowded world, so it is consistent with density dependence of rates as long as enough time has elapsed for diversity to rise and strong competition to come into play.

## Diversity equilibrium

The only clear overall pattern in the data, a flat diversity trajectory, could be explained by several dynamic mechanisms that are worth considering:

1. A secular decline in speciation or a secular increase in extinction. Gilinsky and Bambach (1987) found that origination rates typically decline within major marine animal groups, but extinction rates do not, even though they seem to decline when all groups are considered at once. This kind of a tradeoff between rates might explain the logistic growth patterns seen by Sepkoski (1979, 1984) in his Phanerozoic familial diversity data, even in the absence of density dependence. A symmetrical decline of both rates instead would allow exponential growth to continue.

2. Cross-correlations of rates. A positive correlation between origination and extinction rates could generate a diversity equilibrium regardless of density dependence, with extinction pulses being countered by radiations and vice versa (Webb 1969; Mark & Flessa 1977).

3.  Density dependence of rates. If origination is a negative function of diversity, extinction is a positive function of diversity, or both rates show these properties, then an equilibrium will be reached at some point (Sepkoski 1978).

None of these hypotheses are directly supported by the analyses. However, models involving correlations are supported by marina data (Alroy 2008) and are more likely to be correct here because any strong trends through time would have been easy to spot. Furthermore, unless noise is a problem, the failure to recover any correlations might be a catch-22: the stronger an equilibrium, the less variation in turnover rates there is to correlate. Additionally, Foote (2000) argued that density dependence of both origination and extinction could simultaneously generate a cross-correlation between them and logistic growth. In other words, cross-correlations could be an effect and not a cause of equilibrium. Thus, direct density dependence seems like the most parsimonious interpretation.

Regardless of this dilemma, the evidence for a constrained diversity trajectory per se is fairly strong. The model used to test for a constraint (Fig. 16.4) accounts for biases that might create excessive variation in simulated diversity curves, and the true diversity curve (Fig. 16.1) is about as robust as could be hoped for. In contrast, evidence for density dependence is mixed in lineage through time studies: it is absent in primates (Purvis *et al.* 1995), but present in birds (Nee *et al.* 2002) and plausible in amphibians (Roelants *et al.* 2007). There are also serious technical difficulties with such analyses that might create the appearance of logistic growth, such as failure to discriminate newly evolved species (Purvis *et al.*, this volume).

If density dependence is real, we can make some inferences about the tempo of recovery from the current mass extinction (Alroy 2008). Given that the variance in diversity explained by saturation is 53% and that this saturation is caused by a relationship between the net diversification rate and standing diversity, it is easy to back-compute the slope of a regression between these variables. The slope and the respective means imply an intercept of 0.849 spp/sp/Myr, i.e. the rate when there is only one species. This value is more than three times higher than the background origination rate, much less normal net diversification rates. It means that a single species should give rise to 90% of the regional equilibrium number in about 18 Myr, and that the same level should be attained about 9 Myr after a 50% extinction pulse. A much faster recovery occurred after the Cretaceous–Tertiary boundary mass extinction (Fig. 16.1), but it seems safe to predict that any recovery from the current disruption will be fast on a geological time scale and extremely slow on a human time scale. Furthermore, a large radiation starting with a handful of species, some of them invasive and others closely related to each other, will do nothing to replace any phylogenetic, morphological, functional, trophic, or biogeographic diversity that will be lost.

The exact, population-level mechanism governing the statistically significant lack of variation through time in diversity cannot be pinpointed. Furthermore, one might wonder whether this pattern relates to real chances of speciation and extinction, or to chances that species will increase above or fall below the levels of abundance and geographical breadth required to be sampled in the fossil record. This difference is somewhat academic, and it stands to reason that the mechanism creating an equilibrium is primarily ecological, just as ecological factors such as abundance and reproductive rate govern variation among mammalian clades in speciation rates (Isaac *et al.* 2005). For example, competition for resources may cause average population sizes to fall as diversity increases, an inevitable outcome at some point. Small populations are more likely to go extinct, and must as population sizes limit on one, so an equilibrium eventually must result.

A more detailed mechanistic interpretation would be desirable, and recent work that predicts speciation rates based on explicit assumptions about population-level processes and their relationship to metabolism (Allen *et al.* 2006) shows great promise. However, this model would seem to imply a decline in origination rates because it posits that speciation rates scale to environmental temperature, and there was a strong overall decline through the Cenozoic in global temperatures (Zachos *et al.* 2001). There is no such decline in rates (Fig. 16.3).

A key assumption of this new theory is that although only species speciate, speciation rates need to be expressed in terms of per-individual, not per-species, rates. Moreover, the model predicts that not just per-individual but therefore per-species rates will be diversity-independent, if other factors such as total community abundance and temperature are constant. Thus, if any equilibrium exists, it must relate to extinction rates. However, because again species become extinct and individuals do not, extinction rates can vary only if species-level properties such as geographic range and population size do. This contradiction calls into question whether per-species rates are needless. The metabolic model (Allen *et al.* 2006) also does not address the mathematical certainty that constant total speciation rates will eventually produce more species than individuals, which could only lead to a diversity equilibrium.

Both fossil (Alroy 1998b) and molecular (Nee *et al.* 1992; Roelants *et al.* 2007) data show that the diversity of vertebrates does not change randomly, much less increase exponentially. This fact suggests that more process-oriented work needs to be done, although it need not be entirely reductionistic. Witnessing limits to diversity is just a matter of time: an infinite evolutionary process must produce them. Thus, the open issue is not whether limits exist, but rather whether they are approached quickly on a geological time scale (Sepkoski 1978, 1979, 1984; Benton 1995). In the case of North American mammals, the answer is yes.

## Acknowledgements

I thank Jon Bridle, Roger Butlin, and Dolph Schluter for the opportunity to contribute to this volume, and the British Ecological Society for allowing me to participate in the related Symposium. This work was conducted while a Center Associate at the National Center for Ecological Analysis and Synthesis, a Center funded by NSF (Grant #DEB-0553768), the University of California, Santa Barbara, and the State of California. I thank NSF's Division of Earth Sciences for recognizing the need for the Paleobiology Database to serve its discipline, and an anonymous donor for funding that effort. This is Paleobiology Database publication number 69.

## References

Allen, A. P., Gillooly, J. F., Savage, V. M. and Brown, J. H. (2006) Kinetic effects of temperate on rates of genetic divergence and speciation. *Proceedings of the National Academy of Sciences of the United States of America* **103**, 9130–9135.

Alroy, J. (1992) Conjunction among taxonomic distributions and the Miocene mammalian biochronology of the Great Plains. *Paleobiology* **18**, 326–343.

Alroy, J. (1996) Constant extinction, constrained diversification, and uncoordinated stasis in North American mammals. *Palaeogeography, Palaeoclimatology, Palaeoecology* **127**, 285–311.

Alroy, J. (1998a) Cope's rule and the dynamics of body mass evolution in North American mammals. *Science* **280**, 731–734.

Alroy, J. (1998b) Equilibrial diversity dynamics in North American mammals. In: *Biodiversity Dynamics: Turnover of Populations, Taxa, and Communities* (ed. M. L. McKinney and J. A. Drake), pp. 232–287. Columbia University Press, New York.

Alroy, J. (1999) The fossil record of North American mammals: evidence for a Paleocene evolutionary radiation. *Systematic Biology* **48**, 107–118.

Alroy, J. (2000) New methods for quantifying macroevolutionary patterns and processes. *Paleobiology* **26**, 707–733.

Alroy, J. (2002) How many named species are valid? *Proceedings of the National Academy of Sciences of the United States of America* **99**, 3706–3711.

Alroy, J. (2008) Dynamics of origination and extinction in the marine fossil record. *Proceedings of the National Academy of Sciences of the United States of America* **105**, 11536–11542.

Alroy, J., Aberhan, M., Bottjer, D. J., *et al.* (2008) Phanerozoic trends in the global diversity of marine invertebrates. *Science* **321** 97–100.

Alroy, J., Koch, P. L. and Zachos, J. C. (2000) Global climate change and North American mammalian evolution. Deep time: paleobiology's perspective (ed. D. H. Erwin and S. L. Wing), pp. 259–288. *Paleobiology* **26** (supplement).

Alroy, J., Marshall, C. R., Bambach, R. K., *et al.* (2001) Effects of sampling standardization on estimates of Phanerozoic marine diversification. *Proceedings of the National Academy of Sciences of the United States of America* **98**, 6261–6266.

Benton, M. J. (1995) Diversification and extinction in the history of life. *Science* **268**, 52–58.

Brett, C. E. and Baird, G. C. (1995) Coordinated stasis and evolutionary ecology of Silurian to Middle Devonian faunas in the Appalachian Basin. In: *New Approaches to Speciation in the Fossil Record* (ed. D. H. Erwin and R. L. Anstey), pp. 285–315. Columbia University Press, New York.

Bush, A. M. and Bambach, R. K. (2004) Did alpha diversity increase during the Phanerozoic? Lifting the veils of taphonomic, latitudinal, and environmental biases. *Journal of Geology* **112**, 625–642.

Bush, A. M., Markey, M. J. and Marshall, C. R. (2004) Removing bias from diversity curves: the effects of spatially organized biodiversity on sampling standardization. *Paleobiology* **30**, 666–686.

Clyde, W. C. and Gingerich, P. D. (1998) Mammalian community response to the latest Paleocene thermal maximum; an isotaphonomic study in the northern Bighorn Basin, Wyoming. *Geology* **26**, 1011–1014.

Connolly, S. R. and Miller, A. I. (2001) Joint estimation of sampling and turnover rates from fossil databases: capture-mark-recapture methods revisited. *Paleobiology* **27**, 751–767.

Crampton, J. S., Foote, M., Beu, A. G., *et al.* (2006) The ark was full! Constant to declining Cenozoic shallow marine biodiversity on an isolated midlatitude continent. *Paleobiology* **32**, 509–532.

Foote, M. (1993) Discordance and concordance between morphological and taxonomic diversity. *Paleobiology* **19**, 185–204.

Foote, M. (1994a) Morphological disparity in Ordovician-Devonian crinoids and the early saturation of morphological space. *Paleobiology* **20**, 320–344.

Foote, M. (1994b) Temporal variation in extinction risk and temporal scaling of extinction metrics. *Paleobiology* **20**, 424–444.

Foote, M. (1999) Evolutionary radiations in Paleozoic and post-Paleozoic crinoids. *Paleobiology Memoir* **25**, 1–115.

Foote, M. (2000) Origination and extinction components of taxonomic diversity: Paleozoic and post-Paleozoic dynamics. *Paleobiology* **26**, 578–605.

Foote, M. (2003) Origination and extinction through the Phanerozoic: a new approach. *Journal of Geology* **111**, 125–148.

Foote, M. and Raup, D. M. (1996) Fossil preservation and the stratigraphic ranges of taxa. *Paleobiology* **22**, 121–140.

Foote, M. and Sepkoski, J. J., Jr. (1999) Absolute measures of the completeness of the fossil record. *Nature* **398**, 415–417.

Fox, D. L. and Koch, P. L. (2004) Carbon and oxygen isotopic variability in Neogene paleosol carbonates: constraints on the evolution of the C-4-grasslands of the Great Plains, USA. *Paleogeography, Palaeoclimatology, Palaeoecology* **207**, 305–329.

Gilinsky, N. L. (1994) Volatility and the Phanerozoic decline of background extinction intensity. *Paleobiology* **13**, 427–445.

Gilinsky, N. L. and Bambach, R. K. (1987) Asymmetrical patterns of origination and extinction in higher taxa. *Paleobiology* **13**, 427–445.

Gotelli, N. J. and Colwell, R. K. (2001) Quantifying biodiversity: procedures and pitfalls in the measurement and comparison of species richness. *Ecology Letters* **4**, 379–391.

Hulbert, R. C., Jr. (1993) Taxonomic evolution in North American Neogene horses (subfamily Equinae): the rise and fall of an adaptive radiation. *Paleobiology* **8**, 159–167.

Isaac, N. J. B., Jones, K. E., Gittleman, J. L. and Purvis, A. (2005) Correlates of species richness in mammals: body size, life history, and ecology. *American Naturalist* **165**, 600–607.

Jablonski, D., Roy, K., Valentine, J. W., Price, R. M. and Anderson, P. S. (2003) The impact of the Pull of the Recent on the history of bivalve diversity. *Science* **300**, 1133–1135.

Kirchner, J. W. and Weil, A. (2000) Correlations through time in fossil extinctions and originations. *Proceedings of the Royal Society of London, Series B*, **267**, 1301–1309.

Kosnik, M. A. (2005) Changes in Late Cretaceous-early Tertiary benthic marine assemblages: analyses from the North American coastal plain shallow shelf. *Paleobiology* **31**, 459–479.

Kowalewski, M., Kiessling, W., Aberhan, M., *et al.* (2006) Ecological, taxonomic, and taphonomic components of the post-Paleozoic increase in sample-level species diversity of marine benthos. *Paleobiology* **32**, 533–561.

Krug, A. Z. and Patzkowsky, M. E. (2004) Rapid recovery from the Late Ordovician mass extinction. *Proceedings of the National Academy of Sciences of the United States of America* **101**, 17605–17610.

Magallon, S. and Sanderson, M. J. (2001) Absolute diversification rates in angiosperm clades. *Evolution* **55**, 1762–1780.

Mark, G. A. and Flessa, K. W. (1977) A test for evolutionary equilibria: Phanerozoic brachiopods and Cenozoic mammals. *Paleobiology* **3**, 17–22.

Miller, A. I. and Foote, M. (1996) Calibrating the Ordovician radiation of marine life: implications for Phanerozoic diversity trends. *Paleobiology* **22**, 304–309.

Nee, S., Holmes, E. C., May, R. M. and Harvey, P. H. (1994) Extinction rates can be estimated from molecular phylogenies. *Philosophical Transactions of the Royal Society Series B* **344**, 77–82.

Nee, S., Mooers, A. O. and Harvey, P. H. (1992) Tempo and mode of evolution revealed from molecular phylogenies. *Proceedings of the National Academy of Sciences of the United States of America* **89**, 8322–8326.

Newman, M. E. J. and Eble, G. J. (1999) Decline in extinction rates and scale invariance in the fossil record. *Paleobiology* **25**, 434–439.

Olszewski, T. D. (2004) A unified mathematical framework for the measurement of richness and evenness within and among multiple communities. *Oikos* **104**, 377–387.

Paul, C. R. C. (1982) The adequacy of the fossil record. In: *Problems of Phylogenetic Reconstruction* (ed. K. A. Joysey and A. E. Friday), pp. 75–117. Academic Press, London.

Peters, S. E. (2004) Evenness of Cambrian-Ordovician benthic marine communities in North America. *Paleobiology* **30**, 325–346.

Peters, S. E. and Foote, M. (2001) Biodiversity in the Phanerozoic: a reinterpretation. *Paleobiology* **27**, 583–601.

Powell, M. G. and Kowalewski, M. (2002) Increase in evenness and sampled alpha diversity through the Phanerozoic: comparison of early Paleozoic and Cenozoic marine fossil assemblages. *Geology* **30**, 331–334.

Purvis, A., Nee, P. and Harvey, P. H. (1995) Macroevolutionary inferences from primate phylogeny. *Philosophical Transactions of the Royal Society Series B* **260**, 329–333.

Raup, D. M. (1976) Species diversity in the Phanerozoic: an interpretation. *Paleobiology* **2**, 289–297.

Raup, D. M. (1977) Stochastic models in evolutionary paleontology. In: *Patterns of Evolution as Illustrated by the Fossil Record* (ed. A. Hallam), pp. 59–78. Elsevier Scientific, Amsterdam.

Raup, D. M. (1979) Biases in the fossil record of species and genera. *Bulletin of the Carnegie Museum of Natural History* **13**, 85–91.

Raup, D. M. (1985) Mathematical models of cladogenesis. *Paleobiology* **11**, 42–52.

Raup, D. M. and Sepkoski, J. J., Jr. (1982) Mass extinctions in the marine fossil record. *Science* **215**, 1501–1503.

Roelants, K., Gower, D. J., Wilkinson, M., *et al.* (2007) Global patterns of diversification in the history of modern amphibians. *Proceedings of the National Academy of Sciences of the United States of America* **104**, 887–892.

Sepkoski, J. J., Jr. (1978) A kinetic model of Phanerozoic taxonomic diversity I. Analysis of marine orders. *Paleobiology* **4**, 223–251.

Sepkoski, J. J., Jr. (1979) A kinetic model of Phanerozoic taxonomic diversity II. Early Paleozoic families and multiple equilibria. *Paleobiology* **5**, 222–252.

Sepkoski, J. J., Jr. (1982) A compendium of fossil marine families. *Milwaukee Public Museum Contributions in Biology and Geolology* **51**, 1–125.

Sepkoski, J. J., Jr. (1984) A kinetic model of Phanerozoic taxonomic diversity III. Post-Paleozoic families and mass extinctions. *Paleobiology* **10**, 246–267.

Sepkoski, J. J., Jr. (1997) Presidential address: Biodiversity: past, present, and future. *Journal of Paleontology* **71**, 533–539.

Sepkoski, J. J., Jr. (1998) Rates of speciation in the fossil record. *Philosophical Transactions of the Royal Society Series B* **353**, 315–326.

Sepkoski, J. J., Jr. (2002) A compendium of fossil marine animal genera. *Bulletins of American Paleontology* **363**, 1–560.

Simpson, G. G. (1944) *Tempo and Mode in Evolution.* Columbia University Press, New York.

Smith, A. B. (2001) Large-scale heterogeneity of the fossil record: implications for Phanerozoic biodiversity studies. *Philosophical Transactions of the Royal Society Series B* **356**, 351–367.

Smith, T., Rose, K. D. and Gingerich, P. D. (2006) Rapid Asia-Europe-North America geographic dispersal of earliest Eocene primate *Teilhardina* during the Paleocene-Eocene Thermal Maximum. *Proceedings of the National Academy of Sciences of the United States of America* **103**, 11223–11227.

Stanley, S. M. (1979) *Macroevolution: Pattern and Process.* W.H. Freeman, San Francisco.

Van Valen, L. (1973) A new evolutionary law. *Evolutionary Theory* **1**, 1–30.

Vrba, E. S. (1985) Environment and evolution: alternative causes of the temporal distribution of evolutionary events. *South African Journal of Science* **81**, 229–236.

Vrba, E. S. (1992) Mammals as a key to evolutionary theory. *Journal of Mammalogy* **73**, 1–28.

Wagner, P. J. and Erwin, D. H. (1995) Phylogenetic tests of speciation hypotheses. In: *New Approaches to Speciation in the Fossil Record* (ed. D. H. Erwin and R. L. Anstey), pp. 87–122. Columbia University Press, New York.

Wagner, P. J., Kosnik, M. A. and Lidgard, S. (2006) Abundance distributions imply elevated complexity of post-Paleozoic marine ecosystems. *Science* **314**, 1289–1292.

Wang, X., Tedford, R. H. and Taylor, B. E. (1999) Phylogenetic systematics of the Borophaginae (Carnivora, Canidae). *Bulletin of the American Museum of Natural History* **243**, 1–659.

Webb, S. D. (1969) Extinction-origination equilibria in late Cenozoic land mammals of North America. *Evolution* **23**, 688–702.

Zachos, J., Pagani, M., Sloan, L., Thomas, E. and Billups, K. (2001) Trends, rhythms, and aberrations in global climate 65 Ma to present. *Science* **292**, 686–693.

# Index